D0824639

ECOLOGICAL RESTORATION: A GLOBAL CHALLENGE

Ecological restoration projects seek to recover the natural functioning of degraded ecosystems, most often areas disturbed by human large scale projects such as agriculture developments, road building, mining and urban sprawl. While scientists focus on the ecological basis of habitat repair, practitioners, governmental and non-governmental organizations, and local people, tend to worry about social, economic, cultural and political aspects. The value of ecological restoration is discussed here with examples including tropical forests in Vietnam and Australia, degraded environments in China as a paradigm of global issues, the restoration of wetlands and the coastal zone and how to proceed with urban developments. The author also uniquely assesses how ecological restoration can be used against the impacts of climate change. In addition to strategies for extending ecological restoration on a global scale, it provides useful ideas and tools for the everyday work of practitioners, professionals, researchers and students.

FRANCISCO A. COMÍN is an ecologist with background experience on ecosystem analysis and development of management tools. He has been Professor of Ecology at University of Barcelona (Spain) for twenty-five years and visiting professor of CINVESTAV-IPN (Mérida, Yucatan, Mexico) for fifteen years. Books published include X. Rodo & F. A. Comín (2003), Global Climate. Current Research and Uncertainties in the Climate System; F. A. Comín & T. Northcote (1990), Saline Lakes. Developments in Hydrobiology 59; F. A. Comín, J. A. Herrera, J. Ramírez (2000). Limnology and Aquatic Birds. Monitoring, Modelling and Management Univ. Autónoma del Yucatán Publ.. He has been performing research and practical work on ecological restoration for twenty years. His aim in his work is to integrate scientific, technical, economic and social aspects of ecological restoration. Topics in the frontiers between different sciences are a special subject of interest, as is shown in this book. He is member of SER International, Intecol and RIACRE.

ECOLOGICAL RESTORATION

A Global Challenge

Edited by

FRANCISCO A. COMÍN

CAMBRIDGE
UNIVERSITY PRESS

CAMBRIDGE UNIVERSITY PRESS
Cambridge, New York, Melbourne, Madrid, Cape Town, Singapore,
São Paulo, Delhi, Dubai, Tokyo

Cambridge University Press
The Edinburgh Building, Cambridge CB2 8RU, UK

Published in the United States of America by Cambridge University Press, New York

www.cambridge.org
Information on this title: www.cambridge.org/9780521877114

© Cambridge University Press 2010

This publication is in copyright. Subject to statutory exception
and to the provisions of relevant collective licensing agreements,
no reproduction of any part may take place without the written
permission of Cambridge University Press.

First published 2010

Printed in the United Kingdom at the University Press, Cambridge

A catalogue record for this publication is available from the British Library

ISBN 978-0-521-87711-4 Hardback

Cambridge University Press has no responsibility for the persistence or
accuracy of URLs for external or third-party Internet websites referred to
in this publication, and does not guarantee that any content on such
websites is, or will remain, accurate or appropriate.

To the extent permitted by applicable law, Cambridge University Press is not liable for direct
damages or loss of any kind resulting from the use of this product or from errors or faults
contained in it, and in every case Cambridge University Press's liability shall be limited to the
amount actually paid by the customer for the product.

This book is dedicated to previous scientists and restoration practitioners who inspired our ideas and to the future generations, in the hope that they will use our planet wisely and will extend the practice of ecological restoration on a global scale.

Contents

List of contributors		*page* xi
Foreword		xv
Preface		xvii
Acknowledgements		xxii
List of abbreviations		xxiii
Part I	**Global perspectives for ecological restoration**	1
1	The challenges of humanity in the twenty-first century and the role of ecological restoration	3
	Francisco A. Comín	
	1.1 A changing world	3
	1.2 The challenges to the Earth in the twenty-first century	5
	1.3 Global changes versus local actions	8
	1.4 Scientific approaches to our environmental problems	10
	1.5 Socio-ecological approaches to solving environmental problems	11
	1.6 The role of ecological restoration in solving our problems	13
	References	17
2	The global carbon cycle: current research and uncertainties in the sources and sinks of carbon	21
	Dario Papale and Riccardo Valentini	
	2.1 Introduction	21
	2.2 Magnitude: "slow in, fast out"?	21
	2.3 Methodology: how to investigate carbon cycle patterns and processes	26
	2.4 Variability: how variable is the terrestrial part of the carbon cycle?	34
	2.5 Ecological restoration and the Kyoto Protocol	38
	References	39

3   Using international carbon markets to finance forest restoration   45
    Johannes Ebeling, Malika Virah-Sawmy and Pedro
    Moura Costa
    3.1   Introduction   45
    3.2   Forestry and international carbon markets   46
    3.3   How do forestry projects work in carbon markets?   55
    3.4   Opportunities for ecological restoration through forest carbon
          markets   60
    3.5   Restoration through carbon forestry in practice   65
    3.6   Challenges for integrating carbon forestry and ecological
          restoration   66
    3.7   Conclusions and outlook   70
          References   73
4   The value of a restored Earth and its contribution to a sustainable
    and desirable future   78
    Robert Costanza
    4.1   Introduction   78
    4.2   Ecosystem services and natural capital   79
    4.3   Ways to restore the Earth's natural capital   83
    4.4   Estimates of the value of a restored Earth   84
    4.5   Paths to a restored Earth   88
    4.6   Conclusions   89
          References   89
5   Focal restoration   91
    Eric Higgs
    5.1   Introduction   91
    5.2   Effective, efficient and engaging restoration   91
    5.3   Cultural variation   93
    5.4   Technological restoration   95
    5.5   Focal restoration   97
          References   98
6   Ethical dimensions of ecological restoration   100
    Rebecca L. Vidra and Theodore H. Shear
    6.1   Introduction   100
    6.2   Definitions and expectations of ecological restoration   102
    6.3   Restoration as a process: an art, practice or science?   103
    6.4   Restoration as the acid test: the next big lie?   105
          References   109

**Part II Towards the practice of ecological restoration
on a global scale** 113

7 Undertaking forest restoration on a landscape scale
in the humid tropics: matching theory and practice in developed
and developing countries 115
David Lamb
  7.1 Introduction 115
  7.2 Reforestation at a particular site 116
  7.3 Reforestation of degraded sites in the wet tropics of Australia
and Vietnam 119
  7.4 Forest restoration across landscapes 124
  7.5 Implementing forest landscape restoration 127
  7.6 Discussion and conclusions 134
  References 136

8 Land degradation and ecological restoration in China 140
*Bojie Fu, Dong Niu, Yihe Lu, Guohua Liu
and Wenwu Zhao*
  8.1 Introduction 140
  8.2 Land degradation and related impacts 142
  8.3 Ecological restoration 152
  8.4 Concluding remarks 165
  References 166

9 Conservation, restoration and creation of wetlands: a global
perspective 175
William J. Mitsch
  9.1 Introduction 175
  9.2 Wetland definitions and global extent 176
  9.3 Wetland losses 177
  9.4 A more optimistic approach: creating and restoring wetlands
and watersheds 178
  9.5 Wetland restoration on a global scale 186
  9.6 Conclusions 187
  References 187

10 Uses, abuses and restoration of the coastal zone 189
Francisco A. Comín, Jordi Serra and Jorge A. Herrera
  10.1 Introduction 189
  10.2 Types of coastal ecosystems and their restoration 191
  10.3 Uses of the coastal zone 201
  10.4 Abuses in the coastal zone 206

10.5   Restoration of the coastal zone: what is being done and what
       is necessary   210
10.6   Conclusions   215
       References   217

11   Spatial ecological solutions to mesh nature and people: Boston
     suburb, Barcelona region and urban regions worldwide   225
     Richard T. T. Forman
11.1   Introduction   225
11.2   Local unit in the Boston region   226
11.3   Greater Barcelona region   230
11.4   Urban regions compared   239
       References   242

12   The role of ecological modelling in ecosystem restoration   245
     Sven Erik Jørgensen
12.1   Ecological modelling as a tool for ecological restoration   245
12.2   Models applied in ecosystem restoration   247
12.3   Case studies   249
12.4   Conclusions   262
       References   262

13   Restoration as a bridge for cooperation and peace   264
     Amos Brandeis
13.1   Introduction   264
13.2   The Alexander River Restoration Project   265
13.3   The Lake Bam Restoration Project   268
13.4   Lessons learned about how restoration projects become
       bridges for cooperation   271
13.5   Conclusions   283
       References   287

Index   289

Colour plates appear between pages 112 and 113

Contributors

Brandeis, Amos
Architecture, Urban & Regional Planning, and Project Management Ltd.
36 Eshkol St.
Hod-Hasharon. Israel.

Comín, Francisco A.
Pyrenean Institute of Ecology-CSIC Av. Montañana 1005.
Zaragoza. Spain.

Costanza, Robert
Gund Institute of Ecological Economics, Rubenstein School of Environment
and Natural Resources, The University of Vermont.
590 Main Street.
Burlington.Vermont. USA.

Ebeling, Johannes
EcoSecurities Ltd.
1st Floor, Park Central, 40/41 Park End Street.
Oxford. United Kingdom.

Forman, Richard T. T.
Graduate School of Design, Harvard University. Cambridge, USA.
48 Quincy Street.
Cambridge, MA. USA.

Fu, Bojie
State Key Laboratory of Urban and Regional Ecology, Research Centre
for Eco-Environmental Sciences, Chinese Academy of Sciences.
No.18 Shuangqing Road.
Haidian District. Beijing. China.

Gann, George
The Institute for Regional Conservation.
22601 S.W. 152 Avenue.
Miami. Florida. USA.

Herrera, Jorge A.
CINVESTAV-IPN.
Carretera Antigua a Progreso km 6.
A.P. 73. CORDEMEX.
Merida. Yucatan. Mexico.

Higgs, Eric
School of Environmental Studies, University of Victoria.
132A Sedgewick Building.
Victoria. British Columbia. Canada.

Jørgensen, Sven Eric
Department of Environmental Chemistry. Institute of Analytical
and Pharmaceutical Chemistry. Copenhagen University.
Universitetsparken 2.
Copenhagen. Denmark.

Lamb, David
Rainforest Cooperative Research Center and School
of Integrative Biology.
University of Queensland.
Brisbane. Australia.

Liu, Guohua
State Key Laboratory of Urban and Regional Ecology, Research
Centre for Eco-Environmental Sciences, Chinese
Academy of Sciences.
No.18 Shuangqing Road.
Haidian District. Beijing. China.

Lu, Yihe
State Key Laboratory of Urban and Regional Ecology,
Research Centre
for Eco-Environmental Sciences, Chinese Academy of Sciences.
No.18 Shuangqing Road.
Haidian District. Beijing. China.

Mitsch, William J.
Heffner Wetland Research Building. The Olentangy River Wetland Research Park.
The Ohio State University.
352 W Dodridge Street.
Columbus. Ohio. USA.

Moura Costa, Pedro
EcoSecurities Ltd.
1st Floor, Park Central, 40/41 Park End Street.
Oxford. United Kingdom.

Niu, Dong
State Key Laboratory of Urban and Regional Ecology, Research Centre
for Eco-Environmental Sciences, Chinese Academy of Sciences.
No.18 Shuangqing Road.
Haidian District. Beijing. China.

Papale, Dario
Department of Forest Environment and Resources –DISAFRI
University of Tuscia
via S. Camillo de Lellis.
Viterbo. Italy.

Serra, Jordi
Dept. Estratigrafía i Paleontología. Facultat de Geologia. Universitat de Barcelona.
c/ Martí i Franqués s/n.
Barcelona. Spain.

Shear, Theodore H.
Department of Forestry and Environmental Resources.
North Carolina State University.
Raleigh. USA.

Valentini, Riccardo
Department of Forest Environment and Resources –DISAFRI
University of Tuscia
via S. Camillo de Lellis.
Viterbo. Italy.

Vidra, Rebecca L.
University Writing Program. Duke University.
Box 90025.
Durham. NC. USA.

Virah-Sawmy, Malika
Oxford University Centre for the Environment, School of Geography and the Environment, University of Oxford.
South Parks Road.
Oxford. United Kingdom.

Zhao, Wenwu
Key Laboratory of Environmental Change and Natural Disaster, Ministry of Education. Institute of Resources Management, Beijing Normal University, Beijing. China.

Foreword

The time for complacency and "business as usual" with regard to the management of our home, planet Earth, is in the past. We are now entering a period in which massive action must be taken to halt and reverse damage to the environment. For the last several millennia we humans have expanded our geographic reach and accelerated our uses of the environment and its resources. The effects of this on the biosphere are well documented and have been thoroughly discussed. The second half of the nineteenth century bore witness to a shift in our thinking as a species, as the appreciation of nature and wild places became points of public discussion. The early twentieth century saw the development of the concepts of conservation and resource management, while the second half saw the rise of modern environmentalism. These concepts focus primarily on halting environmental damage and the sustainable use of natural resources. The idea of repairing environmental damage, or ecological restoration, and its associated scientific discipline, restoration ecology, developed concurrently with modern environmentalism, but were until recently less well known.

While environmental exploitation has benefited relatively few throughout history, the collapse of colonialism following World War II and the increasing pressure on governments to foster economic development for the masses has accelerated the utilization of the Earth's finite resources. What became clear before the close of the twentieth century, however, was that the unbridled use of the planet's resources would not bring prosperity to the poor, but rather would ultimately harm everyone. Thus, the idea of sustainability gained traction within the global political community, culminating in the World Commission on Environment and Development's 1989 definition of sustainability: "[to meet] the needs of the present without compromising the ability of future generations to meet their own needs."

Unfortunately, not everyone yet understands the seriousness of our situation as a species and the affects of our actions on the planet and our fellow inhabitants. The Bush administration's decision to prevent the United States from full participation in international agreements such as the Kyoto Protocol and the Convention on Biological Diversity put the global environment at higher risk and illustrates how old ideas die hard. Ironically, it is failed economic policies in the first decade of the twenty-first century that now provide us with a critical opportunity for change. The global meltdown in 2008 may take social scientists years or even decades to understand, but what is clear is that the world will never be the same.

Ecological Restoration: A Global Challenge could not come at a better time. Massive change is on its way and ideas like sustainability and ecological restoration are poised for increased prominence. For twenty years, the Society for Ecological Restoration International and its members have worked to develop the techniques, philosophy and science behind environmental repair and rehabilitation. Now it is time to move beyond the limited conservation actions of the past and put this knowledge to work at the planetary scale. Ecological restoration is not a utopian idea: it is a powerful tool, and if implemented on a large scale, it could change the planet for the better by transforming the way humans relate to it and to each other. Ecological restoration can help protect biodiversity, mitigate climate change, sequester carbon, reunite indigenous peoples with their landscapes and cultures, and restore a healthy relationship between people and nature. Ecological restoration also complements other allied efforts such as rare species conservation, natural landscaping, reconciliation ecology, organic agriculture, the restoration of natural capital, environmental justice and the elimination of armed conflict. If implemented properly, it can help alleviate poverty and assist in the equitable sharing of resources among the Earth's inhabitants.

Around the world, local communities are picking up their shovels and restoring degraded ecosystems, in many cases, with little or no funding. These Herculean efforts point to an increasing public awareness that restoring "green infrastructure" is the most efficient and desirable way to secure goods and services derived from nature. In many cases, these are irreplaceable and their economic value is immeasurable: how can we replace the environment's role in protecting us from droughts, floods, and storms or in providing water and air purification or supporting agriculture and fisheries? What is needed now is more funding and political commitment to ecological restoration at the local, national and international level. In my mind, there is an urgency to get to work and there is a tremendous amount to do. This book provides the background and stimulus to do just that.

George D. Gann
Chair
Society for Ecological Restoration International (1997–1999, 2007–2009)

Preface

The state of the environment at the end of the twentieth century was diagnosed as rather negative (Starke, 2000). Evidence of the direct relationship between greenhouse gas emissions caused by human activity and global warming of the Earth's atmosphere was proven (IPCC, 2007), and the loss of habitats was still very high. At the same time, never before had there been so much interest in improving the conservation of natural resources and the state of the environment, from governmental and non-governmental entities, from individuals and global-scale associations. It seemed that the world was becoming focused on changing the paradigm in the relationship between human beings and nature, and that the principles of caution and nature conservation were to be integrated into development planning. A new kind of human-nature relationship, embodying conservation and a rational use of the ecosystem, had become necessary, as well as a new form of socioeconomic development which integrated this rational use of the ecosystem and involved local populations.

However, the twenty-first century has hardly began to consider these experiences or this diagnosis (MEA, 2005; WorldWatch Organization, 2009). Far from it, human demand for resources and natural services has long exceeded the Earth's capacity to provide them (Ewing *et al.*, 2008). The globalization of terrorism, war and the countless cases of financial, political and social irresponsibility, on all scales from global to local, have maximized the separation of the factors regulating economic development and human well-being in general, and are still clearly contributing to the degradation of natural systems. The increase in human migrations, in the international traffic of exploited human beings, in the gap between urban and rural worlds, is not independent of environmental degradation and the loss of habitats and natural resources. The confirmation of the connection between atmospheric global warming and environment-degrading human activities, together with the forecast of the impact of climate change on the ecosystem, should make us adopt measures for mitigation and adaptation to these phenomena quickly.

There is a growing interest in this direction. Environment-improving campaigns are being developed by small local groups and by whole governments, with many diverse interests, from species conservation to the purchase of land and its sustainable management from strict nature protection to improvements in people's social and economic status. However, the general state of the world does not seem to improve, but worsen. Although the number and extension of protected areas increase, at regional and global scale, natural habitats are still being lost, and changes in land use are identified as the cause for 50 percent of the greenhouse gas emissions in the world (World Watch, 2009). Sustainability has become a concept difficult to put into practice, although there are significant small-scale cases proving that it is achievable (Gulf *et al.*, 2006; Munasinghe, 2009).

Under these circumstances, the restoration of degraded ecosystems stands out as an imperative activity in order to improve the state of the Earth. It is not enough to conserve or protect, it is necessary to restore, and to do so on a global scale, since degradation also happens on a global scale. Such is the inspiration and main thesis of this book. It is not possible to imagine now the final form and state of the planet that will be attained through global ecological restoration. However, the concept is beginning to develop. More and more scientists are contributing their ideas along these lines, and more practitioners their experience. There are increasingly more organizations participating in restoration projects, and agencies contributing material goods or finance for this activity (Clewell and Aronson, 2008).

Nevertheless, practice is currently more advanced than theory, despite the fact that the last decade has seen a proliferation of publications on restoration ecology and ecological restoration. The scientific journal *Restoration Ecology*, sponsored by the Society for Ecological Restoration International, has published scientific research works on restoration ecology for nearly twenty years. Since 1981, the journal *Ecological Restoration* (formerly *Restoration & Management Notes*), has published a relevant combination of practical articles, restoration experiences, and scientific information. The journal *Ecological Management & Restoration* has been similarly publishing since 2000, focusing on the Australian continent. Many other journals have published articles on restoration in the last decade. Both small-scale experiments on ecological restoration and large-scale practical applications offer excellent opportunities for two complementary aspects of scientific and social progress: extending knowledge and gaining practical experience.

Moreover, a growing number of books, as well as other types of documents, are beginning to form a formal corpus on restoration ecology. A group of these books is promoted by the Society for Ecological Restoration International (available at Island Press); each deals with a specific topic or type of ecosystem. Others are well-structured, general books (e.g., Perrow and Davy, 2002; Mitsch and Jorgensen, 2003; Van Andel and Aronson, 2006). These books constitute an excellent collection which shows the significant quality attained by ecological restoration projects and its contribution to the progress of ecology. However, ecological restoration includes, as does ecology, important social and economic aspects which confer a wider dimension than that of restoration ecology. These aspects are considered in all these publications, which markedly demonstrates how ecological restoration juxtaposes various approaches and integrates different disciplines.

Thus, this volume aims at conveying two main ideas: (i) the above-mentioned need to develop ecological restoration practices on a global scale in order to improve the state of our world, and (ii) the need for global-scale ecological restoration to positively integrate and contribute to socioeconomic development. For the first idea to be efficiently disseminated it is necessary to base ecological restoration practices on the foundations proved by experts and scientists, whose advice is to develop restoration actions at ecosystem-scale, that is, at the scale of the functional relations among the components of an ecologic system (SER International Science & Policy Working Group, 2004; Clewell *et al.*, 2005). Consequently, it is not so much a matter of achieving the establishment of a predetermined physical or biological structure, as a matter of re-initiating adequate functional relations (e.g., geomorphological and biogeochemical processes, the renewal of key biological populations) and allowing ecosystems to develop within the framework of general environmental conditions. It is important that research work and restoration practices take into account the dynamic, changing character of the ecosystems. Otherwise, there is a risk of considering fixed, invariable structures, either physical or biological, as the aim of restoration, whereas they are something alien to natural ecosystems and to ecological restoration aims, and tend to restrict the ecosystems' self-organization capacity and their potential for adaptation to changing environmental and climatic conditions.

The second idea stems from the common experience acquired with the implementation of restoration projects. Social and economic factors are as important as scientific and technical factors, or maybe more, and need to be integrated into restoration projects (Comin *et al.*, 2005). Neglecting these aspects may lead to the failure of a restoration project. This extends to ethical considerations that are included in restoration practices; these are beginning to be formalized by scientists, as well as by practitioners and philosophers interested in these issues.

In this sense, ecological restoration can become a global tool for cooperation and development in situations of war and hostility between groups. Without losing any scientific-technical rigor or expert basis, restoration projects can be agglutinants of interests that motivate and strengthen relations between groups of persons and entities and also between governmental organizations, be their interests boundary-oriented or based on peaceful coexistence and mutual respect. The capacity of ecological restoration to integrate scientific-technical, economic and social concerns offers opportunities for the betterment of the human condition.

Consideration of all these issues motivated the content structure of this book. Thus, the contents are grouped in two main parts. The first part includes more generic content, related to the challenges faced in this century by humankind, the uncertainties concerning the carbon cycle as the engine for change in our ecosystem the Earth, and the mitigation of carbon emissions to the atmosphere made possible through ecological restoration. The second part includes global-scope restoration discussions, either related to key ecosystems of the world, or to specific cases which could constitute examples to be developed on a global scale.

This book neither is, nor pretends to be, an exhaustive treatise on ecological restoration techniques. As was mentioned above, there are already many books and manuals on the restoration of specific types of ecosystems. This book collects together the most usual ideas and practices of ecological restoration, considering both the practicalities and the design and implementation of the projects, and tries to further the expansion of ecological restoration to the whole world, contributing its scientific-technical rigor, its practical experience, and its ability to assist with socioeconomic development and cooperation. In summary, this book's purpose is to further the aim of ecological restoration to leave the Earth at the end of the twenty-first century in a better state than at the beginning of the century.

References

Clewell, A. and Aronson, J. (2008). *Ecological Restoration: Principles, Values, and Structure of an Emerging Profession*. Washington DC: Island Press.

Clewell, A., Rieger, J. and Munro, J. (2005). *Guidelines for Developing and Managing Ecological Restoration Projects*. Tucson AZ: Society for Ecological Restoration International.

Comín, F. A., Menendez, M., Pedrocchi, C. *et al*. (2005). Wetland restoration: integrating scientific-technical, economic and social perspectives. *Ecological Restoration*, **23**: 182–186.

Ewing B., Goldfinger, S., Wackernagel, M. *et al*. (2008). *The Ecological Footprint Atlas 2008*. Oakland: Global Footprint Network.

Gulf, C, Newton, A. and Gerber, L. (2006). *Desarrollo sostenible: Conceptos y ejemplos de buenas prácticas en Europa y América Latina*. Munster: Waxman Verlag.

IPCC-Intergovernmental Panel on Climate Change (2007). *Climate Change 2007: Synthesis Report. Contribution of Working Groups I, II and III to the Fourth Assessment*. Report of the Intergovernmental Panel on Climate Change [Core Writing Team, Pachauri, R. K. and Reisinger, A. (eds.)]. Geneva: IPCC.

MEA-Millenium Ecosystem Assessment (2005). *Ecosystems and Human Well-being: Synthesis*. Washington DC: Island Press.

Mitsch, W. J., and Jorgensen, S. E. (2003). *Ecological Engineering and Ecosystem Restoration*. New York: Wiley.

Munasinghe, M. (2009). *Sustainable Development in Practice. Sustainomics Methodology and Applications*. Cambridge University Press.

Perrow, M. R., and Davy, A. J. (2002). *Handbook of Ecological Restoration*. Cambridge University Press.

SER-Society for Ecological Restoration International Science & Policy Working Group (2004). *The SER International Primer on Ecological Restoration*. Tucson AZ: Society for Ecological Restoration International.

Starke, L. (ed.) (2000). *State of the World 2000*. Washington DC: The Worldwatch Institute.

Van Andel, J. and J. Aronson (eds.) (2006). *Restoration Ecology: The New Frontier*. Malden: Blackwell Publishing.

WorldWatch Organization (2009). *State of the World 2009: Into a Warming World*. Washington DC: The WorldWatch Institute.

Acknowledgements

A book of this nature owes much to the authors' efforts. Their commitment and dedication must be acknowledged here, for they have contributed their wide perspective and sound experience to the contents of this book. The authors' varied geographical origins, together with their internationalization, are another token of this book's global character. We also would like to express our recognition to Mª Paz Errea for her dedicated revision of the graphical materials, to Susana Artieda for her sound reviewing of the text edition, and to Mercedes Garcia for her efficient assistance during the process of editing the book. We wish to acknowledge Cambridge University Press, for their encouraging reception of this book, Dominic Lewis in special for his great professionalism, and the positive vision of the Publications Department of the Consejo Superior de Investigaciones Científicas (CSIC), which made this joint publication possible.

Francisco A. Comín

Abbreviations

ARRA	Alexander River Restoration Administration
ARRP	Alexander River Restoration Project
BAU	business as usual
BD	buffer discount
BMZ	German Federal Ministry for Economic Cooperation and Development
CBD	Convention on Biological Diversity
CDM	Clean Development Mechanism
CER	Certified Emission Reduction
CERN	Chinese Ecosystem Research Network
CFERN	China Forest Ecosystem Research Network
CNSCN	China Network of Soil Conservation Monitoring
COP	United Nations Climate Change Conference
CREP	Conservation Reserve Enhancement Program
CSIRO	Commonwealth Scientific and Industrial Research Organisation (of Australia)
CSR	corporate social responsibility
ENSO	El Niño Southern Oscillation
ER_{net}	net emission reduction
$ER_{project}$	project emission reduction
$ER_{baseline}$	baseline emission reductions
$EO_{project}$	other project emission
ET	Emission Trading
EU ETS	European Emission Trading Scheme
FAO	Food and Agricultural Organization
FAPAR	fraction of photosynthetically absorbed active radiation
GDP	Gross Domestic Product
CGP	Grain for Green Program

GHG	greenhouse gas
GIS	Geographic Information System
GOALS	Global Ocean-Atmosphere-Land System
GPP	gross primary production
GTOS	Global Terrestrial Observing System
GUMBO	Global Unified Metamodel of the Biosphere
HIV-AIDS	Human Immunodeficiency Virus-Acquired Immunodeficiency Syndrome
IAP	Institute of Atmospheric Physics
ICZM	integrated coastal zone management
ILTER	International Long Term Ecological Research Network
IPCC	Intergovernmental Panel for Climate Change
IRF	International River Foundation
IUCN	International Union for Nature Conservation
JI	Joint Implementation
L	leakage
LAI	leaf area index
LASG	Laboratory of Atmospheric Sciences and Geophysical Fluid Dynamics
LBRP	Lake Bam Restoration Project
LCA	Louisiana Coastal Area
LPJ	Lund-Potsdam-Jena (dynamic global vegetation model)
LULUCF	Land-Use, Land-Use Change and Forestry
MEA	Millenium Ecosystem Assessment
MOM	Missouri-Ohio-Mississippi (river basin)
NBP	net biome productivity
NDVI	normalized difference vegetation index
NFPP	Natural Forest Protection Program
NGO	non governmental organization
NEP	net ecosystem production
NPP	net primary production
ODA	Official Development Assistance
ORCHIDEE	Organizing Carbon and Hydrology in Dynamic Ecosystems (terrestrial biosphere model)
PFT	plant functional type
Ra	autotrophic respiration
RE	restored Earth
REDD	reduced emissions from deforestation and degradation
REW	relative extractable water (in soil)

Rh	heterotrophic respiration
RIEMS	Regional Integrated Environmental Model System
RUE	radiation use efficiency
SDM	structurally dynamic model
SER(I)	Society for Ecological Restoration International
TER	total ecosystem respiration
TGDP	Three Gorges Dam Project
TGR	Three Gorge Reservoir
TOPEX	Topography Experiment for Ocean Circulation
VCS	Voluntary Carbon Standard
VOC	volatile organic compound
UNCCD	United Nations Convention to Combat Desertification
UNEP	United Nations Environmental Program
UNFCCC	United Nations Framework Convention on Climate Change
WWF	World Wildlife Fund

Part I

Global perspectives for ecological restoration

1

The challenges of humanity in the twenty-first century and the role of ecological restoration

FRANCISCO A. COMÍN

1.1 A changing world

Humanity is facing its most critical time in recent history. Humans as a population recognize the intricate network that biotic and abiotic components of the Earth constitute. This is not new, as it has been the essential interest of ecology as a science since its beginning, early in the industrial era (Haeckel, 1866; Forbes, 1887; Forel, 1892; Warming, 1909). We acknowledged the benefits of maintaining a healthy environment long ago and are aware of the impacts caused to the environment by our unsustainable way of living. What is new is the intensity of the negative impacts that humanity has caused and the global scale of this impact. What is also new is the global awareness of the fact that humans are just a population interacting all around the world with other biotic and abiotic components of our unique ecosystem, Planet Earth. Earth is our habitat but we share this habitat with millions of other species. Human life has cultural and economic aspects and not only basic environmental aspects, and thus we have the unresolved problem of humanity being separated into groups that compete for resources instead of collaborating to use them in a sustainable way, which would be expected from a rational species (Lipschutz, 2004).

Human history can be interpreted as a series of stages initiated by the technical changes which occur with the development of knowledge. The discovery of fire, the extraction of metals, the invention of the wheel, etc. have shaped living conditions for humans. In modern times, after the industrial era and the general development of production based on mechanized procedures, lifestyles have changed for most people, but this has also initiated a massive exploitation of natural resources on a global scale. It can be argued that the industrial era has generally increased human lifespans and improved the average standard of living in most countries of the world. However, the costs of the impacts to the Earth's ecosystems and to humanity itself have been very high, leading to major global concern about our unsustainable way of life (MEA, 2005).

Ecological Restoration: A Global Challenge, ed. Francisco A. Comin. Published by Cambridge University Press.
© Cambridge University Press 2010.

This simple view of the cultural evolution of humanity is useful to emphasize the major challenges we face at this time. The speed of new scientific-technological steps pushing the cultural evolution of the human population is increasing as predicted (Kurzweil, 2005). However, the problem of how we are going to mitigate the impact on the Earth has not been solved yet, as populations and consumption of resources continue to grow. It has been estimated that we already need the equivalent of 1.3 Planet Earths to produce and digest the present global human demand (Global Footprint Network, 2006). However, the heterogeneous spatial distribution of the human footprint for any variable we may consider (continents, demography, culture) proves the irregular distribution of the benefits obtained from the exploitation of natural resources (Plate 1.1). This situation, together with many other environmental, historical and cultural variables, shows also that humans have not found yet a way to cooperate in the optimization of the use of natural resources and the application of new technologies for the common good. Some signs of progress can be observed in terms of both the environment, and human health and culture (World Health Report, 2007). Nevertheless, this progress does not compensate for the increasing global demand for food and the accelerating impacts on the environment (Ramankutty *et al.*, 2002). Not surprisingly, the planet's ecosystem is taking some time to show the consequences of the human impacts caused during the Anthropocene (Crutzen and Stoermer, 2000). Both space and time compose the framework of the ecological theater for the evolutionary play (Hutchinson, 1965). In an ecosystem, a disturbance in one or more components or processes on a large enough scale may provoke changes in the entire network in the long term. In the case of the recent global disturbances to biodiversity, land use, land cover change and biogeochemical cycles, the changes will last for decades or even centuries, and they will have an impact on human well-being on a global scale, even if we reduce the impacts now (IPCC, 2007). Paradoxically, it seems that having more wealth does not necessarily increase one's well-being (Stutz and Mintzer, 2006), that is, being richer is not much better, or at least it is not much better for most of the population (Goklany, 2002).

Therefore, it is necessary to adopt a different strategy for human development within the biosphere, as has been claimed insistently for many years (IUCN, UNEP, WWF, FAO and Unesco, 1980; United Nations, 1987). A new economy is required to solve the environmental problems of the Earth (Hawken, 1983), or rather, a new development model for the whole world (Costanza, 2008). Furthermore, a new political system is required whose purpose would be the progress of people, since those based on production maximization and consumption of materials do not produce genuine progress (Talberth *et al.*, 2006). New perspectives on social and cultural development on a global scale are also required. Ecological restoration, the process of assisting the recovery of an ecosystem that has been degraded, damaged, or destroyed (Society for Ecological Restoration International Science and Policy Working Group, 2004), needs to take place on a global scale if sustainable development and environmental goals are to be met during the twenty-first century. This is the major thesis of this book. Ecological restoration on a global scale must be implemented as soon as possible if we are to achieve sustainable, desirable development. But ecological restoration is not just an end in itself. Ecological restoration must be a tool to mitigate the environmental problems of this world and to develop a happier human society living in the biosphere in a sustainable way. So, the challenge is to put into practice ecological restoration on a global scale in order to improve the socio-ecological conditions of the Earth, the habitat of the human population. There is a wide consensus that we must develop and apply tools which integrate the social, economic and cultural aspects of human life and the use of natural resources, and this will be one of the major challenges for humanity during this century (NAS *et al.*, 1997; MEA, 2005). Ecological restoration is one such tool.

1.2 The challenges to the Earth in the twenty-first century

Various issues were discussed during the last part of the twentieth century regarding challenges to the future of humanity. The classical Malthusian versus anti-Malthusian debate on the limits to human population growth based on the resources available and environmental degradation (Meadows *et al.*, 1972; Meadows *et al.*, 2004) has been reopened following demands from rapidly changing societies. In particular the Chinese are following the production-consumption model of development and triggering large land-use changes in other parts of the world, for example in Brazil (Lowe and Bowlby, 1992; Brown, 2005).

Globally, the world will have the capacity to produce enough food to feed the human population in the short term (FAO, Economic and Social Department, 2002). However, critical stresses are foreseen in both local and global ecosystems following the long-term trends of population increase and rise in per capita consumption (Vitousek *et al.*, 1997). It is clear that in order to provide enough food for a growing human population, it is necessary to have not only the capacity to produce enough food, which could be improved by scientific and technical advancement, and the availability of enough agronomic land (Ramankutty *et al.*, 2002), but also to distribute food in a more equitable way and decrease the environmental costs (Daily *et al.*, 1998). This includes reducing biodiversity losses derived from conventional and bioengineering agricultural and fisheries practices.

The inability to produce enough food for the global population is a problem of spatial heterogeneity, related to a lack of coordination and interest in providing enough food at the right time. So, basically, it is a cultural problem. Some parts of the world lack the basic food production to feed their population due to several environmental and social factors, while other parts of the world have a food production surplus caused by factors such as general climatic conditions, meteorological and environmental phenomena, structural organization, techno-logical capacity, or trade and cultural organization. In fact, the millions of tons of food transported and traded around the world constitute a very lucrative multinational business, which is one of the major contributors to greenhouse gas emissions.

So, we have the capacity to produce enough food to feed the world, as well as to transport this food to those communities currently experiencing deficits (Vitousek *et al.*, 1986). In addition, we generally have knowledge in advance about where there will be food production shortfalls and where the structural weak points are located. The drain on the Earth's resources is due, in part, to cultural aspects of development, such as over-consumption of water, wood, soil, land, fossil fuels, etc., in an attempt to further improve our quality of life.

Raising our living standards is a laudable objective and should be encouraged globally. However, problems arise when the quality of life for a particular popula-tion group is undermined because of unfair land rights and unequal access to natural resources.

The use and abuse of energy is another major challenge for humankind. It constitutes another example of the imbalance in development and distribution of benefits on regional and global scales, as well as within national boundaries. Providing enough energy is a daily challenge for governments and energy compa-nies. It is not a challenge on a global scale, as the Earth can provide enough energy for all. It is a cultural problem, since the spatial distribution of energy sources is extremely heterogeneous, and physical and cultural barriers make it difficult to distribute equitably.

Two major challenges arise when energy is regarded as the basis of human development. The first one is the development of new technologies and energy sources which do not exacerbate environmental degradation while providing enough energy for our needs. New sources and technologies (wind, solar, hydrogen, etc.) can be very promising if they incorporate an appropriate use of natural resources, includ-ing the territory. The exploitation of those energy sources that create global environ-mental and social problems must not be considered as a good alternative. The production of biomass as a renewable energy source is currently taking place in agricultural areas that were once diverse ecosystems, like forests or grasslands. This situation is similar to the land use changes in the tropical rainforest and mangroves of developing countries, which are taking place in order to produce food for humans and livestock.

The second energy challenge is common to any other natural resource required for human survival. A large difference is observed between the average *per capita* demand for resources in the developed countries and that in the developing countries. Again, a tremendous imbalance associated with various historical and cultural aspects has been established across scales. In any case, much more energy than the Earth can provide is being demanded, and much more garbage and residues than the planet can digest and recycle through its biogeochemical cycles is being generated. As new technologies develop to counteract this trend, we need to educate people in order to decrease their impacts and to help others not to follow the same route. If all the people on Earth lived like the average US person, three times the global land area would be needed to assimilate and recycle all the human demands and waste. Again, huge differences exist between countries (Plate 1.2), but also between social groups within a country.

The major challenge arising at the beginning of this century is how to achieve food security and improve the standard of living for those in the developing world without worsening the problems mentioned above (food availability, energy and natural resources consumption, environmental degradation, loss of natural habitats). Globalization is all around but the negative effects (increased consumption, cultural homogenization, unnecessary movements of goods, rise in crime, etc.) tend to overshadow the positive (improved communication and information, products and cultural exchange, human association).

Inequities in society are often the origin of violence. Cultural and religious differences can also exacerbate tensions between groups. This is one of the many challenges that humans face during the twenty-first century. All of them are related to the use and management of natural resources, environmental degradation, and the consequences of global changes.

In the second half of the past century, scientists and philosophers discussed the limits to growth and the Earth's capacity to provide enough resources to maintain the levels of expected population growth. During the last decade, the discussion has focused on the need to change the prevailing development model, including the dominant economic system driving current socio-environmental conditions. The growth rate of human society is unsustainable and may lead to global collapse. It is time to focus the discussion on the methods that will allow us to achieve sustainable and desirable development. Most probably, it will be necessary to adopt new social, economic and ecological paradigms and practices to deal with what are, in essence, cultural problems.

Ecological restoration at a global scale will be critical, since the Earth's capacity to sustain us has been diminished significantly by degradation of the environment. Of course, human conditions have improved in recent times, as is shown by various indicators (life expectancy, etc.). As one global society, we must live in a sustainable world with a more equitable resource distribution that would allow for an average theoretical carrying capacity of 5 billion people, with a theoretical maximum of 14 billion (Cohen, 1995), although major uncertainties for human demography are also ahead (Cohen, 2003). Obviously, this figure may change depending on the variables used for its calculation, including the technological improvements assumed and the average standard of living. Human history tells us that global crises, including wars, arise frequently where or when there are large differences in the standards of living between social groups. They also emerge where or when cultural minorities are discriminated against, instead of being acknowledged as enriching our diverse global cultural heritage. Another major challenge for the twenty-first century is to develop more stable and effective systems of communication and collaboration.

1.3 Global changes versus local actions

Today, there is increasing attention paid to the consequences of climate change at local, national and global level. However, it is important to remember that the present climate change is the result of an accelerated increase in the amount of greenhouse gases (GHG) being pumped into the atmosphere during the last 150 years. As a result mostly of the burning of fossil fuels such as oil, gas and coal, we are now experiencing global climate changes (Figure 1.1).

Globally, the intensive use of fossil fuels, the changes in the biogeochemical cycles, and land use and cover changes during the industrial era have altered the physical composition of the Earth's ecosystems and their biogeochemical linkages (Vitousek, 1994; Houghton, 2008). Fossil fuel consumption accounts for 58 percent of the global annual emissions of anthropogenic gases (approx. 50 Gt CO_2-eq per year in 2005). Agriculture, forestry (including deforestation) and water resource management account for almost 34 percent of the total anthropogenic GHG emissions, while urbanization accounts for 8 percent (IPCC, 2007). However, all these activities are interlinked in a globalized world. For example, as Brazil becomes a world leading food exporter and China a world leading food importer (Brown, 2005), petroleum-based fertilizers are used to produce grains and soya beans, deforestation takes place at record rates, biodiversity decreases in tropical forests, and fossil fuel use increases to transport Brazilian food products and Chinese manufactured goods. Moreover, the global impact of these changes is represented by the increase in greenhouse gases and changes in the ocean–atmosphere heat exchange, which may be related to major changes in the general ocean circulation (Stocker, 2003; Collins *et al.*, 2005).

There are many examples of ecosystem degradation at local level. Some of the causes have been discussed above. Urban sprawl is one of them. The most obvious consequence is the loss of habitats, the primary cause of species loss around the

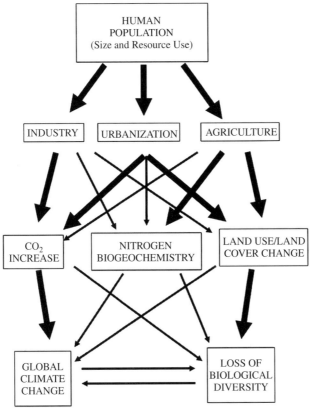

Fig. 1.1. Network of major relationships between global processes that caused global changes and global climate change in the industrial era.

world. In the tropics, deforestation has the greatest impact on habitat and species loss. Both are related, as wood from virgin forests is primarily used in urbanized areas. Agricultural expansion in tropical and temperate zones is also a major factor in ecosystem degradation: directly, due to land use changes, and indirectly, due to the nutrient pollution of aquatic ecosystems by urban and industrial agriculture sources. Invasive, non-indigenous species are also having severe negative impacts on biodiversity. These land cover changes in rural areas are driven by processes controlled by a population increasingly concentrated in urban zones.

The point here is that although the saying "think global and act local" (Eblen and Eblen, 1994; Keyes, 2006) is still effective in increasing participation in daily actions to improve the environment, the opposite statement "think local and act global" can be operative now, as globalization is an arena where actions developed by an individual or group in one part of the world can be matched by similar actions with the same objective in other parts of the world, fostering a participatory global network.

1.4 Scientific approaches to our environmental problems

The current human capacity to solve environmental problems is no match for our capacity to destroy and degrade the environment. For centuries, humans have exploited the Earth's natural resources, both above and below ground, without concern for the state of the original ecosystem or the impact of our footprint on nature. During the last part of the twentieth century, the "rule of the 3 Rs" (reduce, reuse, recycle) has gained popularity on different scales, mostly local, and many private and public businesses flourish through activities based on these practices. However, this is not enough. It is necessary to incorporate another "R", restore, to our usual land management and extractive industry practices. From the very beginning of a change in land use or extractive endeavor, it is necessary to plan for restoration. Otherwise, the natural capital decreases and may cause a permanent loss of the natural resources and the associated benefits provide by their interrelated biotic and abiotic components. It is a similar approach to that used in some forests, where wood extraction is planned in such a way that the tree population is renewed at the same rate with which trees are felled. Restoration takes into account the whole ecosystem – not only one species, but all the habitat components and the ecological processes networking them. This way of using natural resources provides more benefits for humans, since many ecosystem services that are not commodified present a direct benefit to the whole biosphere. This was referred to by the Ancient Greek philosophers, and Hardin (1968) depicted "the tragedy of the commons" as the conflict between individual and community interests in terms of nature's benefits. Now is the time to reclaim the benefits that the commons can provide (Barnes, 2006).

Creating environmental problems can take just seconds in comparison with the time required to repair them, which must include research and practice. The root causes of these problems may range from synthetic molecules and materials dispersed in the environment to disturbances in the absolute and relative abundance of plants and animals living in an ecosystem, as well as in the functioning of biogeochemical processes, in addition to direct alterations of their physical features. These are the causes of the current climate change, and they have impacts on a global scale that will require action on a global scale to absorb and buffer them. Worldwide forecasts include changes in the spatial distribution of ecosystems both in latitude and longitude, changes in species networks that are a substantial part of these ecosystems, and changes in the functioning of biogeochemical processes. It is important to take into account these forecasts in order to manage present ecosystems and bring them towards the desired functional level (Walker and Steffen, 1997; Adger *et al.*, 2007). The establishment of ecological corridors in natural or protected areas is an adaptive strategy which could be implemented on a global scale

(Simberloff *et al.*, 1992; Council of Europe, 2007). Although there are pros and cons (Rosenberg *et al.*, 1997), ecological corridors established in accordance with the scale of ecosystem and landscape fragmentation and disturbance are highly recommended in order to recover vital ecosystem goods and services (Taylor *et al.*, 1993; Bennett, 1999).

This new evolutionary scenario created by global changes may require adjustments in the structural and functional characteristics of our ecosystems. A fifth "R" should be added, that is, "rethink" the methods by which we harvest and consume natural resources. This is a major challenge which requires considering cultural perspectives and our impact on the biosphere.

1.5 Socio-ecological approaches to solving environmental problems

It is clear that the array of problems we face at the beginning of the twenty-first century have originated from the characteristics of the recent relationship between humankind and nature. As this is a global issue, an effective response requires a global cultural approach. Human development has been largely based on the exploitation and degradation of natural resources and ecosystems, and has created great imbalances between countries and social groups. Cultural aspects, such as religion and food habits, are part of this system and have been used to divide rather than to unite humankind. The challenge is not a single one – for example, reducing the number of people dying per year, increasing the availability of potable water, achieving a greater proportion of land under protected status, or obtaining a balanced mix of energy sources. The challenge is huge: adopting a sustainable way of development at all levels as soon as possible. Otherwise, the risks of social, economic and political instability will increase substantially and the negative impacts of global changes will affect everyone, particularly the more weak and vulnerable.

Using models to simulate future scenarios helps us to elucidate the advantages and disadvantages of alternative strategies. It is a difficult task because the results depend on the amount and quality of data. It may also be unrealistic or impossible to represent future scenarios accurately. However, modelling is very useful when making decisions on what not to do or which condition is more favorable for our purposes (Jørgensen, this volume). In general, people like to assume that the Earth has unlimited resources and that technology can solve most problems but this is unrealistic (Costanza, 2000). Skepticism is part of this vision, as it is one that relies less on technological change and more on social and community development, even though technologies can be used to achieve social goals. This perspective could make us behave as if any problem in this world could be solved by developing and using technologies. However, even though technologies are valuable and useful, they also produce unintended impacts and some of them may create more problems than they solve, especially if they are used incorrectly or at the expense of others.

There are many scenarios for human civilisation which range from a globally connected society with free market and liberalized trade that maintains a reactive approach towards environmental problems to a society based on regional structures that strengthen local institutions and helps maintain a proactive approach towards environmental problems. This last scenario, as defined by the Millenium Ecosystem Assessment (2005) is called the *Adapting Mosaic* scenario. Following the model of development as suggested by this scenario would yield the greatest benefits to society, improving all types of ecosystem services (provisioning, regulating, and cultural). As foreseen by the Millennium Ecosystem Assessment, other models of development may result in some positive services but also negative ones. The *TechnoGarden* model (globally connected world relying on technologies) improves material well-being, health and security, but it also reduces the cultural services of ecosystems (spiritual and religious, aesthetic, recreational), which have been referred to as the essence of existence. Either a globally connected world (*Global Orchestration* scenario) or a regionalized one (*Order from Strength* scenario) with a reactive approach to environmental problems and little respect for the commons will result in highly degraded ecosystem services by 2050, particularly in the developing countries.

The beauty of Planet Earth is that it is unique in the Universe; life on Earth is extremely diverse as the result of millions of years of evolution. Ecological processes take place in the channel provided by evolution (Margalef, 1997). At the same time, evolutionary conditions are the framework for ecological processes. So, although it is difficult to imagine the details of a future network of ecological, economic and social relationships, it is clear that adopting cautious principles and adaptive strategies will give us more advantage than continuing with development strategies based on the intensive exploitation and degradation of natural resources and ecosystems, and on dividing our society along economic and cultural lines, rather than on collaborating for a more even distribution of social and economic benefits while maintaining cultural diversity.

Adopting an adaptive management strategy for human development, as stated in the *Adapting Mosaic* scenario, will contribute to a more even distribution of benefits which may reduce the risks of social failures or crisis. This is also a desirable part (Costanza, 2003) of sustainable development, which has been repeatedly mentioned as a global objective for humans (United Nations, 1987). Sustainable development will be an essential part of the strategy required for true human progress, which should no longer be measured or interpreted as a simple set of statistical indexes, related to the market economy such as employment rates, interest rates, housing levels and other assorted indicators, which do not consider or integrate human (health, happiness), social (education, relationships, confidence), and natural capital (ecosystem services and values). These are real components of everyday life and must be considered in order that we can correctly evaluate the progress and quality of life (Costanza *et al.*, 2007; Talberth *et al.*, 2006).

All these aspects, (not those used in conventional indices of progress), must be incorporated both as part of the scenario and as part of the tools needed to keep within the evolutionary channel and to achieve the ultimate objectives of human actions on any scale (from individual to global) (Anielski and Rowe, 1999; Easterly, 2007).

1.6 The role of ecological restoration in solving our problems

Degraded ecosystems are the origin of a significant portion of the environmental problems in the world. Restoring degraded ecosystems can provide substantial benefits by counteracting or buffering climate change impacts (Table 1.1). In addition, biodiversity loss and ecosystem degradation will continue at high rates under any current development scenario, due to the momentum of global climate change impacts and continued extraction and consumption of natural resources. So, the ecosystem approach to restoration is critical in any global strategy leading to a desirable habitat for all species.

Ecological restoration is the process of assisting the recovery of an ecosystem that has been degraded, damaged, or destroyed (SER International Science & Policy Working Group, 2004). The exact and/or complete recovery of an ecosystem so that all past characteristics are present is acknowledged to be impossible, since nature is continually subjected to evolutionary forces. In fact, this issue is frequently raised when discussing the ethics of ecological restoration (Vidra and Shear, this volume). However, extending ecological restoration to the global scale presents a huge challenge for the recovery and improvement of ecosystem and cultural services (Costanza, this volume).

There is no universal formula for restoring ecosystems. Every ecosystem has unique characteristics and the objectives of restoration tend to be subjective and related to economic and social factors and technical possibilities (Comín *et al.*, 2005). However, some generalizations may be useful groupings of ecosystems known as ecotypes. For example, there are many types of coastal zones, but general rules to recover essential forcing functions within these, such as sea currents, the interplay between continental and marine flows, and the sand exchange between dune systems and beach zones can be stated (Comín *et al.*, this volume). Using facilitation relationships between species and establishing heterogeneous patterns for revegetating damaged forests could be standard rules for restoration, instead of planting trees of the same species in more or less regularly spaced zones. These standard rules can be applied in the recovery of the complex network of functions and values of any forest for the long term, rather than just the production of a certain volume of wood for a certain number of years before harvesting. Planning and managing urban growth with some sensitivity to ecological

Table 1.1. *Potential net carbon storage activities related to land use management and restoration included in Article 3.4 of the Kyoto Protocol (AI- Annex I countries, NAI- non-Annex I countries).*

Activity	Group*	Area (10⁶ ha)	Adoption/Conversión (percent of Area)		Rate C Gain (tC ha⁻¹ yr⁻¹)	Potential (Mt C yr⁻¹)	
			2010	2040		2010	2040
Improved management within a land use							
Cropland	AI	589	40	70	0.32	75	132
	NAI	700	20	50	0.36	50	126
Rice paddies	AI	4	80	100	0.10	<1	<1
	NAI	149	50	80	0.10	7	12
Agroforestry	AI	83	30	40	0.50	12	17
	NAI	317	20	20	0.22	14	28
Grazing land	AI	1297	20	20	0.53	69	137
	NAI	2104	20	20	0.80	168	337
Forest land	AI	1898	10	50	0.53	101	503
	NAI	2153	10	30	0.31	69	200
Urban land	AI	50	5	15	0.30	1	2
	NAI	50	5	15	0.30	1	2
Land-use change							
Agroforestry	AI	0	0	0	0	0	0
	NAI	630	20	30	3.1	391	586
Restoring severely degraded land	AI	12	5	15	0.25	<1	1
	NAI	265	5	10	0.25	3	7
Grassland	AI	602	5	10	0.8	24	48
	NAI	855	2	5	0.8	14	34
Wetland restoration	AI	210	5	15	0.4	4	13
	NAI	20	1	10	0.4	0	1
Off-site Carbon storage							
Forest products	AI					210	210
	NAI					90	90
Totals	AI					497	1063
	NAI					805	1422
	Global					1302	2485

*AI: Annex I countries; NAI: non-Annex I countries.
Source: Sampson and Scholes (2000).

processes (preserving the functional role of streams, hill slopes, groundwater flows) is a general approach which can be used for creating new urban sites, as well as for revitalizing degraded ones (Cunningham, 2002; Cunningham, 2008).

The real challenge of global ecological restoration is not simply achieving a defined goal. Objectives and scenarios can be established but nature, including humankind, imposes its own rules. Ecological restoration is not an objective in itself, but a tool to improve the environment and thus improve human well-being, defined not just as economic prosperity but also as healthy social relationships between groups (Higgs, this volume), and the integration of culture and nature. The challenge lies in incorporating the practice of ecological restoration in the activities of local communities, in regional administrations as part of their usual management planning, special interest groups or volunteers, and business enterprises and corporations and in using it for networking opportunities between social groups, communities and countries. With these aims in mind, the practice of ecological restoration is progressively developing a knowledge base increasingly rich in theories and experience, all based on restoration ecology, the science that provides the fundamentals for the practice of ecological restoration (Perrow and Davy, 2002; Van Andel and Aronson, 2006). The major strength of ecological restoration is its comprehensive approach towards nature. Even if the primary focus of a study or project is a single species, ecological restoration is concerned with recovering the degraded functions of the ecosystem that will allow the recovery of that species. It is based on the relationships between forcing functions, and abiotic and biotic characteristics. Ecological restoration efforts are primarily focused on assisting the recovery process in order to foster healthy communities and relationships, rather than directly trying to establish a number of individuals of a defined number of species or a community structure. This is a major difference between ecological restoration and other approaches and technologies adopted to solve environmental problems. The fields of bioengineering, ecological engineering, hydroecology, and conventional engineering, which are very useful at certain scales, can be combined with ecological restoration for the common objective of improving the status of degraded ecosystems, habitats and sites (Mitsch and Jørgensen, 2003).

In spite of its brief history, the scientific fundamentals provided by restoration ecology have contributed strong guidelines for the practice of ecological restoration. Three ecological restoration key issues are:

- identifying a reference system for comparison
- defining the trajectory for recovery
- using indicators to evaluate the results of the restoration.

Identifying an appropriate reference system to compare with a restored one may entail the search for historical information, comparisons with similar contemporaneous systems in good or degraded states, and the manipulation of an experimental system to assess the response to restoration actions (Aronson *et al.*, 1995; White and Walker, 1997). The validation of reference sites and systems may require the adoption of a flexible framework that can handle disturbances occurring on different spatial and temporal scales while global climate changes may alter the present or future character-istics of the reference systems. This framework can be provided by the theory of dynamic equilibrium as applied to ecosystems (Von Bertalanffy, 1968; Ritter, 2006) or by establishing a broad landscape structure that contains ecosystems connectivity and fluctuations (Harker *et al.*, 1999; Jentsch, 2007).

This is also the case when studying and evaluating the trajectory followed by a restored ecosystem which may not follow the temporal pattern of evolution exemplified by the reference systems because of the impact of large disturbances, such as climate change (Harris *et al.*, 2006). Ecological restoration has inherited the background of classical ecology in terms of secondary succession (Margalef, 1974; Egan and Howell, 2001) but it goes further, since ecological restoration is based on restoration ecology and it is used as an acid test to ensure that ecological processes are studied and restored in a holistic way (Bradshaw, 1987). It is contained in a system framework that may change in space, time and concept, as the reference and interest of the research study or practice changes (Winterhalder *et al.*, 2004).

It is impossible to predict all the characteristics of a restored ecosystem. Simply, there are too many, compared to those that are measurable in a restoration ecology research study or an ecological restoration project. But there are many indicators that can be used to estimate the success of restoration (Ruiz-Jaén and Aide, 2005). They range from the species level (Williams, 1993) to the community level (Rader *et al.*, 2008), and from site to habitat and landscape scales (Lirman and Millar, 2003; Baustian and Turner, 2005; Jentsch, 2007). They usually require long-term mon-itoring (Comín *et al.*, 2004; Klein *et al.*, 2007). Ecological restoration also involves social and economic aspects that are associated with the cultural characteristics of the place and the stakeholders (Higgs, this volume). So, assessing the results of ecological restoration projects must also take into account all these aspects (Comín *et al.*, 2005).

With the scientific and technical fundamentals provided by restoration ecology, ecological restoration also involves the social and economic aspects referred to above as key components for the implementation of practical projects. Topics related to the ethics of ecological restoration actions are currently being discussed (Light and Higgs, 1996; Vidra and Shear, this volume). Ecological restoration can also be a tool to establish cooperation between groups of people living in different countries, confronting long-term conflicts (Brandeis, this volume), and contributing to the reduction of poverty and inequality among groups (Maathai, 2004). Establishing ecological restoration as a current practice around the world represents a major global challenge for this century. If successful, it will contribute to a sustainable and more desirable Earth.

References

Adger, W. N., Agrawala, S., Mirza, M. M. Q. *et al.* (2007). Assessment of adaptation practices, options, constraints and capacity. Climate Change 2007: Impacts, Adaptation and Vulnerability. *Contribution of Working Group II to the Fourth Assessment Report of the Intergovernmental Panel on Climate Change*, ed. M. L. Parry, O. F. Canziani, J. P. Palutikof, P. J. van der Linden and C. E. Hanson. Cambridge University Press.

Anielski, M. and Rowe, J. (1999). *The Genuine Progress Indicator: 1998 Update*. San Francisco, CA: Redefining Progress.

Aronson, J., Dhillion, S., Le Floc'h, E. (1995). On the need to select an ecosystem of reference, however imperfect: A reply to Pickett and Parker. *Restoration Ecology*, **3** (1): 1–3.

Barnes, P. (2006). *A Guide to Reclaiming the Commons*. San Francisco: Berrett-Koehler Publishers.

Baustian, J. J. and Turner, R. E. (2005). Restoration success of backfilling canals in coastal Louisiana marshes. *Restoration Ecology*, **14** (4): 636–644.

Bennett, A. F. (1999) *Linkages in the Landscape: The Role of Corridors and Connectivity in Wildlife Conservation*. Gland, Switzerland and Cambridge: IUCN-International Union for Conservation of Nature.

Bradshaw, A. D. (1987). Restoration: An acid test for ecology. In *Restoration Ecology – A Synthetic Approach to Ecological Restoration*, ed. W. R. Jordan, M. E. Gilpin, and J. D. Aber. Cambridge University Press, pp. 23–30.

Brown, L. R. (2005). *Outgrowing the Earth: The food security challenge in an age of falling water tables and rising temperatures*. New York: Earth Policy Institute, W. W. Norton & Co.

Cohen, J. E. (1995). Population growth and earth's human carrying capacity. *Science*, **269**: 341–346.

Cohen, J. E. (2003). Human population: The next half century. *Science*, **302**: 1172–1175.

Collins, M. and the CMIP Modelling Groups (2005). El Niño or La Niña-like climate change? *Climate Dynamics*, **24**: 89–104.

Comín, F. A., Menéndez, M. and Herrera, J. A. (2004). Spatial and temporal scales for monitoring coastal aquatic ecosystems. *Aquatic Conservation: Freshwater and Marine Ecosystems*, **14**: 5–17.

Comín, F. A., Menendez, M., Pedrocchi, C. *et al.* (2005). Wetland restoration: Integrating scientific-technical, economic and social perspectives. *Ecological Restoration*, **23** (3): 182–186.

Costanza, R. (2000). Visions of alternative (unpredictable) futures and their use in policy analysis. *Conservation Ecology*, **4** (1): 5. (www.consecol.org).

Costanza, R. (2003). A vision of the future of science: Reintegrating the study of humans and the rest of nature. *Futures*, **35**: 651–671.

Costanza, R. (2008). Stewardship for a "Full" World. *Current History*, **107**: 30–35.

Costanza, R., Fisher, B., Ali, S. *et al.* (2007). Quality of life: An approach integrating opportunities, human needs, and subjective well-being. *Ecological Economics*, **61**: 267–276.

Council of Europe (2007). Sixth Ministerial Conference: "Environment for Europe." Belgrade, Serbia (10–12 October 2007). *The Pan-European Ecological Network: Taking Stock* (submitted by the Council of Europe through the Ad Hoc Working Group of Senior Officials). Background Document United Nations.

Crutzen, P. J. and Stoermer, E. F. (2000). The "Anthropocene." *Global Change Newsletter*, **41**: 12–13.

Cunningham, S. (2002). *The Restoration Economy: The Greatest New Growth Frontier*. San Francisco: Berrett-Koehler Publishers.

Cunningham, S. (2008). *Rewealth!* Columbus: McGraw-Hill.

Daily, G. *et al.* (1998). Global food supply: Food production, population growth, and the environment. *Science*, **281**: 1291–1292.

Easterly, W. (2007). Inequality does cause underdevelopment: Insights from a new instrument. *Journal of Development Economics*, **84**: 755–776.

Eblen, R. A. and Eblen, W. (1994). *The Encyclopedia of the Environment*. Boston: Houghton Mifflin Company.

Egan, D. and Howell, E. (2001). *The Historical Ecology Handbook. A Restorationist's Guide to Reference Ecosystems*. Washington DC: Island Press.

FAO-Food and Agriculture Organization, Economic and Social Department (2002). *World agriculture:towards 2015/2030*. Summary report. Rome: FAO.

Forbes, S. A. (1887). The lake as a microcosm. *Bulletin of the Scientific Association (Peoria, IL)*, 77–87. (Reprinted in 1925, *Bulletin of the Illinois State Natural History Survey*, **15**: 537–550).

Forel, F. A. (1892). *Lac Léman: Monographie Limnologique*. Lausanne: Rouge.

Global Footprint Network (2006). *Annual Report*. Oakland: Global Footprint Network.

Goklany, I. (2002). The globalization of human well-being. *Policy Analysis*, **447**: 1–20.

Haeckel, E. (1866). *Generelle Morphologie der Organismen*. Berlin: Reimer.

Hardin, G. (1968). The tragedy of the commons. *Science*, **162**: 1243–1248.

Harker, D., Libby, G., Harker, K., Evans, S. and Evans, M. (1999). *Landscape Restoration Handbook*. Boca Raton: Lewis Publications.

Harris, J. A., Hobbs, R. J., Higgs, E. and Aronson, J. (2006). Ecological restoration and global climate change. *Restoration Ecology*, **14** (2): 170–176.

Hawken, P. G. (1983). *The Next Economy*. New Cork: Henry Holt and Company.

Houghton, R. A. (2008).Carbon flux to the atmosphere from land-use changes: 1850–2005. In *TRENDS: A Compendium of Data on Global Change*. Oak Ridge: Carbon Dioxide Information Analysis Center, Oak Ridge National Laboratory, US Department of Energy. (http://cdiac.ornl.gov/trends/landuse/houghton/)

Hutchinson, G. E. (1965). *The Ecological Theater and the Evolutionary Play*. New Haven: Yale University Press.

IPCC-Intergovernmental Panel on Climate Change (2007). *Climate Change 2007: Synthesis report. Contribution of Working Groups I, II and III to The Fourth Assessment. Report of the Intergovernmental Panel on Climate Change* [Core Writing Team, Pachauri, R. K and Reisinger, A. (eds.)]. Geneva: IPCC.

IUCN-International Union for Conservation of Nature, UNEP-United Nations Environment Programme, WWF-World Wide Fund for Nature; FAO-Food and Agriculture Organization, and Unesco (1980). *World Conservation Strategy: Living Resource Conservation for Sustainable Development*. Gland: IUCN-UNEP-WWF-FAO-Unesco.

Jentsch, A. (2007). The challenge to restore processes in face of nonlinear dynamics – On the crucial role of disturbance regimes. *Restoration Ecology*, **15** (2): 334–349.

Keyes, R. (2006). *The Quote Verifier*. New York: Simon & Schuster.

Klein, L. R., Clayton, S. R., Alldredge, J. R. and Goodwin, P. (2007). Long-term monitoring and evaluation of the Lower Red River Meadow Restoration Project, ID, USA. *Restoration Ecology*, **15** (2): 223–239.

Kurzweil, R. (2005). *The Singularity Is Near*. New York: Viking.

Light, A. and Higgs, E. S. (1996). The politics of ecological restoration. *Environmental Ethics*, **18**: 227–247.

Lipschutz, R. D. (2004). *Global Environmental Politics. Power, Perspectives and Practice*. Washington DC: CQ Press.

Lirman, D. and Millar, M. W. (2003). Modeling and monitoring tools to assess recovery status and convergence rates between restored and undisturbed coral reef habitats. *Restoration Ecology*, **11** (4): 448–456.

Lowe, M. S. and Bowlby, S. R. (1992). Population and environment. In *Environmental issues in the 1990s*, ed. A. M. Mannion and S. R. Bowlby. Chichester: J. Wiley, 117–130.

Maathai, W. (2004). *The Green Belt Movement: Sharing the Approach and the Experience*. Herndon: Lantern Books.

Margalef, R. (1974). *Ecología*. Barcelona: Ed. Omega.

Margalef, R. (1997). *Our Biosphere. Excellence in Ecology 10*. Oldendorf/Luhe: Ecology Institute.

Meadows, D. H., Meadows, D. L., Randers, J. and Behrens, W. W. III. (1972). *The Limits to Growth*. New York: Universe Books.

Meadows, D. H., Randers, J. and Meadows, D. L. (2004). *Limits to Growth: The 30-Year Update*. White River Junction, VT: Chelsea Green Publishing Company.

MEA-Millenium Ecosystem Assessment (2005). *Ecosystems and Human Well-being: Synthesis*. Washington DC: Island Press.

Mitsch, W. J. and Jørgensen, S. E. (2003). *Ecological Engineering and Ecosystem Restoration*. Hoboken: J. Wiley & Sons Inc.

NAS-National Academy of Sciences, National Academy of Engineering, Institute of Medicine and National Research Council (1997). *Preparing for the 21st Century, The Environment and The Human Future*. Washington DC: National Academy of Sciences.

Perrow, M. R. and Davy, A. J. (eds.) (2002). *Handbook of Ecological Restoration*. Cambridge University Press.

Rader, R. B., Voelz, N. J. and War, J. V. (2008). Post-flood recovery of a macroinvertebrate community in a regulated river: Resilience of an anthropogenically altered ecosystem. *Restoration Ecology*, **16**: 24–33

Ramankutty, N., Foley, J. A. and Olejniczak, N. J. (2002). People on the land: Changes in global population and croplands during the 20th century. *Ambio*, **31**: 251–257.

Ritter, M. E. (2006). *The Physical Environment: An Introduction to Physical Geography*. (www.uwsp.edu/geo/faculty/ritter/geog101/textbook/title_page.html).

Rosenberg, D. K., Noon, B. R. and Meslow, E. C. (1997). Biological corridors: Form, functions, and efficacy. *BioScience*, **47**: 677–687.

Ruiz-Jaén, M. C. and Aide, T. M. (2005). Restoration success: How is it being measured? *Restoration Ecology*, **13** (3): 569–577.

Sampson, R. N. and Scholes, R. J. (2000). Additional human induced activities. In *Land use, land-use change and forestry (A special report of the IPCC)*, ed. R. L. Watson, I. A. Noble, B. Bolin, N. H. Ravindranath, D. J. Verardo and D. J. Dokken. Cambridge University Press, 181–281.

Simberloff, D., Farr, J. A., Cox, J. and Mehlman, D. W. (1992). Movement corridors: Conserving bargains or poor investments. *Conservation Biology*, **6** (4): 493–504.

SER-Society for Ecological Restoration International Science & Policy Working Group (2004). *The SER International Primer on Ecological Restoration*. Tucson, AZ. (www.ser.org).

Stocker, T. F. (2003). Changes in the global carbon cycle and ocean circulation on the millenium time scale. In *Global Climate: Current Research and Uncertainties in the Climate System*, ed. X. Rodó and F. A. Comín. Berlin: Springer, 129–152.

Stutz, J. and Mintzer, E. (2006). *The Affluence Paradox: More Money Is Not Making Us Happier. A Review of Statistical Evidence*. Boston: Tellus Institute.

Talberth, J., Cobb, C. and Slattery, N. (2006). *The Genuine Progress Indicator 2006*. Oakland, CA: Redefining Progress.

Taylor, P. D., Fahrig, L., Henein, K. and Merriam, G. (1993). Connectivity as a vital element of landscape structure. *Oikos*, **68**: 571–73.

United Nations Documents Cooperation Circles (1987). *Report of the World Commission on Environment and Development. General Assembly Resolution 42/187, 11 December 1987*. New York: United Nations.

Van Andel, J. and Aronson J. (eds.) (2006). *Restoration Ecology*. Malden: Blackwell.

Vitousek, P. M. (1994). Beyond global warming: Ecology and global change. *Ecology*, **75**: 1861–1876.

Vitousek, P. M., Erlich, P. R., Erlich, A. H. and Matson, P. A. (1986). Human appropriation of the products of photosynthesis. *BioScience*, **36**: 368–373.

Vitousek, P. M., Mooney, H. A., Lubchenco, J. and Melillo J. M. (1997). Human domination of Earth's ecosystems. *Science*, **277**: 494–499.

Von Bertalanffy, L. (1968). *General System Theory: Foundations, Development, Applications*. New York: George Braziller.

Warming, E. (1909). *Oecology of Plants, an Introduction to the Study of Plant-Communities*. Oxford: Clarendon Press.

Walker, B. and Steffen, W. (1997). An overview of the implications of global change for natural and managed terrestrial ecosystems. *Conservation Ecology*, **1** (2): 2. (www.consecol.org/vol1/iss2/art2/)

White, P. S. and Walker, J. L. (1997). Approximating nature's variation: Selecting and using reference information in Restoration Ecology. *Restoration Ecology*, **5** (4): 338–349.

Williams, K. S. (1993). Use of Terrestrial Arthropods to Evaluate Restored Riparian Woodlands. *Restoration Ecology*, **1** (2): 107–116.

Winterhalder, K., Clewell, A. F. and Aronson, J. (2004). Values and science in ecological restoration. A response to Davis and Slobodkin. *Restoration Ecology*, **12** (1): 4–7.

World Health Report (2007). *A Safer Future: Global Public Health Security in the 21st Century*. Geneva: World Health Organization.

2

The global carbon cycle: current research and uncertainties in the sources and sinks of carbon

DARIO PAPALE AND RICCARDO VALENTINI

2.1 Introduction

The plant life of the Earth acts as a carbon sink, which is currently absorbing carbon from the atmosphere at a rate of about 2.9 Gt C/yr through photosynthesis. This sink is of similar magnitude and opposite in sign to deforestation, particularly in the tropical regions, which is releasing carbon to the atmosphere at an annual rate of 2.0 Gt C/yr (Gruber *et al.*, 2004). There is, however, no scientific evidence that the Earth's flora will continue to absorb carbon at the same rate in the future. In this respect, any measure taken by means of enhanced terrestrial sinks to offset fossil fuel emissions will have a temporary effect which will allow us to buy between 15 and 100 years in which to develop alternative longer-term CO_2 reduction strategies, depending on the mean residence time of the protected sinks. After that, any measure for emission reduction will have to be permanent and achieved by technological means.

Despite the temporary nature of biological carbon sequestration, ecological restoration through enhancement of ecosystem services can have a positive effect on climate stabilization. Indeed, compared to reforestation and afforestation, human-induced natural regeneration and/or conservation and improvement of existing vegetation can be more effective in enhancing both carbon sequestration and biological diversity, and protecting soil against degradation and desertification. However biological carbon sequestration is also affected by climate and it is influenced by its variability. Also, disturbance regimes (e.g., fires, wind storms, pests, etc.), although often dominated by human activities, are triggered by climate variability and extremes. In this chapter we discuss the current carbon cycle variability and the different methods and approaches available to quantify the terrestrial ecosystem carbon budget.

2.2 Magnitude: "slow in, fast out"?

The carbon cycle is essential to the Earth system, being inextricably coupled with climate, the water cycle, nutrient cycles, and the production of biomass by

Ecological Restoration: A Global Challenge, ed. Francisco A. Comin. Published by Cambridge University Press.
© Cambridge University Press 2010.

photosynthesis on land and in the oceans. This production ultimately sustains the entire animal kingdom, including humans through their dependence on food and fibre. Hence, a proper understanding of the global carbon cycle is a central component both for understanding the environmental history of our planet and its human inhabitants, and for predicting and steering their joint future (Falkowski *et al.*, 2000; IPCC, 2001).

The Vostok ice core record (Figure 2.1) illustrates the speed and severity of the current changes in the global carbon cycle. For around the last half-million years up to about 200 years ago, the climate system has operated within a relatively constrained range of temperatures and concentrations of atmospheric carbon dioxide (CO_2) and methane (CH_4). In the pre-industrial world, atmospheric CO_2 concentrations oscillated in roughly 100,000 year-cycles between 180 and 280 ppmV (parts per million by volume), as the CO_2 climate system pulsed between glacial and interglacial states. Today we have about 100 CO_2 ppmV more than the maximum registered in the interglacial periods. The ice core record also shows clearly that atmospheric composition and climate (especially temperature) are closely linked.

Comparison of the Vostok record with contemporary measurements of atmospheric CO_2 concentration reveals that we are currently far from this regular domain of glacial-interglacial cycling. Atmospheric CO_2 concentrations are now nearly 100 ppmV higher than at the interglacial maximum, and the increase rate has been at least 10 and possibly 100 times faster than at any other time in the last 420,000 years. Concentrations of other greenhouse gases, including methane

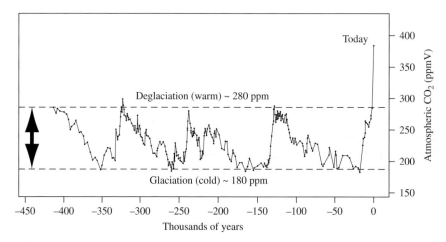

Fig. 2.1. Atmospheric CO_2 concentration in ppmV in the past 450,000 years from the Vostok ice core record. Modified from Petit *et al.*, 1999.

and nitrous oxide, are increasing at comparable rates. These increases are unquestionably due to human activities, and we are already experiencing the consequences on climate: a temperature record for the past millennium indicates that the contemporary climate system is responsive to the changing greenhouse gas concentrations in the atmosphere. Far greater changes are predicted over the next century, with a degree of confidence that has increased substantially between the different Assessments of the Intergovernmental Panel on Climate Change (IPCC, 2001; IPCC, 2007).

These changes mean that the Earth system is now well outside the operating range of the carbon cycle over the past half million years. The change has been unidirectional and the change rate unprecedented. Thus, we have pushed the Earth system into uncharted territory.

The role of humans in the carbon cycle is not new, as we have influenced it for thousands of years through agriculture, forestry, trade and energy use in industry and transport. However, only over the last two or three centuries have these activities become sufficiently widespread and far-reaching to match the great forces of the natural world. Moreover, human societies are not unidirectional drivers of change: they are also impacted by changes in the carbon cycle and climate, and they respond to these impacts in ways that have the potential to feed back on the carbon cycle itself. Social, cultural, political and economic institutions provide the context in which this complex human-environment system is evolving, and in which attempts will be made to alter the future direction of the carbon cycle. A key present example is the effort to manage greenhouse gas emissions into the global atmospheric commons, through global governance initiated by the United Nations Framework Convention on Climate Change (UNFCCC).

On time scales of up to a few thousand years, atmospheric CO_2 is controlled by exchanges with the carbon reservoirs of the oceans and the terrestrial biosphere. Considering the terrestrial ecosystem carbon sink of about 2.9 Gt C/yr and the ocean uptake that accounts for 2.4 Gt C/yr on one side, and the carbon emissions to the atmosphere due to tropical deforestation (2.2 Gt C/yr) and fossil fuel use and cement production (6.3 Gt C/yr) on the other side, we are able to confirm the current growth rate of carbon accumulation in the atmosphere (3.2 Gt C/yr) as measured by long-term atmospheric stations.

Although land use change emissions and uptake rates are comparable in magnitude to the ocean uptake rate, their timing is radically different. Indeed, fossil fuel and land use change (mainly deforestation) emissions are fast processes, occurring in time scales of the order of a few days, while land and ocean uptakes occur on a time scale of many years and centuries. This is clearly indicated by tropical deforestation, where destroying the forest by fire or logging can take a few days, with a large release of carbon stored in the biomass,

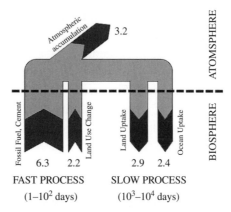

Fig. 2.2. The slow in – fast out paradigm. The difference in carbon emissions and sinks between biosphere and atmosphere is not only in the quantities: processes (units as Gt C/yr) like fossil fuel and land use change emissions occur with speeds three to four times greater than the uptake processes.

and the recovery of the same carbon per unit of land would require a much longer time (Figure 2.2).[1]

The carbon exchange between the terrestrial biosphere and the atmosphere is a key driver of the current carbon cycle. The carbon flows between the pools and the pools contents themselves have rich spatial and temporal structures, affecting strongly the global carbon budget and climate forcing.

In terms of quantities, the gross primary production (GPP) of an ecosystem is defined as the gross uptake of CO_2 that is used for photosynthesis. The synthesis of new plant tissue from CO_2, water and nutrients and the maintenance of living tissues are energy-demanding processes where some photo-assimilated compounds are lost from the ecosystem as autotrophic respiration (Ra). The amount of carbon that is not used for respiration and is available for other processes is defined as net primary production (NPP):

$$NPP = GPP - Ra$$

The bulk of NPP is allocated to the production of biomass in several ecosystem components: foliage, wood, root and reproductive organs.

The residence time of carbon, the time between fixation in photosynthates and the return to the atmosphere through the respiration process differs among NPP components. Carbon incorporated in wood has a residence time varying from years to centuries, whereas the carbon deposited in foliage and fine roots has residence times

---

[1] While this book was in preparation a new assessment of the global carbon budget for the years 2000–2008 was made available (Le Querie **et al.**, 2009). The current carbon budget reveals an uptake of vegetation equal to 2.7 + −1.0 Gt C/yr, ocean uptake of 2.3 + −0.5 Gt C/yr and emissions by land use change of 1.4 + −0.7 Gt C/yr and from fossil fuel and cement of 7.7 + −0.5 Gt C/yr.

varying from months to years. Each year, part of the standing biomass dies and becomes part of carbon pools (with different residence times) in the litter and/or soil layers. These carbon pools are subjected to decomposition by microbial activity, a process defined as heterotrophic respiration (Rh). The difference between NPP and Rh is the net ecosystem production (NEP):

$$NEP = NPP - Rh$$

The sum of Rh and Ra represents the total ecosystem respiration (TER), which includes a fraction coming from the soil (soil respiration, which is both hetero-trophic and autotrophic because it comes from microbial decomposition and roots respiration) and a fraction coming from the above ground components (mainly autotrophic respiration). NEP is thus determined by the difference between the two large components GPP and TER, generally on an annual base, and is considered to be the primary source of the observed interannual variability in atmospheric accu-mulation of CO_2 (Peylin *et al.*, 2005)

$$NEP = GPP - TER$$

NEP represents the net ecosystem carbon sink or source due to physiological processes. This balance, however, differs from the long-term carbon balance mainly because of CO_2 losses by disturbances and management, such as forest fires, harvests and/or erosion (Luyssaert *et al.*, 2007). Therefore, these CO_2 losses should be accounted for to obtain the long-term ecosystem carbon balance. The net biome productivity (NBP) (Schulze *et al.*, 2000) is defined as:

$$NBP = NEP - CO_2 \; losses \; due \; to \; disturbances$$

Tropical forests contain the largest carbon pool of terrestrial biota, and also have the largest carbon net primary productivity (Table 2.1). Boreal forests, although extended over a larger area, represent a less significant carbon pool and have lower net primary production. They are also greatly affected by climate-induced disturbances. Temperate forests show high net carbon ecosystem uptake rates (NEP), with a significant latitudinal trend, depending on forest management and structure. Tropical savannas and grasslands cover the largest percentage of the Earth's land mass and they represent a substantial fraction of the global NPP, but they are also experiencing their largest change in land coverage, due to human activities. Cultivated land is almost neutral in terms of carbon balance, compared to forest and savannas. Most of the carbon fixed is harvested and respired elsewhere. However, any conversion from forest area to cultivated land is a major source of carbon loss to the atmosphere.

The net ecosystem production (NEP) of the tropical forests has been the subject of debate in recent years. Indeed, canopy forest fluxes indicate rates of uptake ranging from 100 to 700 g C/m^2 per year. Globally, these fluxes could mean an assimilation

Table 2.1. *Distribution of terrestrial carbon pools and net primary productivity.*

Biome	Area (million km^2)	Total carbon pool (Gt C))	Total C NPP (Gt C/yr)
Tropical forests	17.5	340	21.9
Temperate forests	10.4	139	8.1
Boreal forests	13.7	57	2.6
Arctic tundra	5.6	2	0.5
Mediterranean shrublands	2.8	17	1.4
Crops	13.5	4	4.4
Tropical savannas and grasslands	27.6	79	14.9
Temperate grasslands	15	6	5.6
Deserts	27.7	10	3.5
Ice	15.5	–	–
Total	149.3	652	62.6

Source: Saugier *et al.* (2001); Sabine *et al.* (2004).

of 1.75–12.25 Gt C/yr. However, there are ecological issues influencing the ultimate destination of this carbon, such as biotic and abiotic disturbances which are not captured by inventories and reduce the net carbon sequestration rates of tropical rainforest. Possible carbon lateral transport and degassing from rivers is another source of carbon usually not included in carbon budgets of tropical biomes. Thus the overall net carbon sink of tropical forest biomes is still not well quantified.

Tropical deforestation releases 1–2 Gt C/yr, and is one of the major driving forces for the increase in atmospheric CO_2. Carbon is also lost from terrestrial vegetation in the form of non-CO_2 hydrocarbons such as those of biogenic VOCs (volatile organic compounds). Globally VOC emissions can be equivalent to 0.2 to 1.4 Gt C/yr (Kesselmeier *et al.*, 2002). For grasslands and cultivated land, large emissions of other greenhouse gases such as N_2O, CH_4 also occur and these emissions are strongly affected by land use changes.

2.3 Methodology: how to investigate carbon cycle patterns and processes

To understand, quantify and predict global carbon cycle magnitudes and patterns, a large array of tools and methods used by different research communities exists. For example, tools such as satellite data, air sampling networks, atmospheric transport models and inverse numerical methods ("top-down" approaches) allow us to study the strength and location of the global-scale and continental-scale carbon sources and sinks. Surface monitoring, process studies and biogeochemical models ("bottom-up" approaches) provide estimates of land–atmosphere and ocean–atmosphere carbon fluxes at finer spatial scales, and help us to examine the mechanisms that control fluxes at these regional and ecosystem scales.

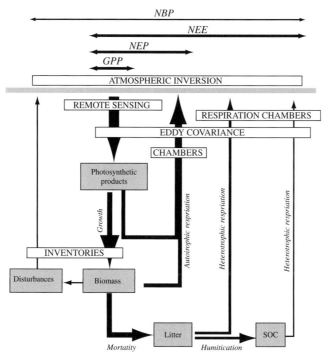

Fig. 2.3. Processes involved in the carbon cycle and methods available for assessing the different components. Modified from Schulze *et al.*, 2000.

A further set of investigations is concerned with activities which humans initiate purposefully and which are geared to manipulating components of the carbon cycle, including methods of maximizing and tracking the net sequestration of carbon in the terrestrial biosphere, technologies for ocean sequestration, and the monitoring and minimization of the massive emissions from fossil fuel combustion. In the search for an integrated understanding of the carbon cycle and its current changes, each of these tools provides a piece of the puzzle. However, no single existing technique, approach, discipline or region can explain the whole picture (Figure 2.3). Hereafter, we describe briefly the different methods and techniques available to measure or assess the carbon cycle and budgets at different scales, stressing particularly uniqueness and limits.

2.3.1 Chambers

Chambers are used commonly to measure soil respiration fluxes, although they are also used to measure the CO_2 fluxes of plant components, particularly stems, branches and leaves. The method of use is based on the change in the CO_2 concentration in a known volume or the flux due to the release (respiration) or absorption (photosynthesis) by the component measured. The data acquired with this method are "point measurements" and for this reason, are difficult to use to assess carbon balance at ecosystem level, particularly for soil respiration due to the high spatial heterogeneity. This method is, however, the only one available for measuring component fluxes directly and, for this reason, it is very useful in model structure definition and parameterization.

Chambers are also used to assess the two soil efflux components: autotrophic and heterotrophic respirations. Various methods have been proposed (Hanson *et al.*, 2000; Kuzyakov, 2006), for example living roots are removed from a sample of soil which is then put in a root-exclusion bag and replaced in the ground; these bags do not allow the passage of roots, but they do not interfere with the movement of water and nutrients. After a period of adaptation, needed because an external disturbance is introduced in the system, it is possible to measure total (from control plots) and heterotrophic (from the soil into the bag) soil respirations, and then calculate the autotrophic respiration.

2.3.2 Inventories

Inventory assessments of carbon budgets are based on well-established techniques of conventional inventory, particularly for forest ecosystems. The assessment of stock changes is accomplished by applying available, consecutive forest inventory data of a region (Ney *et al.*, 2002) or by deducting drain from increment (Penman *et al.*, 2003).

In most of the European and North American countries these inventories are carried out with a sample-based approach, whereby currently over 400,000 plots in Europe are measured at intervals of five or ten years. These regional or national forest inventories have been designed primarily to measure traditional forest vari-ables like forest area, stem volume of growing stock and net annual stem wood increment. To derive carbon budgets from forest inventory data, allometric relation-ships, expansion factors and average carbon densities for each tree species or forest type are used to convert the measured stem wood volume into the corresponding above- and below-ground biomass and carbon pools (Kauppi *et al.*, 1992; Schroeder *et al.*, 1997; Löwe *et al.*, 2000; Watson *et al.*, 2000; Coomes *et al.*, 2002; Linder *et al.*, 2004). European forest inventories are, however, not yet harmonized and the applied inventory and carbon conversion methods may vary between countries and, sometimes, even between administrative regions within countries, especially for the biomass expansion factors needed to convert stem volumes into dry weights for the different biomass components (Köhl *et al.*, 2000; Löwe *et al.*, 2000).

Further completion of the inventory-based carbon budget is achieved by measur-ing or modelling the soil carbon pool. Although several countries have made soil carbon assessments based on national forest soil surveys (Baritz and Strich, 2000; Pignard *et al.*, 2000) the heterogeneity of the soil and the long time required for measurable changes in the soil carbon content to occur make the sampling approach difficult and uncertain. Another possibility is to apply soil models to estimate the soil carbon budget (Liski *et al.*, 2002; Karjalainen *et al.*, 2003) but this clearly introduces new uncertainties, and the results are strongly dependent on the correctness of the model assumptions.

In the case of grasslands and croplands, the inventory approach has fewer difficulties than in the case of forest ecosystems, although the uncertainty in the final budget remains quite large (Belelli-Marchesini *et al.*, 2007).

2.3.3 Eddy covariance

The eddy covariance method, a micrometeorological technique, has proven to be a valuable direct measure of net carbon and water fluxes between ecosystems and the atmosphere (Baldocchi *et al.*, 1988; Aubinet *et al.*, 2000) over short and long timescales (hours to years). Currently, more than 350 eddy covariance towers are acquiring data around the world (Baldocchi, 2008), covering different climate conditions, land use and land cover changes: some of them have been running continuously for more than 10 years.

The eddy covariance technique is based on high-frequency (10–20 Hz) measurements of wind speed and direction, as well as CO_2 and H_2O concentrations at a point over the canopy using a three-axis sonic anemometer and a fast-response infrared gas analyzer (Aubinet *et al.*, 2000; Aubinet *et al.*, 2003). Assuming perfect turbulent mixing these measurements are typically integrated over periods of half an hour (Goulden *et al.*, 1996) building a basis for calculating carbon and water balances from daily to annual time scales.

The area contributing to the measurements (the footprint) is a function of the wind speed, wind direction and the difference between the measurement point (top of the eddy covariance tower where the equipment is located) and the height of the canopy height, but generally it is an area varying between a few hundred meters and one kilometer around the tower. For this reason, the eddy covariance technique is the only method available today to measure net carbon exchanges at ecosystem level, and a number of synthesis activities (Valentini *et al.*, 2000; Law *et al.*, 2002) and studies on the effects of climatic gradients on ecosystem production (Reichstein *et al.*, 2007b) have been done based only on eddy covariance data.

Moreover, it is possible to statistically partition the net carbon fluxes measurements into their two major components: gross primary production and ecosystem respiration (Reichstein *et al.*, 2005), thus allowing a better interpretation of the fluxes in terms of ecosystem processes.

Eddy covariance measurements are currently processed using standardized techniques (Papale *et al.*, 2006; Moffat *et al.*, 2007) and used in various synthesis activities; the data registered in sites around the world are the basis for analysis about the global carbon balance and cycle.

Another interesting application of the same technique is the use of an aircraft instead of the tower. The idea of using aircrafts to measure gas exchange between the land surface and the atmosphere was developed more than 20 years ago (Lenschow *et al.*, 1981; Desjardins *et al.*, 1982) and airborne eddy flux observations may give results no less accurate than those from flux-tower operation. These measurements provide an almost instantaneous picture of the fluxes over an area of a few square kilometres that can be very useful in particular for model output validation. Despite initial successes (Oechel *et al.*, 1998) and continuous technological improvements, such a novel technique still requires large-scale validation so that we can better understand its reliability in regional flux estimates (Doran *et al.*, 1992; Desjardins *et al.*, 1995; Crawford *et al.*, 1996; Desjardins *et al.*, 1997; Oechel *et al.*, 1998).

2.3.4 *Remote sensing*

Remote sensing data from satellites and recently also from airborne sensors provide continuous and detailed observations of the state of terrestrial vegetation that are spatially extensive and relatively uniform in quality. The observed vegetation properties can be related to properties of the terrestrial carbon cycle, either empirically or through vegetation modelling (Linder *et al.*, 2004).

Spectral reflectance differences caused by the spectrally-dependent fractional absorption of light in vegetation, particularly in the red and near-infrared wavelengths, are closely related to the abundance of chlorophyll present, and the structure of leaves in the target (Myneni *et al.*, 1995). The widely used normalized difference vegetation index (NDVI) measures the normalized contrast between red and near-infrared reflectance of solar radiation. It can be converted into the fraction of photosynthetically absorbed active radiation (FAPAR) and to the leaf area index (LAI), using either simplified field-calibrated relations or more complex numerical models of light scattering in vegetation (Myneni and Williams, 1994; Myneni *et al.*, 1997). FAPAR and LAI are important data related to the capacity of the vegetation to fix atmospheric carbon and are used in carbon models (e.g., Nemani *et al.*, 2003).

Satellite observation is also an important tool for following the phenological phases of deciduous vegetation (Reed *et al.*, 1994; Moulin *et al.*, 1997) and this information may be used to constrain the seasonal dynamics of carbon cycle computations. In addition, remote sensing is uniquely suited for spatial mappings of vegetation types and land use patterns. These, as well as the land use and land cover changes, are of primary interest to carbon budget computations.

Applications of remote sensing data for the assessment of the terrestrial carbon cycle mostly employ diagnostic carbon balance simulation models, ranging from simple light use efficiency models to complex biogeochemical models (Fischer *et al.*, 1997). In radiation use efficiency (RUE) models, RUE is the basis for deriving carbon uptake per unit of observed FAPAR, either as synthesized dry matter or as net primary production integrating autotrophic carbon losses (Prince, 1991; Gower *et al.*, 1999; Potter *et al.*, 2003). Some RUE models provide only estimates of vegetation NPP or GPP Assessing NEP over a year requires further estimations of ecosystem respiration and some time integration. Respiration terms can be measured using other methods, such as airborne eddy covariance, or simulated with respiration models (Reichstein *et al.*, 2005).

Radar remote sensing also offers interesting perspectives in the context of carbon balance studies, in particular when used as a unique tool for biomass and carbon stock determination (Le Toan *et al.*, 1992; Ranson *et al.*, 1997; Hyyppä *et al.*, 2000). In fact, this important information is generally available only at plot level and since there is high heterogeneity between the different countries, the use of radar remote sensing can help to create standard and homogeneous global forest biomass maps.

In conclusion, the use of remotely sensed data in the study and determination of regional and global carbon cycle and carbon stocks is important and promising, and new high-resolution satellite sensors become available every year. One of the major examples in recent years is the MODIS sensors that are routinely used to estimate important vegetation characteristics like LAI and FAPAR (Myneni *et al.*, 1997) and NPP (Running *et al.*, 2000) on a global scale.

2.3.5 *Modelling and data integration*

Chamber measurements, inventory data, eddy covariance technique and remote sensing data provide unique information about the terrestrial carbon cycle; however, each of these methods is able to give answers only for a particular sector or component of the global carbon balance. To have a global picture of all the processes and budgets it is crucial to integrate these different methods in a common framework.

Models can use site level measurements and remotely sensed data as constraints or input in carbon cycle simulations. Modelling approaches are differentiated by modelling schemes, assumptions, input used, spatial and time resolutions and parameters.

Generally it is possible to distinguish between two main modelling approaches in carbon balance determination: bottom-up and top-down. Bottom-up models simulate the terrestrial ecosystem carbon cycle starting from some knowledge of ecosystem functioning that can be either a reproduction of the eco-physiological processes or direct measurements of one or more processes. By contrast, top-down modelling is based on carbon dioxide concentration measurements in the atmosphere and the use of atmospheric circulation models to attribute carbon sources or sinks at ground level.

There is a range of bottom-up carbon models that include calculations of carbon exchange between the atmosphere, the terrestrial biosphere, and soils (Melillo *et al.*, 1993; Cramer *et al.*, 1999). These models can operate at very different scales in time and space. Data on climate and soils are the usual input, but some models also need additional information or constraints, such as land use data or phenological observations derived from remote sensing. It is useful to classify bottom-up ecosystem models along a gradient from data-oriented to process-oriented models. Data-oriented models typically do not try to reproduce the different processes involved in the carbon cycle, but more or less directly relate relevant known variables (e.g., vegetation type, meteorological variables, remote sensing data) to desired unknown quantities (e.g., carbon balance components). Typical examples are artificial neural networks, multivariate regression approaches and regression trees. For example, investigators have used artificial neural networks to assess European forest carbon balance using as input meteorological data, land cover maps and a remote sensing vegetation index, and have "parameterized" the model with the relation between input and output eddy covariance measurements collected in sixteen forest sites in Europe (Papale and Valentini, 2003). At the other extreme lie completely process-oriented models that explicitly represent all important state variables and processes of the studied system and only variables that are outside the system boundary are fed into the modelling system. This other extreme is hardly ever reached, since there are always state variables that are either prescribed or held constant in terrestrial ecosystem models, but dynamic global vegetation models come closest (Cramer *et al.*, 2001), where both biogeochemical cycles (mostly carbon, nitrogen, and water) and vegetation distribution are described. Examples of these models are the LPJ (Lund-Potsdam-Jena) dynamic global vegetation model (Sitch *et al.*, 2003; Bondeau *et al.*, 2007; Zaehle *et al.*, 2007), which describes the coupled carbon and water cycles and a distribution of ten plant functional types (PFTs) with different physiological (C3 or C4 photosynthesis), phenological (deciduous, evergreen), and physiognomic (tree, grass) attributes, based on bioclimatic limits for plant growth and regeneration and plant-specific parameters that govern plant competition for light and water, and the ORCHIDEE biosphere model (Viovy, 1996; Krinner *et al.*, 2005), which is a dynamic global vegetation model (DGVM) designed to be coupled with ocean-atmosphere circulation models in order to simulate the global carbon, energy and water fluxes on a half-hourly basis for twelve different plant functional types.

Between the two poles of data- and process-oriented models is a continuum of hybrid data/process-oriented models where internal model state-variables are partly replaced by external inputs, and partly internally represented. An example, on the data-oriented side, are the semi-empirical remote sensing-based models where gross primary production is assessed on the basis of incoming radiation, meteorological

Table 2.2. *Main characteristics of the different bottom-up modelling approaches.*

Approach	Advantages	Major concerns
Data-oriented	Few parameters Few or no assumptions Computational efficiency	Data precision Data representativeness (spatial, thematic)
Process-oriented	Possible attribution of sinks and/or sources to human activity Predictions and/or extrapolations possible	Validity of process assumptions

conditions, FAPAR (derived from remote sensing) and radiation use efficiency coefficients (Veroustraete *et al.*, 2002; Nemani *et al.*, 2003). Process-oriented hybrid models are commonly structured in such a way as to represent most of the state variables except one aspect. Biome-BGC for example is a terrestrial ecosystem model describing the carbon, nitrogen and water cycles (Running and Gower, 1991; Thornton *et al.*, 2002) for seven biomes but does not simulate the distribution and dynamics of vegetation.

Table 2.2 summarizes the main characteristics of the two bottom-up modelling approaches (data-oriented and process-oriented). The main differences lie in the assumptions concerning ecosystem functioning: these are few or not present in the data-oriented models, while in the process-oriented approaches they are important, so that the correctness of these models is a function of the validity of the assumptions; in contrast, data-oriented models are strongly dependent on the representativeness of the dataset used in the model parameterization or application and, as compared to the process-oriented models, are hardly usable for extrapolation in time or space.

The top-down modelling approach, also referred to as inverse modelling, can give estimates of regional or continental net carbon balance but without information about which ecosystems are contributing to the sink (Janssens *et al.*, 2003). In this approach, small spatial gradients of tropospheric absolute CO_2 concentration (possibly together with isotopic composition of CO_2 and O_2) are used to provide a spatial surface map of CO_2 sources and sinks. Global transport and dispersion models link sources and sinks of CO_2 to atmospheric CO_2 concentrations. When such a model is inverted, CO_2 concentrations are used to calculate the best fitting set of regional CO_2 sources and sinks. To assess the terrestrial ecosystem carbon balance it is, however, necessary to take into consideration and subtract the fossil fuel and disturbance emissions contribution. Anthropogenic emissions are known from governmental statistics or may be estimated from the concentrations of carbon monoxide (CO) or $^{14}CO_2$, while emissions due to ecosystem disturbances can be assessed using other models, statistics or remote sensing (Linder *et al.*, 2004).

The different modelling approaches described above (top-down, bottom-up, process-oriented, data-oriented) have advantages and disadvantages that are quite complementary and this makes the choice of the models to use more difficult. For this reason, it is important to try to integrate the different approaches with the measurements available which can be used for model parameterization and validation of both results and model assumptions. The multiple constraint approach, where bottom-up and top-down modelling are used to assess the carbon balance of the same area, gives allows the investigator to evaluate the uncertainty in each model by comparing the different results obtained. As an example, in Figure 2.4 the approach used in the CarboEuropeIP European project (GOCE-CT-2003–505572) is represented: measurements at different scales (from point to single tree, and at ecosystem and regional levels) and with different time resolutions are integrated within ecosystem models (bottom-up) and atmospheric transport models (top-down) in a common framework. The comparison and integration of model outputs, together with verification activities based on the measured fluxes and processes, makes this scheme an excellent example of a multiple constraint and model-data integration exercise.

2.4 Variability: how variable is the terrestrial part of the carbon cycle?

2.4.1 Global scale variability

The accumulation rate of CO_2 in the atmosphere as documented by atmospheric stations around the globe for the past forty years can vary by a factor of two from one year to the next, that is by up to several Gt C per year. Such changes reflect interannual shifts in the carbon uptake of land and oceans of the same magnitude as the average uptake itself. Changes in fossil fuel emissions tend to be smooth in time: year-to-year variations are less than 4 percent of the total. Both ocean data and global ocean carbon models suggest that air-sea carbon fluxes tend to be stable. The land biosphere may thus be responsible for the majority of observed CO_2 interannual growth rate variation.

Climate variability could be a major driver of changes in the carbon balance of ecosystems. Both photosynthesis and respiration respond to changing radiation, temperature and precipitation. Analysis of temperature and vegetation index data (a satellite measurement that relates to photosynthesis) suggests that different ecosystems respond in different ways, with significant lagged responses. Responses to temperature appear significant in boreal and temperate forests. Direct flux measurements have also shown that a small change in the timing of the growing season, for instance an unusually warm spring, can trigger a large shift in the annual carbon balance of forests. Along the same lines, it has also long been observed that crop yield can vary significantly from one year to the next. In addition to the carbon fluxes of respiration and photosynthesis, the disturbance regime, characterized by

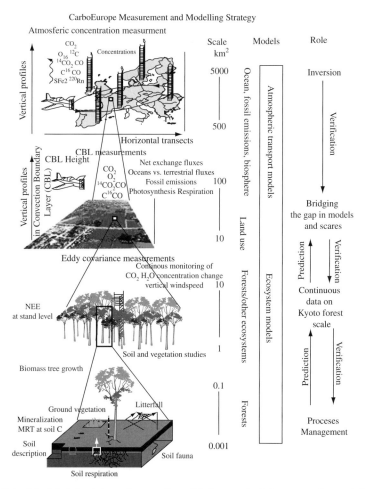

CarboEurope Measurement and Modelling Strategy

Fig. 2.4. Model-data integration and multiple constraint approach used in the CarboEurope project (www.carboeurope.org).

the occurrence of insect infestations, or fires in Mediterranean or boreal forests, is also influenced by changes in climate and in climate variability.

2.4.2 The heat–drought wave of 2003 in Europe: an example of extreme events

Carbon fluxes between terrestrial ecosystems and atmosphere and, as a consequence, the productivity of terrestrial ecosystems, are strongly influenced by temperature and water availability as confirmed by the relationship between average climate and forest productivity (Lieth, 1973; Gholz *et al.*, 1990; Scurlock and Olson, 2002).

At global level, the year 2003 is associated with one of the largest atmospheric CO_2 rises on record (Jones and Cox, 2005). This was particularly significant, as there was no accompanying large El Niño event as is normally the case in years with high CO_2 increases. It has been suggested that drought periods in mid-latitudes of the northern hemisphere caused an additional carbon release to the atmosphere large enough to modify dominant ENSO (El Niño Southern Oscillation) responses in 1998–2002 (Zeng *et al.*, 2005). Atmospheric model inversions have indicated that during these years the northern hemisphere mid-latitudes went from being a sink (0.7 Gt C/yr) to being close to neutral (Vetter *et al.*, 2007).

Western and central Europe experienced extremely hot and dry conditions during the summer of 2003 which caused a decrease in carbon sequestration over large areas (Ciais *et al.*, 2005; Schindler *et al.*, 2006; Reichstein *et al.*, 2007a), although areas normally cooler than the the rest of Europe, such as the Alps, experienced an increase in carbon sequestration (Jolly *et al.*, 2005). Ciais *et al.* (2005) showed in a model study that the carbon flux anomaly was mainly caused by a drop in the gross primary production rather than increased ecosystem respiration, resulting in an anomalous source of 0.5 Gt of carbon to the atmosphere through July–September 2003 relative to the average carbon flux between 1998–2002.

The drought and heat wave that occurred in summer 2003 in Europe was exceptional, both in duration and in its wide distribution across Europe (Ciais *et al.*, 2005) and had significant impacts on European ecosystems, including large losses in agriculture. However, due to global change, more frequent and severe droughts are expected in some regions of the globe, mainly in the northern hemisphere (IPCC, 2007). The effects of such extreme events are poorly documented because of their limited occurrence under past and actual climatic conditions, but since terrestrial ecosystems seem to respond to droughts with an increased carbon flux to the atmosphere, frequent droughts may lead to a faster increase in atmospheric carbon dioxide concentration and accelerate global warming. Thus, understanding the response of ecosystems to large scale drought events is an important issue.

The European 2003 summer heat wave provided a good opportunity to examine the response of a wide range of terrestrial ecosystems to extreme climatic conditions and also to test the ability of the various observing and modelling systems to characterize the biosphere's response to climate anomalies.

Reichstein *et al.* (2007a) conducted an integrated analysis using eddy covariance monitoring sites, spatial remote sensing observations and ecosystem modelling to get complementary views on the effect of the 2003 heat wave on the productivity and carbon balance of the European biosphere. In this analysis they showed, particularly through the use of the eddy covariance data, that the strong negative anomaly in GPP was not primarily due to the high temperatures, but rather to drought stress and also that ecosystem respiration was affected and strongly reduced by the heat–drought wave.

Vetter *et al.* (2007) did a more systematic study to analyze the magnitude of the carbon flux anomaly and ecosystem processes behind it using a modelling approach. In their simulations, they used seven vegetation models of different complexity and types (process-oriented and data-oriented) and a set of common and standardized input data. The model outputs have generally similar patterns in terms of NEP anomalies in Europe (Plate 2.1), with western and central Europe showing a strong NEP negative anomaly, more prominent in western parts than in the central region, and eastern Europe-western Russia, where the anomalous conditions were cold and wet, with a positive anomaly in terms of net primary production.

When it comes to the question of which ecosystem processes are responsible for the NEP anomalies, the different models gave different indications, with some of them indicating an increase of TER as responsible for the reduction in productivity, and others indicating a decrease of GPP as the driving factor. These differences are due principally to different soil structure and hydrology descriptions in the models used, particularly regarding water availability and drought effects, as well as to differences in crop management simulation. The results, however, confirm that by comparing different models one can have a clearer picture of the uncertainty in the predictions. The seven models used in the analysis by Vetter *et al.* (2007) estimated an additional carbon emission from the European terrestrial ecosystem to the atmosphere of 0.27–0.03 Gt C in response to the 2003 drought, which is comparable to the 0.5 Gt C calculated also using the modelling approach by Ciais *et al.* (2005), but for a shorter reference period.

Granier *et al.* (2007) in their analysis based on eddy covariance data from twelve forest sites in Europe confirmed the wide spatial distribution of drought stress over Europe, with a maximum intensity within a large band extending from Portugal to north-east Germany. They found a relation between relative extractable water in soil (REW = actual extractable water (mm) / potential maximum extractable water (mm)) and water vapor fluxes and net ecosystem production. Both the NEP components, gross primary production and total ecosystem respiration, decreased when REW dropped below 0.4 and 0.2 for GPP and TER respectively. Scientists have analaysed the effect of drought on tree growth at three sites and they found that growth reduction was more pronounced in 2004, the year following the drought, particularly in beech trees, one of the species most affected by heat waves. Such lag effects on tree growth should be considered an important feature in forest ecosystems, which may enhance vulnerability to more frequent climate extremes.

Extreme climatic events like this are unique and interesting opportunities for study and analysis, with the aims of increasing understanding and improving our knowledge and ability to predict the possible effects of climate change on terrestrial ecosystem functioning and their capacity as sinks of atmospheric carbon dioxide in the future.

2.5 Ecological restoration and the Kyoto Protocol

During the United Nations Framework Convention on Climate Change COP6bis meeting in Bonn and the subsequent COP7 meeting in Marrakesh, both held in 2001, a historic consensus on the Kyoto Protocol rules was achieved. One of the most controversial issues was how much credit developed countries could receive towards their Kyoto targets through the use of sinks. The meeting agreed that the eligible activities for Annex I countries would include not only afforestation, reforestation and deforestation (Article 3.3), but also management of forests, croplands and grazing lands, and revegetation. A net-net accounting approach was agreed for croplands and grazing lands, while individual country quotas for forest management were agreed. As a result of these decisions, sinks will account for a fraction of the emission reductions that can be counted towards the Kyoto targets.

On the other hand, the current rates of deforestation and associated fluxes of GHG which amount to about 2.2 Gt C/yr, were not considered in the Kyoto Protocol as potential targets for carbon credits (Figure 2.5). In fact, halting deforestation in tropical forests alone would be equivalent to a global saving of more than the amount of emission reductions required by the Kyoto protocol by 2012.

The result of such agreements with different levels of regulation is the creation of a complex landscape made of different patches where different policies apply. Scientifically, this makes the accounting very difficult and almost impossible to verify, since the atmosphere (the final target of such agreements) is an integrator of all the possible combinations of natural and human processes, both Kyoto and non-Kyoto relevant. This situation may also create possible perverse incentives, for example, to deforest a particular area of land while reforesting another piece of land in a developing country and claiming carbon credits only for the latter. Recently, negotiation has started to include avoided deforestation in the carbon market, but a final agreement has not yet been reached.

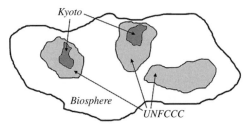

Fig. 2.5. Different domains of the UNFCCC which include a bigger proportion of the land than the Kyoto Protocol.

Furthermore, the Kyoto Protocol, by setting country-specific caps on forest management carbon credits and not allowing developing countries to benefit by such activities (only afforestation and reforestation is allowed through the Clean Development Mechanism (CDM), prevents ecological restoration from being an effective method also in the context of climate mitigation.

However, ecological restoration can be a really effective tool for enhancing carbon sequestration on land and to improve several other ecosystem services, such as conservation of biodiversity and mitigation of land degradation and desertification processes.

From this perspective, it is absolutely necessary that the principles of ecological restoration should be included both in Article 3.3 of the Kyoto Protocol (afforestation and reforestation) and in Article 3.4 (forest management and revegetation). For the former, enhancement and promotion of natural regeneration of forest in degraded lands can be more effective than artificial plantation, where in most cases, invasive species and fertilizer loads impacting the local environment. In the latter case, restoration of pristine forests, their conservation and sustainable management can be very effective in tropical, temperate and boreal zones for the improvement of carbon sequestration. In particular, avoiding tropical deforestation can be very important for reducing the emissions of GHGs in the atmosphere and, at the same time, for the conservation of natural habitats. Fortunately, the developments of the COP13 Bali UNFCC conference held in 2007 have increased the chances of including deforestation avoidance in the mechanisms of carbon credit markets, although there are still a number of unresolved issues related to accounting and financial regulations.

In conclusion, the terrestrial carbon cycle is much more vulnerable and dependent on both human and climate disturbances than the oceanic one. Human activities in particular dominate the carbon cycle on land and make it more unpredictable. In this perspective, climate policy and the future targets of emission reductions should take into consideration the risk of terrestrial carbon outbursts into the atmosphere. The protection and conservation of terrestrial ecosystems is of paramount importance today. Ecological restoration can provide an important contribution to the stabilization and improvement of terrestrial carbon uptake.

References

Aubinet, M., Grelle, A., Ibrom, A. *et al.* (2000). Estimates of the annual net carbon and water exchange of forests: The EUROFLUX methodology. *Advances in Ecological Research*, **30**: 113–175.

Aubinet, M., Clement, R., Elbers, J. A. *et al.* (2003). Methodology for data acquisition, storage and treatment. In: *Fluxes of Carbon, Water and Energy of European Forests*, ed. R. Valentini. Berlin: Springer-Verlag. Ecological Studies 163, pp. 9–36.

Baldocchi, D. (2008). "Breathing" of the terrestrial biosphere: Lessons learned from a global network of carbon dioxide flux measurement system. *Australian Journal of Botany*, **56**: 1–26.

Baldocchi, D. D., Hicks, B. B. and Meyers, T. P. (1988). Measuring biosphere–atmosphere exchanges of biologically related gases with micrometeorological methods. *Ecology*, **69**: 1331–1340.

Baritz, R. and Strich, S. (2000). Forests and the National Greenhouse Gas Inventory of Germany. *Biotechnology, Agronomy, Society and Environment*, **4**: 267–271.

Belelli-Marchesini, L., Papale, D., Reichstein, M., Vuichard, N., Tchebakova, N. and Valentini, R. (2007). Carbon balance assessment of a natural steppe of southern Siberia by multiple constraint approach. *Biogeosciences*, **4**: 581–595.

Bondeau, A., Smith, P. C., Zaehle, S. *et al.* (2007). Modelling the role of agriculture for the 20th century global terrestrial carbon balance. *Global Change Biology*, **13**: 679–706.

Ciais, P., Reichstein, M., Viovy, N. *et al.* (2005). Europe-wide reduction in primary productivity caused by the heat and drought in 2003. *Nature*, **437**: 529–533.

Coomes, D. A., Allen, R. B., Scott, N. A., Goulding, C. and Beets, P. (2002). Designing systems to monitor carbon stocks in forests and shrublands. *Forest Ecology and Management*, **164**: 89–108.

Cramer, W., Kicklighter, D. W., Bondeau, A. *et al.* and "Potsam'95" (1999). Comparing global models of terrestrial net primary productivity (NPP): Overview and key results. *Global Change Biology*, **5**: 1–15.

Cramer, W., Bondeau, A., Woodward, F. I. *et al.* (2001). Global response of terrestrial ecosystem structure and function to CO_2 and climate change: Results from six dynamic global vegetation models. *Global Change Biology*, **7**: 357–373.

Crawford, T. L., Dobosy, R. J., McMillen, R. T., Vogel, C. A. and Hicks, B. B. (1996). Air-surface exchange measurement in heterogeneous regions: Extending tower observations with spatial structure observed from small aircraft. *Global Change Biology*, **2**: 275–285.

Desjardins, R. L., Brach, E. J., Alno, P. and Schuepp, P. H. (1982). Aircraft monitoring of surface carbon dioxide exchange. *Science*, **216**: 733–735.

Desjardins, R. L., MacPherson, J. I., Neumann, H., Denhartog, G. and Schuepp, P. H. (1995). Flux estimates of latent and sensible heat, carbon dioxide, and ozone using an aircraft-tower combination. *Atmospheric Environment*, **29**: 3147–3158.

Desjardins, R. L., MacPherson, J. I.; Mahrt, L. *et al.* (1997). Scaling up flux measurements for the boreal forest using aircraft-tower combinations. *Journal of Geophysical Research*, **102**: 29125–29133.

Doran, J. C., Barnes, F. J., Coulter, R. L. *et al.* (1992). The Boardman Regional Flux Experiment. *Bulletin of the American Meteorological Society*, **73**: 1785–1795.

Falkowski, P. Scholes, R. J., Boyle, E. *et al.* (2000). The global carbon cycle: A test of our knowledge of earth as a system. *Science*, **290**: 291–296.

Fischer, A., Kergoat, L. and Dedieu, G. (1997). Coupling satellite data with vegetation functional models: Review of different approaches and perspectives suggested by the assimilation strategy. *Remote Sensing Review*, **15**: 283–303.

Gholz, H. L., Ewel, K. C. and Teskey, R. O. (1990). Water and forest productivity. *Forest Ecology and Management*, **30**: 1–18.

Goulden, M. L., Munger, J. W., Fan, S. M., Daube, B. C. and Wofsy, S. (1996). Measurements of carbon sequestration by long-term eddy covariance: Methods and a critical evaluation of accuracy. *Global Change Biology*, **2**: 169–182.

Gower, S. T., Kucharik, C. J. and Norman, J. M. (1999). Direct and indirect estimation of leaf area index, fapar, and net primary production of terrestrial ecosystems. *Remote Sensing of Environment*, **70**: 29–51.

Granier, A., Reichstein, M., Breda, N. *et al.* (2007). Evidence for soil water control on carbon and water dynamics in European forests during the extremely dry year 2003. *Agricultural and Forest Meteorology*, **143**: 123–145.

Gruber, N., Friedlingstein, P., Field, C. B., *et al.* (2004). The vulnerability of the carbon cycle in the 21st century: An assessment of carbon-climate-human interaction. In: *The Global Carbon Cycle*, ed. C. B. Field and M. R. Raupach. Washington DC: Island Press, pp. 45–76.

Hanson, P. J., Edwards, N. T., Garten, C. T. and Andrews, J. A. (2000). Separating root and soil microbial contributions to soil respiration: A review of methods and observations. *Biogeochemistry*, **48**: 115–146.

Hyyppä, J., Hyyppä, H., Inkinen, M., Engdahl, M., Linko, S. and Zhu, Y. H. (2000). Accuracy comparison of various remote sensing data sources in the retrieval of forest stand attributes. *Forest Ecology and Management*, **128**: 109–120.

IPCC-Intergovernmental Panel on Climate Change (2001). Climate change 2001: The Scientific Basis. *Contributions of Working Group I to the Third Assessment Report of the Intergovernmental Panel on Climate Change*, ed. J. T. Houghton *et al.* Cambridge University Press.

IPCC-Intergovernmental Panel on Climate Change (2007). *Climate Change 2007: Synthesis Report. Summary for Policymakers*. (www.ipcc.ch).

Janssens, I., Freibauer, A., Ciais, P. *et al.* (2003). Europe's terrestrial biosphere absorbs 7 to 12% of European anthropogenic CO_2 emissions. *Science*, **5625**: 1538–1542.

Jolly, W. M., Dobberlin, M., Zimmermann, N. E. and Reichstein, M. (2005). Divergent vegetation growth responses to the 2003 heat wave in the Swiss Alps. *Geophysical Research Letters*, **32**: L18409, doi:10.1029/2005GL023252.

Jones, C. D. and Cox, P. (2005). On the significance of atmospheric CO_2 growth-rate anomalies in 2002–03. *Geophysical Research Letters*, **32**: L14816, doi:10.1029/2005GL023027.

Karjalainen, T., Pussinen, A., Liski, J, *et al.* (2003). Scenario analysis of the impacts of forest management and climate change on the European forest sector carbon budget. *Forest Policy and Economics*, **5**: 141–155.

Kauppi, P. E., Mielikäinen, K. and Kuusela, K. (1992). Biomass and carbon budget of European forests, 1971 to 1990. *Science*, **256**: 70–74.

Kesselmeier, J., Ciccioli, P., Kuhn, U. *et al.* (2002). Volatile organic compound emissions in relation to plant carbon fixation and the terrestrial carbon budget. *Global Biogeochemical Cycles*, **16**: 73(71)-73(79).

Köhl, M., Päivinen, R. and Traub, B. (2000). Harmonisation and standardisation in multi-national environmental statistics – mission impossible? *Environmental Monitoring and Assessment*, **63**: 361–380.

Krinner, G., Viovy, N., de Noblet-Ducoudré, N. *et al.* (2005). A dynamic global vegetation model for studies of the coupled atmosphere-biosphere system. *Global Biogeochemical Cycles*, **19**: GB1015, doi:10.1029/2003 GB002199.

Kuzyakov, Y. (2006). Sources of CO_2 efflux from soil and review of partitioning methods. *Soil Biology & Biochemistry*, **38**: 425–448.

Law, B. E., Falge, E., Gu, L. *et al.* (2002). Environmental controls over carbon dioxide and water vapor exchange of terrestrial vegetation. *Agricultural and Forest Meteorology*, **113**: 97–120.

Le Toan, T., Beaudoin, A., Riom, J. and Guyon, D. (1992). Relating forest biomass to SAR data. *IEEE Transactions on Geoscience and Remote Sensing*, **30**: 403–411.

Le Quéré, C., Raupach, M. R., Canadell, J. G., Marland, G. *et al.* (2009). Trends in the sources and sinks of carbon dioxide. *Nature Geoscience*, **2**: 831–836.

Lenschow, D. H., Pearson Jr, R. and Stankov, B. B. (1981). Estimating the ozone budget in the boundary layer by use of aircraft measurements of ozone eddy flux and mean concentration. *Journal of Geophysical Research*, **86**: 7291–7297.

Lieth, H. (1973). Primary production: Terrestrial ecosystems. *Human Ecology*, **1**: 303–332.

Linder, M., Lucht, W., Bouriaud, O. *et al.* (2004). *Specific study on forest greenhouse gas budget. CarboEurope-GHG Concerted Action – Synthesis of the European Greenhouse Gas Budget.* Report 8/2004. Specific study 1. Viterbo, Italy.

Liski, J., Korotkov, A. V., Prins, C. F. L. and Karjalainen, T. (2002). Increasing carbon stocks in the forest soils of western Europe. *Forest Ecology and Management*, **169**: 159–175.

Löwe, H., Seufert, G. and Raes, F. (2000). Comparison of methods used within Member States for estimating CO_2 emissions and sinks according to UNFCCC and EU Monitoring Mechanism: Forest and other wooded land. *Biotechnology, Agronomy, Society and Environment*, **4**: 315–319.

Luyssaert, S., Inglima, I., Jung, M. *et al.* (2007). CO_2 balance of boreal, temperate, and tropical forests derived from a global database. *Global Change Biology*, **13**: 2509–2537.

Melillo, J. M., McGuire, A. D., Kicklighter, D. W., Moore, B. I., Vorosmarty, C. J. and Schloss, A. L. (1993). Global climate change and terrestrial net primary production. *Nature*, **363**: 234–240.

Moffat, A. M., Papale D., Reichstein M. *et al.* (2007). Comprehensive comparison of gap-filling techniques for eddy covariance net carbon fluxes. *Agricultural and Forest Meteorology*, **147**: 209–232.

Moulin, S., Kergoat, L., Viovy, N. and Dedieu, G. G. (1997). Global-scale assessment of vegetation phenology using NOAA/AVHRR satellite measurements. *Journal of Climate*, **10**: 1154–1170.

Myneni, R. B., and Williams, D. L. (1994). On the relationship between FAPAR and NDVI. *Remote Sensing of Environment*, **49**: 200–211.

Myneni, R. B., Hall, F., Sellers, P. and Marshak, A. (1995). The interpretation of spectral vegetation indices. *IEEE Transactions on Geoscience and Remote Sensing*, **33**: 481–486.

Myneni, R. B., Nemani, R. R. and Running, S. W. (1997). Estimation of global leaf area index and absorbed PAR using radiative transfer models. *IEEE Transactions on Geoscience and Remote Sensing*, **35**: 1380–1393.

Nemani, R. R., Keeling, C. D., Hashimoto, H. *et al.* (2003). Climate-driven increases in global terrestrial net primary production from 1982 to 1999. *Science*, **300**: 1560–1563.

Ney, R. A., Schnoor, J. L. and Mancuso, M. A. (2002). A methodology to estimate carbon storage and flux in forestland using existing forest and soils databases. *Environmental Monitoring and Assessment*, **78**: 291–307.

Oechel, W. C., Vourlitis, G. L., Brooks, S. B., Crawford, T. L. and Dumas, E. J. (1998). Intercomparison between chamber, tower, and aircraft net CO_2 exchange and energy fluxes measured during the Arctic system sciences land-atmosphere-ice interaction (ARCSS-LAII) flux study. *Journal of Geophysical Research*, **103**: 28 993–29 003.

Papale, D. and Valentini, R. (2003). A new assessment of European forests carbon exchanges by eddy fluxes and artificial neural network spatialization. *Global Change Biology*, **9**: 525–535.

Papale, D., Reichstein, M., Aubinet, M. *et al.* (2006). Towards a standardized processing of Net Ecosystem Exchange measured with eddy covariance technique: Algorithms and uncertainty estimation. *Biogeosciences*, **3**: 571–583.

Penman, J., Gytarsky, M., Hiraishi, T. *et al*. (eds). (2003). *Good Practice Guidance for Land Use, Land-Use Change and Forestry*. Hayama, Kanagawa: IPCC-The Institute for Global Environmental Strategies for the IPCC and IPCC National Greenhouse Gas Inventories Programme (www.ipcc-nggip.iges.or.jp/public/gpglulucf/gpglulucf_contents.html).

Petit, J. R., Jouzel, J.; Raynaud, D. *et al*. (1999). Climate and atmospheric history of the past 420,000 years from the Vostok ice core, Antarctica. *Nature*, **399**: 429–436.

Peylin, P., Bousquet, P., Le Quéré, C. *et al*. (2005). Multiple constraints on regional CO_2 flux variations over land and oceans. *Global Biogeochemical Cycles*, **19**: GB1011, doi:10.1029/2003 GB002214.

Pignard, G., Dupouey, J. L., Arrouays, D. and Loustau, D. (2000). Carbon stocks estimates for French forests. *Biotechnology, Agronomy, Society and Environment*, **4**: 285–289.

Potter, C., Klooster, S., Myneni, R. B., Genovese, V., Tan, P. N. and Kumar, V. (2003). Continental-scale comparisons of terrestrial carbon sinks estimated from satellite data and ecosystem modeling 1982–1998. *Global and Planetary Change*, **39**: 201–213.

Prince, S. D. (1991). A model of regional primary production for use with course resolution satellite data. *International Journal of Remote Sensing*, **12**: 1313–1330.

Ranson, K. J., Sun, G., Lang, R. H., Chauhan, N. S., Cacciola, R. J. and Kilic, O. (1997). Mapping of boreal forest biomass from spaceborne synthetic aperture radar. *Journal of Geophysical Research*, **102**: 29 599–29 610.

Reed, B. C., Brown, J. F., VanderZee, D., Loveland, T. R., Merchant, J. W. and Ohlen, D. O. (1994). Measuring phenological variability from satellite imagery. Journal of Vegetation *Science*, **5**: 703–714.

Reichstein, M., Falge, E., Baldocchi, D. *et al*. (2005). On the separation of net ecosystem exchange into assimilation and ecosystem respiration: Review and improved algorithm. *Global Change Biology*, **11**: 1424–1439.

Reichstein, M., Ciais, P., Papale, D. *et al*. (2007a). Reduction of ecosystem productivity and respiration during the European summer 2003 climate anomaly: A joint flux tower, remote sensing and modelling analysis. *Global Change Biology*, **13**: 634–651.

Reichstein, M., Papale, D., Valentini, R. *et al*. (2007b). Determinants of terrestrial ecosystem carbon balance inferred from European eddy covariance flux sites. *Geophysical Research Letters*, **34**: L01402, doi:10.1029/2006GL027880.

Running, S. W., and Gower, S. T. (1991). Forest-BGC, a general model of forest ecosystem processes for regional applications II. Dynamic carbon allocation and nitrogen budgets. *Tree Physiology*, **9**: 147–160.

Running, S. W., Thornton, P. E., Nemani, R. R. and Glassy, J. M. (2000). Global terrestrial gross and net primary productivity from the Earth Observing System. In *Methods in Ecosystem Science*, eds. O. E. Sala, R. B. Jackson, H. A. Mooney and R. W. Howarth. New York: Springer Verlag, pp. 44–57.

Sabine, C. L., Heiman, M., Artaxo, P. *et al*. (2004). Current status and past trends of the carbon cycle. In *The Global Carbon Cycle: Integrating Humans, Climate, and the Natural World*, eds. C. B. Field and M. R. Raupach. Washington DC: Island Press, pp. 17–44.

Saugier, B., Roy, J. and Mooney, H. A. (2001). Estimations of global terrestrial productivity: Converging toward a single number? In *Terrestrial Global Productivity*, eds. J. Roy, B. Saugier and H. A. Mooney. San Diego: Academic Press, pp. 541–555.

Schindler, D., Türk, M. and Mayer, H. (2006). CO_2 fluxes of a Scots pine forest growing in the warm and dry southern upper Rhine plain, SW Germany. *European Journal of Forest Research*, **125**: 201–212.

Schroeder, P., Brown, S., Mo, J., Birdsey, R. and Cieszewski, C. (1997). Biomass estimation for temperate broadleaf forests of the United States using inventory data. *Forest Science*, **43**: 424–434.

Schulze, E. D., Wirth, C. and Heimann, M. (2000). Managing forests after Kyoto. *Science*, **289**: 2058–2059.

Scurlock, J. M. O. and Olson, R. (2002). Terrestrial net primary productivity – A brief history and a new worldwide database. *Environmental Reviews*, **10**: 91–109.

Sitch, S., Smith, B., Prentice, I. C. *et al.* (2003). Evaluation of ecosystem dynamics, plant geography and terrestrial carbon cycling in the LPJ dynamic vegetation model. *Global Change Biology*, **9**: 161–185.

Thornton, P. E., Law, B. E., Gholz, H. L. *et al.* (2002). Modelling and measuring the effects of disturbance history on carbon and water budgets in evergreen needleleaf forests. *Agricultural and Forest Meteorology*, **113**: 185–222.

Valentini, R., Matteucci, G., Dolman, A. J. *et al.* (2000). Respiration as the main determinant of carbon balance in European forests. *Nature*, **404**: 861–865.

Veroustraete, F., Sabbe, H. and Eerens, H. (2002). Estimation of carbon mass fluxes over Europe using the C-Fix model and Euroflux data. *Remote Sensing of Environment*, **83**: 376–399.

Vetter, M., Churkina, G., Jung, M. *et al.* (2007). Analyzing the causes and spatial pattern of the European 2003 carbon flux anomaly in Europe using seven models. *Biogeosciences*, **5**: 561–583.

Viovy, N. (1996). Interannuality and CO_2 sensitivity of the SECHIBA-BGC coupled SVAT-BGC model. *Physiological Chemistry Earth*, **21**: 497–498.

Watson, R. T., Noble, I. R., Bolin, B., Ravindranath, N. H., Verardo, D. J. and Dokken, D. J. (2000). *Land Use, Land-Use Change, and Forestry. Special Report of the Intergovernmental Panel on Climate Change.* Cambridge University Press.

Zaehle, S., Bondeau, A., Carter, T. R. *et al.* (2007). Projected changes in terrestrial carbon storage in Europe under climate and land-use change, 1990–2100. *Ecosystems*, **10**: 380–401.

Zeng, N. Quian, N. Rödenbeck and C.Heimann, M. (2005). Impact of 1998–2002 midlatitude drought and warming on terrestrial ecosystem and the global carbon cycle. *Geophysical Research Letters*, **32**: L22709.

3

Using international carbon markets to finance forest restoration

JOHANNES EBELING, MALIKA VIRAH-SAWMY AND
PEDRO MOURA COSTA

3.1 Introduction

The rapid growth of the global economy during the last century, coupled with a tripling of the world's population, has taken a heavy toll on the natural environment. Although it is difficult to quantify the current extent of degraded land globally, some estimate this to include around one third of the total global land area (Eswaran *et al.*, 2001). It is clear that land degradation has been occurring on a scale which makes it one of the most pressing environmental problems of our time. In the same way, the need for ecological restoration is not confined to isolated sites but applies to vast areas around the world.

Deforestation, often followed by soil erosion, is a key contributor to land degradation on a global scale. At the current rate, 13 million hectares of forest are lost annually, and this consists almost entirely of deforestation in tropical developing countries (FAO, 2006). There is thus an urgent need for active ecological restoration through reforestation in developing countries. A combination of factors, including poor environmental regulation and enforcement, limited governance capacity, constrained financial resources, and lack of scientific information, make the problem of land degradation in general, and deforestation in particular, especially pressing for developing countries.

Arguably, the majority of the degradation of natural ecosystems occurs because the environmental services they provide to the global community are not valued in monetary terms. There have been frequent calls by researchers, as well as politicians and activists, for people to realize the value of these ecosystem services and ensure their protection; however, this has rarely led to the translation of these global benefits into tangible incentives for local actors who often face strong economic incentives to engage in unsustainable land-use practices. An international financial mechanism exists that could help enable the restoration of formerly forested lands through private sector funds, which is not the case for many other forms of land degradation. International carbon markets could, at least in principle, bridge the notorious gap

Ecological Restoration: A Global Challenge, ed. Francisco A. Comin. Published by Cambridge University Press.
© Cambridge University Press 2010.

between global climatic and conservation benefits and local opportunity costs of sustainable land use through the monetary valuation of the carbon sequestered by growing forests. Project-based mechanisms under the framework of the Kyoto Protocol and within voluntary carbon markets contain concrete opportunities for financing ecological restoration through reforestation. Such carbon forestry projects have been in the making for over a decade (Stuart and Moura Costa, 1998) and, despite temporary setbacks, they are now growing in importance in carbon markets.

This chapter begins by providing a brief background to existing carbon markets and mechanisms for funding carbon forestry activities in developing countries. This includes an overview of the current status of forestry in the regulatory and voluntary carbon markets. The main part of the chapter evaluates the match between the aims and mechanisms of carbon forestry and those of ecological restoration in theory and in practice. The impact of the requirements of carbon standards for forestry projects and the market incentive structure is analyzed with regard to their impact on ecological restoration aims. Potential synergies between carbon crediting and ecological restoration are explored and illustrated with practical examples, and these are contrasted with challenges posed by the economics of carbon markets, regulations under the Kyoto Protocol, and real-world limitations. The concluding section gives some considerations regarding the realistic potential for achieving ecological restoration goals through carbon forestry while suggesting ways to enhance synergies. Throughout, the focus is on carbon forestry under the Clean Development Mechanism (CDM), voluntary carbon markets and, where relevant, emerging non-Kyoto compliance markets. While most of the contents of this chapter would also apply to the use of carbon finance for avoided deforestation projects, or "REDD" (reduced emissions from deforestation and degradation), this is not the subject of this chapter.

3.2 Forestry and international carbon markets

This first section lays out the underlying principles and framework of international carbon markets. It then provides some background on how forestry has been integrated into these markets since they were first established, before summarising the current market status of the carbon forestry sector.

3.2.1 Background on carbon markets

There are two fundamentally different, yet related, types of carbon markets: voluntary, and compliance or regulatory markets. Voluntary funding for carbon offset projects (not yet real "markets") has existed for about two decades, albeit on a small scale, and has provided companies and individuals with an opportunity to offset some of the greenhouse gas (GHG) emissions they produce. These markets are

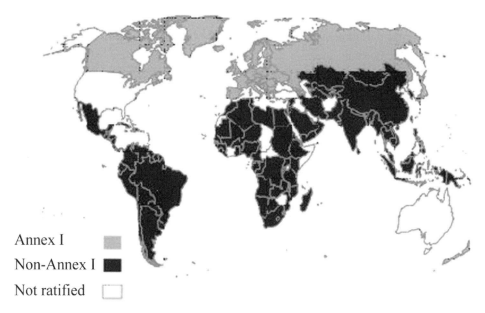

Annex I

Non-Annex I

Not ratified

Fig. 3.1. Parties to the Kyoto Protocol, UNFCCC. Elaboration by EcoSecurities.

discussed below. Regulatory or compliance markets, on the other hand, have been created based on national law or international agreements.

The basis for the main international regulatory market for emission reductions was laid in 1997 when most of the world's nations signed the Kyoto Protocol (Figure 3.1) and became parties to it. This agreement established quantified emission reduction obligations for the industrialized countries which had previously signed the United Nations Framework Convention on Climate Change (UNFCCC), the so-called "Annex I countries" of that convention. Most developing countries are similarly Parties to the UNFCCC and the Kyoto Protocol. However, they do not have any emission reduction targets and are referred to as "Non-Annex I countries" (UNFCCC, 1998). The Kyoto Protocol also established three flexible mechanisms. These were intended to allow for the implementation of emission reductions where it is most economically efficient, which is often in developing countries with low levels of energy efficiency, while the ensuing carbon credits can then be purchased and used by Annex-I countries.

The most important of these flexible mechanisms is the Clean Development Mechanism (CDM), which allows projects implemented in developing countries to generate internationally tradable emission reductions (Certified Emission Reductions, or CERs). CDM credits, including from forestry projects, can be used by developed countries to meet their emissions reductions targets (UNFCCC, 1998; UNFCCC, 2005).

The CDM has proven to be successful in many ways since the Kyoto Protocol finally entered into force in 2005. With double-digit growth rates since that year, the

CDM market reached about US$ 13 billion in 2007. At the same time, overall regulatory carbon markets transacted about US$ 64 billion, including the CDM and the European Emissions Trading Scheme (EU ETS) (Capoor and Ambrosi, 2008). The respective numbers for 2008 markets are projected to be US$ 20 billion for the CDM and US$ 118 billion for overall regulatory markets (New Carbon Finance, 2009). The commodity traded on these markets is measured in tonnes of avoided CO_2 emissions, the main anthropogenic greenhouse gas, and they are therefore generally referred to as carbon markets.

Even before the Kyoto Protocol was signed, and in parallel with the creation of regulatory markets, voluntary carbon markets have emerged. Whereas Kyoto markets are fundamentally compliance markets, created and shaped by governmental regulation, voluntary initiatives are not driven by any legal obligation. Individuals, but also corporations and other organizations without reduction obligations, have the option to purchase carbon credits voluntarily through these markets and to use them as "offsets" for their own emissions. Growing awareness regarding emissions caused by individuals, particularly concern about individual air travel, along with a growing sense of corporate social responsibility (CSR) have fuelled the voluntary markets. There is now an ever increasing demand by organizations for reducing their carbon footprint or even to become "carbon neutral." A growing number of project developers are implementing carbon projects, many of them in developing countries, to create offset credits for the voluntary markets. Voluntary markets have been rapidly growing in the last few years and are thought to have reached a volume of US$ 330 million in 2007 (Hamilton *et al.*, 2008).

The other two flexible mechanisms are Emission Trading (ET), which allows for the trading of emission allowances between Annex-I governments, and Joint Implementation (JI), which allows crediting of emission reduction projects implemented in other Annex-I countries.

3.2.2 Brief history of forestry in international carbon markets

Forestry projects were very much at the focus of early carbon offset projects and climate change mitigation efforts. In many ways, they helped define the concept of carbon offsets as such and prepared the ground for the CDM. Indeed, the requirements for independent certification of carbon credits under the CDM is based on work done for forestry projects and schemes (Moura Costa *et al.*, 1997; Moura Costa *et al.*, 2000). Forestry offsets are still considered by many as typical offset projects. In 1989, years before the Kyoto Protocol or even the UNFCCC were agreed upon, AES Corporation, a US electricity supplier, initiated the first corporate carbon offset project (Faeth *et al.*, 1994). The project, which focused on community forestry and agro-forestry interventions in Guatemala and was aimed at reducing deforestation pressures, helped set the stage for the development of forestry offset projects. Throughout the 1990s, forestry continued to play a central role in the development of the carbon offset concept. At least thirty Land-Use, Land-Use Change and Forestry (LULUCF) offset projects were developed during this period by a variety of companies under a variety of voluntary programs. The project types included forest conservation, reforestation/afforestation, reduced impact logging and forest management, biomass energy deployment, and projects involving agricultural soils and crops (Moura Costa and Stuart, 1998).

However, the history of forestry in international climate negotiations has been marked by many ups and downs with some proponents hailing its potential enormous co-benefits and others condemning its risks as a sound emissions reduction regime. The discussions which eventually led to the Marrakech Accords in 2001 (UNFCCC, 2001) included a controversial and heated debate about sinks (as opposed to sources) of GHG emissions in the land-use sector, irrespective of their merits. The main criticisms revolved around perceived risks of "market flooding," non-permanence and carbon leakage, as well as measurement and monitoring concerns, perverse incentives, and potential negative social impacts of carbon projects (Ebeling *et al.*, 2008).

The suggestion to include sinks, mainly in the form of forests, was made when total emission targets had already been set by Kyoto Parties. This led to fears that forestry activities would simply dilute the focus on or displace some other mitigation efforts. Instead of leading to a net reduction in emissions, assigning carbon benefits to sinks would further delay the necessary restructuring of our fossil-fuel based economies. The concerns were poignantly expressed by the slogan adopted by the non-governmental organizations (NGOs) involved in the debate: "Don't sink Kyoto!" Projections of vast quantities of cheap forestry credits flooding the carbon markets and depressing the price of tradable emission permits led to similar concerns. Cheap credits, while commercially desirable, would decrease incentives to invest in energy-related emission abatement and crowd out such activities (Ebeling, 2008).

The risk of reversal, or non-permanence, of emission reductions if sinks were destroyed, e.g., by burning or cutting down forests, seemed difficult to tackle. Similarly, leakage, the displacement of emission-generating activities outside the project boundaries without actually reducing them, seemed very difficult to quantify or prevent. For example, leakage would occur if plantations were established on agricultural land and if farmers converted forests elsewhere to regain areas for cultivation (Schwarze *et al.*, 2002). Added to this were uncertainties in monitoring and measuring carbon fluxes from land-use and forestry.

Some stakeholders were also concerned that the CDM would create perverse incentives to cut down existing forests and establish plantations in their stead to gain carbon credits. Finally, sovereignty concerns and social impacts generated controversy: paying poor developing countries for keeping certain areas under forest cover for a long time and thereby restricting other development options on these lands was portrayed by some as a form of expropriation and neo-colonialism (see Fearnside, 2001; Dessai *et al.*, 2005). We discuss below how these concerns were addressed in the design of carbon markets.

3.2.3 Current market status of carbon forestry projects

The main distinction here again is between Kyoto regulatory and voluntary carbon markets, as well as forthcoming US and post-Kyoto carbon markets. After the Kyoto Protocol was signed in 1997, it took several years to define concrete rules for the CDM. Most of these were established by the Marrakech Accords in 2001; however, forestry-related rules were not finalized until several years later. This delayed regulatory clarity, the complicated methodological framework being established for forestry, and, perhaps most importantly, the refusal of the European Union to allow forestry CDM credits into its domestic emissions trading scheme contributed to the fact that forestry projects still account for a very small share of the overall CDM pipeline (Figure 3.2). At the end of 2008, only one forestry project had achieved CDM registration (the final step before the issuance of carbon credits), another seventeen projects had been submitted for validation by independent certifiers (the step preceding registration), and several dozen more projects were under development. The corresponding project numbers for the overall CDM pipeline were 1242 registered projects and 4151 projects at validation (UNEP, 2008). However, there are

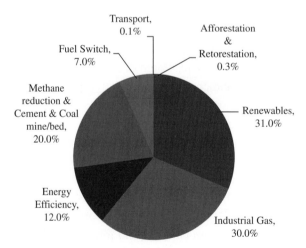

Fig. 3.2. Share of expected carbon credits until 2012 in each CDM project category
Adapted from UNEP, 2008.

signs of increased activity in the CDM forestry sector as more experience is accumu-
lated and carbon buyers become more interested in this asset class.

Most of the first CDM forestry projects in the pipeline were promoted by pilot
funds administered by the World Bank, as well as by several large conservation and
development NGOs. Overall, many of the existing projects rely at least partly on
donor funding (Neeff *et al.*, 2007). This means that at this early stage in the
development of the carbon market, much of the demand for CDM forestry credits
is not actually created by buyers that have to purchase credits for compliance
reasons but rather by multilateral and philanthropic organisations. The registration
and validation of the first CDM forestry projects have lifted the sector beyond the
pilot status and more and more commercial project proponents have started to
become involved. However, interest is still subdued due to the restrictive attitude
of the largest regional carbon market at present, the EU Emission Trading Scheme,
towards forestry (Neeff and Ebeling, 2008). In terms of regional distribution of
CDM forestry project development efforts, Latin America is currently the region
with the highest credit potential in existing projects (with more than half of the
estimated future credits coming from this region), followed by Africa (with about
one third) (Figure 3.3). Similarly, the geographical distribution of projects in the
World Bank's BioCarbon Fund includes a majority of projects being developed in
Latin America, followed by Africa (World Bank, 2006).

In addition to the maturing regulatory (Kyoto) carbon markets, the burgeoning
voluntary carbon markets are fuelling demand for forestry-based carbon offsets.
Indeed, forestry projects have constituted one of the largest sectors in the voluntary

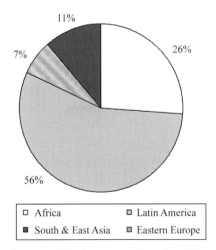

Fig. 3.3. Carbon credit potential of CDM projects in different regions (based on
projected carbon sequestration of CDM projects currently under development).
Modified from Neeff *et al.*, 2007.

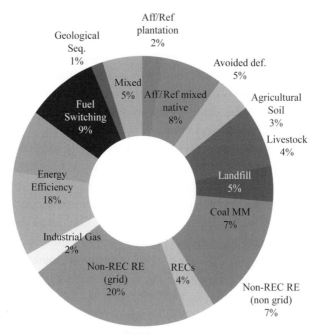

Fig. 3.4. Volume of credits sold by project type on international carbon markets (over-the-counter transactions). After Hamilton *et al*., 2008.

market, with about 18 percent of the volume of all transactions (Figure 3.4) (Hamilton *et al*., 2008). Forestry project types are split between protective and productive reforestation, avoided deforestation, and improved forest management. This high share may decrease somewhat in the future as more projects in other technology sectors are developed.

There are currently clearly more forestry projects under development for the voluntary markets than under the CDM (Neeff and Ebeling, 2008). In the medium term, voluntary markets are also likely to continue to dominate the carbon forestry sector because of their much greater flexibility regarding project types, standards and crediting approaches. In addition, with the United States being increasingly likely to impose cap-and-trade schemes for GHG emissions on a federal level (several US States have already started to implement such schemes), forestry credits may meet significant demand in this market. US buyers have always been very favorable towards carbon forestry in general, and most pre-compliance buyers in the country are likely to flock to the voluntary markets (Ebeling and Fehse, 2008; Hamilton *et al*., 2008). It is expected that any future compliance regime in the US will include a prominent role for forestry credits.

An additional push for the carbon forestry sector in general has been arising from the vigorous debate around the inclusion of Reduced Emissions from Deforestation and Degradation (REDD) under a post-2012 or post-Kyoto climate regime. Some clear indications that such a deal will be reached with a significant role for forestry in general since the announcement of the Bali Roadmap in 2007, have greatly increased the interest of project developers and investors in this activity (Ebeling and Yasue, 2008). Although the REDD discussion initially focused strictly on preventing deforestation, there are increasing signs that a future scheme may in fact move towards a sectoral forestry approach, including gains in forest cover and forest biomass, i.e., reforestation and the restoration of degraded forests (Ebeling *et al.*, 2008). Importantly, there are also signs that future regulatory markets may become more accessible for reforestation projects as such, not only in the US but also under a Kyoto successor agreement. The CDM as a whole is undergoing a review and reform process, and so is the European trading scheme, including current import barriers for forestry credits, which may be eased (Fehse, 2008b).

3.2.4 *Relevance of carbon markets to the financing of forest restoration*

Carbon markets are already transacting billions of Euros each year, and it is evident that they have the potential to bring unprecedented finance into emission reduction projects, including those in the realm of forest restoration. Although the share of carbon forestry in the current regulatory markets is very small, it can be expected to increase greatly in forthcoming international regimes, as well as in regional markets in the US and beyond (Neeff and Ebeling, 2008), and forestry already has a market share of one fifth of the rapidly expanding voluntary carbon markets (Hamilton *et al.*, 2008).

In comparison, funding through Official Development Assistance (ODA) for forestry has been stagnating on a relatively low level for some time and there are few prospects for such funding to significantly increase. Although precise numbers are difficult to establish, current bi- and multilateral government funding for forestry seems to total about US$ 1.1 to 1.5 billion annually during the last decade (Tomaselli, 2006). This number includes investments into the forest industry, and direct expenditures for reforestation or even forest restoration are certainly much lower. Considering the long list of topics on the international environmental and development agenda that are competing for ever scarcer public funding, from poverty relief to fighting HIV-AIDS to providing education and basic sanitation, from combating desertification and biodiversity loss to mitigating and adapting to climate change, it is unlikely that international public funding for forest restoration will receive a major boost. Much hope, therefore, rests on private funding and market-driven initiatives to promote environmental causes (see Scherr *et al.*, 2004). The land-use sector in developing countries includes

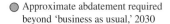

Approximate abdatement required
beyond 'business as usual,' 2030

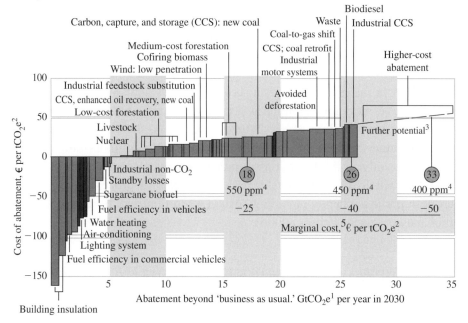

Fig. 3.5. Global cost curve for greenhouse gas abatement measures beyond 'business as usual'. Source: Enkvist *et al.*, 2007.([1]GtCO$_2$e: gigaton of carbon dioxide equivalent; "business as usual" based on emissions growth driven mainly by increasing demand for energy and transport around the world and by tropical deforestation. [2]tCO$_2$e: ton of carbon dioxide equivalent. [3]Measures costing more than €40 a ton were not the focus of this study. [4]Atmospheric concentration of all greenhouse gases recalculated into CO$_2$ equivalents; ppm: parts per million. [5]Marginal cost of avoiding emissions of 1 ton of CO$_2$ equivalents in each abatement demand scenario).

some of the lowest-cost options for carbon emissions abatement on a very significant scale (Enkvist *et al.*, 2007) (Figure 3.5). Against the backdrop of chronically insufficient international donor funding, carbon markets thus hold an immense potential to leverage funding for avoiding degradation and restoring of ecosystems on a global scale.

The Kyoto regime, as well as voluntary carbon markets, created a market value for a less than tangible product, carbon dioxide. The relative ease of measuring different GHGs and assigning a precise value of global warming potential in tons of CO$_2$-equivalent to them has been instrumental in the success of markets for emission reduction certificates. For no ecosystem service other than climate change mitigation has an easily convertible unit been defined to date, although numerous researchers and environmental organizations are attempting to achieve this. A number of local and national payment schemes exist for ecosystem services, such as the regulation of water quality (Jackson *et al.*, 2005) and the provision of biodiversity (Pagiola *et al.*, 2005; Bishop *et al.*, 2008). However, most of these services are not tradable across or even within national borders at present. Even emerging markets for biodiversity offsets, which allow companies to compensate for development impacts in one area by protecting or restoring natural habitat elsewhere, for the most part rely on relative crude area measures without actually measuring their respective biodiversity value (Fehse, 2008a). Schemes such as the US wetland mitigation program have been widely criticized by environmentalists for failing to accurately reflect the habitat value of wetlands that are affected by developments or restored through compensatory measurements (Robertson, 2004). Although far from mature, carbon markets for now remain a uniquely successful attempt at creating an internationally tradable commodity for an environmental service, and they are unrivalled in the scale of resources they leverage.

Apart from payments based on carbon sequestration achieved by growing trees, carbon markets can also be used to monetize other aspects of reforestation activities (Aukland *et al.*, 2002). Wood is receiving increasing attention regarding its potential use as a renewable energy source. The international market for biofuels is expanding rapidly and the CDM and voluntary markets allow for forestry-based bio-energy projects that replace the use of fossil fuels. Such projects can rely on wood residues from timber harvesting and processing or they can directly use wood sustainably harvested from natural or planted forests (Moura Costa and Tippmann, 2003). This significant opportunity has so far received very limited attention from the forestry community and is worth exploring in parallel with more conventional carbon forestry as an additional income source for reforestation projects.

3.3 How do forestry projects work in carbon markets?

Tapping into carbon markets to finance environmental projects can provide resources on a scale far beyond what is available through public financing. However, using carbon finance also means that the requirements of carbon markets have to be met. This refers to eligibility criteria (market entry), as well as buyer preferences (demand). It also means that projects need to consider how to maximize those goods and services that are valued on these markets, i.e., CO_2. In this section, we first give an overview on the requirements that carbon forestry projects need to meet in order to be able to receive carbon finance. We then discuss the economics of the creation of revenue generation for forestry through carbon markets and how these relate to the implementation of projects.

3.3.1 Requirements of carbon forestry projects

Forestry has been a prominent but also one of the most controversial sectors in the design of climate change mitigation schemes and carbon markets. We outlined the main issues in Section 3.2.2, namely concerns around risks of market flooding, non-permanence, leakage, measurement and monitoring, social and environmental impacts, and additionality. All or most of the concerns have been addressed in the meantime, sometimes through painful concessions from the supporters of carbon forestry, and at the price of severely curtailing the potential of forestry to help mitigate climate change. These safeguards are discussed in the following paragraphs, but an overriding limitation of the Kyoto markets is that projects that conserve tropical forests or improve forest management practices cannot gain carbon credits. Instead, only the planting or assisted natural regeneration of forests qualifies under the rules of the CDM, severely reducing the potential that carbon finance could have in reducing the large amount of emissions created by land-use change.

The main requirement of any credible carbon offset projects of any sector is that it needs to be "additional." All carbon projects are in a sense designed to offset emissions that take place in countries with emission reduction targets, in the case of CDM, or emissions from companies or individuals adopting voluntary commitments, i.e., they lead to emission reductions in developing countries *in lieu* of achieving the same result at a higher cost in industrialized, Annex I countries. Each carbon credit issued for a carbon project thus translates into decreased mitigation obligations in developed countries, companies, or households. It is therefore crucial to ensure that offset projects are additional and would not have happened in the absence of carbon finance or other incentives provided through the CDM or voluntary carbon markets (such as visibility and marketing benefits, or political support). If a project were to receive carbon credits but would in fact have gone ahead even in the absence of carbon finance, the net result would be fewer emission reductions globally (Ebeling, 2008).

There are several ways to demonstrate additionality under the CDM framework, and the same approach is part of the voluntary carbon standards. The main approach is to demonstrate that some barrier exists that prevents the proposed project from taking place. For example, project developers can demonstrate that a project would not be sufficiently financially profitable for investors without the added carbon income. Institutional, cultural, technological, or investment barriers could also exist for the CDM or offset activity and plausibly prevent its implementation under a business-as-usual (BAU) scenario. For example, reforestation on a sizable scale may be a novel land-use practice in a region, there may not be any local expertise or desire to reforest, and banks may be unwilling to lend money to such ventures. All of these barriers could prevent the implementation of a reforestation project. Carbon finance helps overcome these barriers.

There are other requirements that are specific to carbon forestry alone. One of these is to address the risk of non-permanence, i.e., reversal of carbon benefits, for example through the felling or burning of trees. Different approaches have been proposed to deal with the issue of non-permanence in forestry projects (Fearnside *et al.*, 2000; Moura Costa and Wilson, 2000). Under the rules of the CDM, for example, forestry projects can only receive temporary carbon credits. Forestry is thus the only project sector in which no permanent credits are issued, and this has proven to be a major bottleneck to creating significant market demand for CDM carbon forestry. Under the temporary crediting approach, the actual carbon stocks in reforested areas have to be reverified periodically, and if a project does not retain the formerly stored carbon, existing temporary CDM forestry credits have to be replaced with emission reduction or sequestration credits from elsewhere (Chomitz and Lecocq, 2003; Pedroni, 2005). Voluntary carbon markets have adopted a different approach and, under the most established standard, the Voluntary Carbon Standard (VCS), a buffer reserve of credits has to be retained. Buffer reserves have been used in some schemes since 1997 (Moura Costa *et al.*, 1997), and under this approach credits from such a pool can be used to compensate for any potential future reversal of carbon benefits. This translates into a powerful financial incentive for project proponents to address non-permanence risks in order to keep the percentage of credits that need to be retained in a buffer reserve low (Ebeling *et al.*, 2008; Voluntary Carbon Standard, 2008).

Another key requirement for forestry activities is accounting for leakage. As mentioned before, activity shifting, i.e., the simple displacement of agriculture or logging to another land area, may lead to deforestation in other areas. This would significantly reduce the carbon sequestration benefits from a carbon project. In the worst case, such emissions may completely offset the climatic benefits achieved through the afforestation or reforestation project. A second type of leakage, market leakage, occurs independently of the direct land-use actors, i.e., does not involve their physical displacement. Rather, changes in supply or demand of products affected by the project leads to increases in production elsewhere (Chomitz, 2002). If agricultural production in an area decreases because of reforestation activities, the diminished supply may lead to increased production elsewhere in order to meet market demand. The reduction in available agricultural land could therefore induce land-use conversion in other, still forested areas (Schwarze *et al.*, 2002; Aukland *et al.*, 2003).

In order for a reforestation project to qualify as a CDM activity, lands to be reforested must not have been forested in 1990. In the language of Kyoto, afforestation, as opposed to reforestation, refers to the establishment of plantations on lands that have not been forested for at least 50 years. The 1990 base-year requirement acts as an efficient safeguard against perverse incentives which might otherwise lead project developers to cut down natural forests before establishing carbon plantations. The corresponding requirement under the Voluntary Carbon Standard is a rolling ten-year threshold, meaning that areas must have been deforested (or not forested) for at least ten years before being reforested (UNFCCC, 2003; Voluntary Carbon Standard, 2008).

Finally, CDM forestry projects (and forward-looking voluntary offset projects) need to seek confirmation by the host government stating that they contribute to sustainable development as defined by that country. This requirement frequently includes a form of social and environmental impact assessment and project developers have to document that the proposed project will not lead to negative environmental impacts, such as ground-water depletion, soil degradation or biodiversity loss. For voluntary offset projects, buyers usually prefer documentation that lays out how a project improves environmental and social conditions apart from climatic benefits. The need to demonstrate how their purchase of offsets will enable such benefits may in fact be the chief motivation of buyers to choose a well-designed forestry project with clear co-benefits (Ebeling and Fehse, 2008).

3.3.2 Economics of carbon forestry plantations

When trees grow, they convert atmospheric carbon into biomass which may be stored for decades or centuries. This is referred to as carbon sequestration and it is one of the services valued in carbon markets. In order to maximize income through emissions trading, project developers face an incentive to maximize the rate of carbon sequestration and eventual volume of carbon in their plantations. The growth rate and biomass of mature trees are therefore important factors in the choice of species for carbon reforestation projects. The net carbon benefits of a carbon project, for which credits can be issued, can be calculated as follows:

$$ER_{net} = (ER_{project} - ER_{baseline} - EO_{project}) \cdot (1 - L) \cdot (1 - BD)$$

where:

ER_{net}	=	net emission reductions
$ER_{project}$	=	project emission reductions
$ER_{baseline}$	=	baseline emission reductions
$EO_{project}$	=	other project emissions (e.g., fuel use)
L	=	leakage (as a fraction of 1)
BD	=	buffer discount (as a fraction of 1) (some standards only)

The project's sequestration benefits through the growth of planted or regenerating trees ($ER_{project}$) has to be adjusted by what would reasonably be expected to occur on the land in the absence of the project, *i.e.*, the baseline or business-as-usual scenario ($ER_{baseline}$) (IPCC, 2003; Pearson *et al.*, 2006). For example, abandoned agricultural lands may regenerate even without any assistance through a carbon project. In addition, it needs to be considered whether the project itself generates emissions such as through the removal of pre-project vegetation ($EO_{project}$). For example, preparing the ground for planting may involve burning or cutting down of existing shrubs or grasslands that would otherwise impede the growth of tree saplings. Similarly, fuel use by machinery or project staff may generate measurable emissions.

Any emissions created through carbon leakage (L), the displacement of activities from the project area, will have to be subtracted. These three latter aspects help explain why it may be easier, and potentially more lucrative, to carry out CDM forestry projects in areas that are in a degraded or degrading state and that are not currently used for agriculture, grazing or collection of wood for fuel.

In order to account for the risk of non-permanence, i.e., the reversal of sequestration benefits through subsequent degradation or destruction of reforested areas, some high-quality carbon standards require the retention of a portion of credits in a buffer reserve (BD) (see Section 3.3.1). The percentage of credits retained in this pool will depend on the apparent risk of non-permanence and these may later be reclaimed if stable carbon stocks can be demonstrated. In contrast, the CDM issues temporary credits for forestry projects which need to be replaced periodically, and there is no buffer discount.

For an indication of the potential carbon revenue reforestation projects can achieve, it is useful to consider the sequestration rates of regenerating forests. Depending on hydrological conditions naturally regenerating tropical and sub-tropical forests typically increase their above-ground dry biomass by approximately 1 to 13 tonnes per year. Assisted regeneration, e.g., through active replanting and removal of weeds, can further increase this unassisted natural sequestration rate, especially in initial years, and so can the choice of faster growing species. The corresponding values for tropical and sub-tropical planted broadleaf forests and assisted natural regeneration are 5 to 18 tonnes per year (IPCC, 2003). Below-ground biomass in roots may add another 27 to 42 percent. Assuming a carbon content of 50 percent in dry biomass (IPCC, 2003), this corresponds to roughly 0.64 to 12.8 tonnes of carbon sequestered per year in above-ground biomass, i.e., 2.6 to 46.9 tonnes of CO_2, obviously a wide range. Assuming a forestry project achieves an annual growth rate of 13 tonnes of CO_2 and valuing this at US$ 2–5 per tonne, carbon finance could then generate US\$ 26 to 65 per hectares per year, or US\$ 260,000 to 650,000 for a 1000-hectares project over a ten-year period. Any emissions created through leakage, baseline sequestration benefits, etc. still need to be deducted in order to obtain the net carbon credit potential. In most cases, carbon finance is likely to cover only a small part of the costs of a good quality forest restoration project, at least at today's relatively low carbon prices. However, it can provide a significant additional income and can enable projects that would otherwise not have been feasible.

A full account of a project's gross sequestration potential would consider the different species used in reforestation and each species' typical above- and below-ground biomass, determined by wood density (determining the relative carbon content in the biomass), the biomass extension factor (indicating the ratio of biomass stored in branches as opposed to the trunk of a tree), and the root:shoot ratio (pointing to the biomass and carbon contained in roots as opposed to the visible parts of the tree). All of these are of course influenced by local climatic, soil, and other growth conditions, as well as management practices. The CDM regulations furthermore stipulate that project developers take into account other carbon sink components which may be affected by the reforestation activities. Besides the biomass components of a standing tree elaborated above, these sinks are deadwood, litter, and soil carbon.

3.4 Opportunities for ecological restoration through forest carbon markets

This section explores how the financial opportunities presented by carbon markets, as well as the particular economics of forest carbon projects, may be harnessed to provide synergies with ecological restoration objectives. This includes an overview of the potential overlap of ecological restoration needs and carbon forestry and also a discussion of settings in which synergies between the aims of ecological restoration and carbon sequestration are particularly evident. Furthermore, this section lays out how the requirements of carbon forestry projects can actually be utilized towards designing robust restoration projects. To illustrate potential synergies, we give examples from existing CDM forestry projects that try to promote ecological restoration. Ecological restoration can be described as an "intentional activity that initiates or accelerates the recovery of an ecosystem to a self-sustaining state with respect to structure, species composition and function, as well as being integrated into the larger landscape" (Clewell *et al.*, 2005). The restoration of formerly forested lands is becoming an increasingly pressing need considering that an estimated 350 million hectares of forests have been lost between 1950 and 2000 and an additional 500 million hectares of forests have been degraded in the same timeframe, both particularly in developing countries (ITTO, 2002). In this context, it is not difficult to see that reforestation, carried out with the help of carbon finance, has the potential to contribute significantly to the aims of global ecological restoration. Most areas of deforestation and severe land degradation (Plate 3.1) can be found in countries that are signatories to the Kyoto Protocol (Figure 3.1), and most carbon offset projects similarly take place in developing countries, simply because these do not have any international emission reduction targets. The potential contribution of carbon finance to global ecological restoration aims is even clearer when considering the very limited public funding and implementation capacity for forest restoration in the developing world.

Carbon finance could support ecological restoration needs mainly in two ways. Firstly, additional funding could enable the restoration of larger areas than might otherwise be possible. Secondly, the added income from carbon finance could make it more feasible to use a wider mix of native species in restorative planting, which can be very expensive. Both aspects, large areas and mixed native species, are attractive also from a carbon market point of view. In the former case, this is because larger areas make it easier to cover non-area dependent, fixed transaction costs. In the latter case, mixed species reforestation projects can be much more attractive to voluntary market buyers of offsets because of their higher biodiversity value and marketing appeal.

Large areas of formerly forested lands remain unused and unproductive in the tropics and do not recover due to a range of factors, including soil degradation, depletion of nutrients, recurring fires, and large distance from intact forest which could otherwise act as seedbanks. A report by the World Bank suggests that with ongoing trends entire ecosystems and countless species in the tropics may be lost, together with the occurrence of widespread changes in water flows, the proliferation of pests, and a decrease in important pollinators (Chomitz *et al.*, 2007). Carbon reforestation projects, either through active planting or through assisted natural regeneration, can support the recovery of a forest ecosystem and generate ecosystem services beyond the sequestration of carbon. Realizing such co-benefits of carbon offset projects is also a near-term opportunity to bring finance towards restoration projects which so far cannot count on any meaningful international markets for biodiversity or water-related services (Ebeling and Fehse, 2008).

Although most existing CDM and voluntary carbon forestry projects have been developed primarily with a view to generating carbon revenues, many do contain design elements to deliver further, non-carbon ecosystem services (co-benefits). Indeed, some of the first carbon forestry projects worldwide aimed at providing a wide range of ecological benefits. No up-to-date review of these multiple benefit projects exists, and the last comprehensive assessment was carried by Landell-Mills and Porras (2002) who, even at this very early stage of carbon markets, identified twenty-eight projects with "bundled" approaches. Such projects combine multiple ecosystem services either into one product or market these services separately. Fehse (2008a) shows that there is great scope for project developers to optimize such synergies in carbon forestry, particularly regarding the restoration of the vegetation cover on degraded and degrading sites that have lost their water and soil retention capacity.

CDM forestry projects need to demonstrate that they achieve a net climatic benefit compared to a baseline situation in which the project activities would not have been implemented. If there are uncertainties about the baseline scenario (e.g., how much carbon would be sequestered in naturally occurring, non-assisted recovery of vegetation), the most conservative outcome, resulting in the lowest carbon benefits, is usually assumed. Indeed, the first methodologies and projects to pass the forestry CDM approval process were based on reforestation of degraded and further degrading soils. This is not due to the co-benefits of these projects but rather to the simplicity of their baseline scenario in which it can be easily demonstrated that there would be no positive changes in the carbon balance if the projects did not take place (Fehse, 2008a). Nevertheless, these projects make a clear contribution to preventing and reversing land degradation.

The apparent contradiction between the need for restoration of natural ecosystems with naturally occurring mixes of species on the one hand and the incentive to use fast-growing species for carbon sequestration on the other hand can be turned into an advantage by drawing on experiences gathered by restoration ecologists. There is an innovative approach in which certain types of fast-growing, non-invasive species are used to create the required initial conditions for colonization by (other) native species in a way that does not compromise the aims of ecological restoration (Feyera et al., 2002; Lamb et al., 2005). For example, in South Africa, ecological restoration projects are using exotic, fast-growing tree species to either act as nurse trees, adding nutrients to the soil, or to provide shade and thereby facilitate natural succession on degraded land. These exotic species are ultimately shaded out by the formation of a secondary native forest or they are actively removed (Geldenhuys, 2004). In this way, fast-growing species sequester carbon before being replaced by non-pioneer, slower-growing species. Removal of the exotic species does not affect the sequestered carbon stocks because they are replaced by secondary vegetation. Obviously, the success of this approach depends on the ability of native species to reach the site during natural succession and on the careful selection of non-invasive species for nurse trees. The replication of this approach in other carbon project will thus depend on the availability of particular ecological expertise and adaptive management practices.

Interestingly, although the requirements for projects regarding eligibility as carbon projects as well as the calculatation of their carbon credit stream may seem very restrictive (see previous section), they can actually be turned into an advantage in the context of ecological restoration. This relates in particular to requirements regarding the accounting for leakage and non-permanence risks. These and other stipulations can greatly enhance risk management and the net ecological benefits of restoration projects.

For example, leakage is not simply a technical requirement with relevance only to carbon crediting. Instead, leakage refers to the general and very real risk that protecting or restoring a piece of land can increase degradation pressures on other areas. Certainly, restoring particularly vulnerable or ecologically valuable land may well be justified even if this takes agricultural land out of production; however, ecologists need to address the fact that reducing land available for agriculture will facilitate degradation and land-use changes elsewhere, unless targeted counter-measures are put in place. In this way, the net benefits of many successful on-site restoration activities may be diminished through harmful off-site effects. In contrast, leakage mitigation measures that are carried out primarily because of the require-ments of a carbon forestry project can enhance overall ecological restoration benefits at a landscape level. Proactive measures to reduce leakage include agricultural intensification in neighbouring areas, and the creation of other income-generating activities for rural communities (Schwarze *et al.*, 2002), reducing land-use pressures not only in the project area.

Similarly, the need to assess and improve the risk profile of carbon forestry projects regarding non-permanence risks can be an advantage for ecological restoration projects. Determining non-permanence risks is a systematic way of identifying threats to the long-term success and resilience of forest restoration endeavours. Once risks have been identified, systematic response strategies, i.e., risk mitigation measures, can be developed and incorporated into the project design (Ebeling, 2008). This can entail direct financial advantages from a carbon crediting perspective (e.g., because of lower discounts if a risk buffer approach is applied), and it can also help to identify and address risks from an ecological restoration point of view early on.

Synergies between ecological and carbon-market oriented restoration aims can also arise regarding the integration of local livelihood benefits. The first CDM forestry projects have been implemented on degraded lands (Fehse, 2008a), partly because degraded agricultural lands and pastures are not very productive, posing a low potential for leakage. As more and more carbon forestry projects are imple-mented on lands which are at least partially in use, there is an increased risk of leakage, requiring that leakage mitigation measures become an integral part of the project design. In some cases, such measures can directly enhance local liveli-hoods. For example, existing cattle grazing in an area that is proposed to be reforested by an offset project could generate a substantial leakage risk because local people may be forced to encroach on neighbouring woodlands. Instead, including tree species that produce suitable fodder for animals, devising silvo-pastoral schemes that allow for continued grazing, reserving a portion of the project area for the production of food and fodder crops, or providing alternative employment and income sources are measures that can diversify livelihood options and prevent leakage.

Ecological restoration practice is also increasingly looking at supporting sustainable livelihoods and is thereby extending its focus beyond purely conservation goals. So far, ecological restoration projects have unfortunately had limited success in compensating for local opportunity costs in the form of land being lost for agricultural production or grazing (Lamb *et al.*, 2005). Integrating approaches devised for carbon forestry projects may help improve this situation.

Although there certainly are a number of challenges in linking the restoration and climate agendas through carbon forestry (discussed in the following section) the potential co-benefits for biodiversity and human development make to the exploration of synergies in the design of projects worthwhile. Instead of merely trying to prevent negative social and environmental impacts, carbon forestry projects can be designed to focus on the enormous potential positive outcomes of restoration (see Table 4.1 in this book). For deliveries of these potential benefits, it is essential that carbon forestry takes into account multiple stakeholders' perspectives. For example, local communities may have an interest in restoration programs using certain types of trees to provide building materials, fruits, firewood, or animal fodder. Ensuring the support and engagement of local land-users is also critical to securing the long-term sustainability and ecological goals of restoration and similarly their carbon benefits.

Finally, it remains important to recognize that carbon finance alone cannot provide all the funding for the multiple benefits of restoration projects. Establishing plantations is generally so costly that in most cases only a small portion of this can be financed through carbon crediting. This is even more so for high-quality restorative tree planting or assisted regeneration. Targeted non-carbon related financial support for the biodiversity and development benefits of reforestation projects can enhance these co-benefits. This could involve direct donor-driven and charitable funding, as well as tapping into markets for other, non-carbon based, ecosystem services that are emerging (Ebeling and Yasue, 2008). There is some indication that forestry projects in voluntary carbon markets that provide clear co-benefits achieve higher prices or meet larger demand (Hamilton *et al.*, 2008). Standards such as the Climate, Community and Biodiversity (CCB) standards aim at providing reliable assurance that projects are designed to deliver measurable net positive benefits to local communities and biodiversity, in addition to credible greenhouse gas reductions (Ebeling and Fehse, 2008).

3.5 Restoration through carbon forestry in practice

Carbon forestry can directly contribute to restoration aims by preventing and reversing degradation of arid and semi-arid lands. An initiative that seeks to secure additional rewards for such projects is the Global Mechanism-EcoSecurities Partnership (Global Mechanism, 2007). The Global Mechanism is a subsidiary body to the UNCCD and is charged with mobilizing finance for implementing that convention. EcoSecurities is a private-sector carbon project developer and environmental finance consultancy. The partnership strives to adopt a synergistic project approach to support the aims of the climate change and desertification conventions (UNFCCC and UNCCD).

One concrete example of the work of the Global Mechanism-EcoSecurities partnership is the Julcuy project in the province of Manabí in coastal Ecuador, as described by Fehse (2008a). This arid region has been suffering for decades from deforestation of the native dry forest and subsequent degradation from overgrazing by goats. This has led to significant soil erosion and contributed to water scarcities in the region. The project seeks to restore around 5000 hectares of the original forest vegetation. In a first step, a mix of seven native tree species will be planted; they were selected to also provide non-timber products to local communities, including fodder for the goats, e.g., the algarrobo (*Prosopis juliflora*) and the palo santo (*Bursera graveolens*). It is envisioned that in the long term the reforested areas will provide the local communities with a sustainable source of timber and fuel wood. The project area also fulfils connects two important coastal nature reserves in the larger Chocó-Manabí Conservation Corridor and will benefit both of them. Furthermore, the project area is of importance for the hydrological supplies of a number of urban centres, which have seen steep population growth in the last three decades. The initially established mix of tree species will create the structural and microclimatic conditions to allow a broader suite of native plant species to colonize through dispersal from nearby forests. In a partnership with Conservation International, EcoSecurities will seek to quantify and market the project's biodiversity benefits and market hydrological benefits to the municipal water companies of nearby cities.

Further examples are carbon projects that rehabilitate degraded mining sites. Legal obligations for companies to rehabilitate decommissioned sites are rarely enforced in most developing countries. Sites therefore often continue to degrade after mining operations have ceased, leading to erosion and related problems of sedimentation or aeolian dust. Any rehabilitation efforts that are made usually only establish a grass cover, even in areas that would naturally be forested. However, although not rehabilitating clearly saves costs in the short term, it can severely damage a company's image, and this is becoming an increasingly important factor for business decisions, especially for international organizations. The additional income obtained through a carbon forestry project can tip the balance in corporate decision-making by turning an environmental cost into an asset. Likewise, without the underlying rehabilitation aims, a carbon project might not take place because carbon credits alone could not provide sufficient financial incentives for restoration.

3.6 Challenges for integrating carbon forestry and ecological restoration

There are, of course a number of challenges to obtaining the benefits of the synergies between carbon forestry and ecological restoration described in the previous section. Many of the potential challenges arise from the very economics of carbon forestry, outlined above. On the other hand, some apparent challenges are simply the result of exaggerated expectations vis-à-vis the carbon markets. We discuss here issues arising from the incentive to maximize carbon sequestration over other ecosystem services, restrictive regulations of the Kyoto framework, weak sustainable development requirements of regulatory carbon markets, ecological constraints, disincentives for fire management and invasive species control, and high transaction costs.

Reforestation projects are probably the carbon project category with the slowest returns on investment in terms of carbon credits produced. This is because trees sequester carbon and thereby reduce emissions relatively slowly compared to the emission reductions achieved more quickly in energy-related or industrial gas project types. Forestry project developers, therefore, often face significant problems in bridging the gap between the necessary upfront investments, ongoing manage-ment costs and eventual generation of credits. This explains the popularity of fast-growing, often exotic, species such as *Pinus* and *Eucalyptus* in commercially oriented carbon forestry projects. Their high growth rate allows for carbon credits to be obtained relatively early after plantation establishment. In addition, timber can be produced from such plantations, in itself an attractive revenue source. However, the introduction of exotic species, in many cases, may clearly not meet the require-ments of ecological restoration or may not appeal to carbon buyers concerned with the ecological impact of their investments.

Early carbon forestry activities, usually promoted by not-for profit NGOs and development organizations, tended to put greater emphasis on providing multiple environmental and socioeconomic benefits as opposed to maximum carbon seques-tration rates. Now, after the successful implementation of many of these initial projects, commercial project proponents have started to become increasingly inter-ested in carbon forestry. What this means will partly depend on the kind of carbon markets these projects target. In regulatory markets, carbon benefits may be the main or sole consideration, whereas in voluntary markets, there may be significant financial advantages to the promotion of ecological co-benefits (Ebeling and Fehse, 2008).

It will remain important to find ways to ensure that the economics of such projects work. Therefore, private project developers in a market-driven environment have to consider carefully both the direct costs of pursuing non-carbon benefits and any potential indirect costs due to reduced carbon sequestration rates, e.g., through planting a mix of slower growing species.

CDM regulations demand that there are no negative environmental and social impacts of afforestation and reforestation activities. The assessment of such impacts is based on the interpretation of the CDM's sustainable development requirement by the respective host countries where such projects are implemented. National regulations may in fact contain concrete stipulations for ecological improvements and many do demand comprehensive environmental impact assessments for proposed projects. Nevertheless, unintended negative impacts can occur in reforestation projects even if such assessments have been carried out. For example, there is a long history of reforestation programs using exotic tree species with negative impacts on the local ecology (Richardson, 1998), although not necessarily in the context of carbon forestry; some of the traits that make these tree species highly suitable for productive forestry (i.e., speed of growth) can also make them potentially invasive. Frequently, negative hydrological impacts of forest plantations are also quoted by organizations opposing carbon forestry.

It is clear that some afforestation and reforestation programs have had undesirable consequences for soil erosion, groundwater levels, biodiversity and local livelihoods (Cossalter and Pye-Smith, 2004). However, it is also important not to associate the shortcomings of poorly designed forest plantations as such with the potential outputs of carbon offset projects. The latter usually involve a careful planning phase and regular environmental monitoring (Ebeling, 2008). It remains true, however, that the economic incentive to maximize sequestration rates presents certain risks. Ecologically desirable reforestation is more costly to manage and may have a lower carbon credit potential.

In general, forest restoration can be conducted in any of the following three main contexts (Clewell *et al.*, 2005):

(a) Recovery of a degraded system
(b) Replacement of a forest system that was entirely destroyed with the same system
(c) Transformation or substitution, i.e., conversion of an ecosystem to a different kind of ecosystem.

The current rules of CDM forestry projects, however, contain severe shortcomings in relation to their potential application to ecological restoration. For example, the restoration of degraded remaining forest (option a) is not allowed under CDM rules since they demand that the area to be planted has been depleted completely of forests since 1990. In this way, Kyoto regulations preclude the possibility of financing the recovery of the more than 500 million hectares of degraded primary and secondary forests that exist worldwide. Restoration of these areas, which may still hold significant biodiversity, provide hydrological services, and support local livelihoods, could be highly beneficial for achieving multiple conservation and sustainable development goals, as well as greatly contributing to mitigating climate change (Ebeling, 2008). Furthermore, restoration of degraded forest is generally more cost efficient and may have a higher success rate than the reforestation of bare lands. This is because degraded forests may still be ecologically functioning i.e., these sites may have sufficient topsoil, nurse trees, and pollinators for successful regeneration and succession. Restoration practitioners can only hope (and lobby) for a reform of pertinent carbon crediting rules in forthcoming post-Kyoto climate regimes. Fortunately, the rapidly growing voluntary carbon markets are much more flexible and do not have such restrictions in place. The restoration of degraded forests as well as sustainable forest management are both eligible under these voluntary schemes. Similarly, preventative actions can be credited, e.g. reducing degrading activities such as extensive logging or collection of fuel wood, or assisting natural regeneration through reducing recurrent disturbances caused by man-made fires, all of which are not eligible for participation under the CDM.

A further Kyoto stipulation is that no CDM forestry project can be implemented on areas deforested after 1990, also severely limiting potential ecological reforestation activities (option b). The 1990 requirement acts as an efficient safeguard against perverse incentives which might otherwise encourage the cutting down of natural forests in order to establish carbon plantations (see Section 3.3.1). However, this restrictive regulation excludes a vast number of potentially beneficial reforestation projects on deforested, degrading lands. This applies to roughly 180 million hectares of land that were deforested from 1990 to 2005 (FAO, 2006). Again, voluntary markets provide a useful alternative here because they have much more flexible eligibility requirements. For example, the Voluntary Carbon Standard uses a rolling 10-year threshold (see above), although even this can still be very restrictive.

Apart from the reforestation of lands deforested before 1990, CDM afforestation projects can also occur in areas that have not been forested at any point in time, or at least not for the last fifty years. This could include sites that once supported functioning non-forested ecosystems, such as grasslands and savannahs, although they may have become unproductive and degraded. Afforestation in these areas could involve substituting one previously existing ecosystem with a very different one (option c). Substitutions, used for a transitional period, can sometimes be useful for restoration aims. For example, certain exotic plant species can be extremely efficiently employed for bioremediation, removing toxic chemicals from soils on mining sites, which then allows the natural systems to recover (Cooke and Johnson, 2002). However, permanent replacement of vegetation cover can be questionable from an ecological restoration point of view, especially if this occurs over large expanses of land, and in systems which could have potentially recovered to their natural ecological state. In these cases, afforestation could interfere with natural succession, especially if non-native species are used that may then dominate the system and change both its physical and biological attributes (Versfeld and van Wilgen, 1986; Richardson, 1998).

In the previous section, we discussed how CDM and voluntary carbon forestry projects can be carried out in regions heavily affected by degradation and desertification (Plate 3.1), including those in arid and semi-arid climatic zones. In practice, however, carbon finance may be easier to obtain in certain environments than in others. Forests planted in the tropics, particularly in humid zones, achieve much higher growth rates and absolute carbon densities, resulting in higher revenues from carbon sales. For example, according to FAO estimates, the average carbon density of forests in Sudan is approximately 6 tons of carbon per hectares, whereas the average value for Malaysia is 102 tons (equalling 374 tons of CO_2) and can be three times this value even in lowland *Dipterocarp* forests (FAO, 2006). Most arid regions may therefore struggle to attract carbon finance because they cannot support high forest biomass levels or growth rates. However, these regions are among the areas most affected by soil degradation and desertification. Managers of restoration projects in these regions may therefore have to rely to a greater extent on additional, non-carbon market-based finance or entirely on public and philanthropic funding.

Another issue that does fit easily into the current carbon forestry framework is fire management. Some tropical dry forest systems require a regime of occasional fires for maintaining natural processes for the regeneration of certain species. This, however, leads to temporary reductions in the level of carbon stored in a sequestration project and also to lower average carbon densities in the long term. A similar example relates to the impact of controlling invasive plant species, often an integral component of restoration programs and considered by some to be one of the "big five" environmental issues of our times (Sala *et al.*, 2000). Removing biomass, while ecologically necessary, would negatively impact the carbon balance of a project at least in the short term.

On a very practical level, transaction costs for designing CDM forestry projects, passing external validation and verification, and obtaining final approval from the CDM Executive Board can be quite high. Similarly, applying high-quality voluntary standards currently under development is likely to entail transaction costs comparable to those of the CDM. Considering that such transaction costs can easily surpass US$ 100,000 per project and do not vary significantly with project size, larger reforestation projects have a definite economic advantage over smaller ones. It may be difficult for many smaller projects to generate carbon revenues that are significantly higher than the transaction costs involved. Given that ecological restoration activities are often carried out in relatively small and patchy areas, they may incur disproportionately high transaction costs, possibly hampering their viability as carbon projects. Some adjustments in documentation required have been made under the CDM in order to lower transaction costs to encourage small-scale projects, defined as having a maximum sequestration potential of 40,000 tons CO_2 per year. However, costs for validation and verification are still expected often to be high. Although the CDM allows the bundling of small projects (the implementation of reforestation on multiple sites under one project), this increases the administrative complexity and may make it difficult to apply consistent baseline and leakage scenarios because the ecological and socioeconomic land-use context on a very local level may vary. Efforts are ongoing to find ways to lower carbon market transaction costs for smaller projects, for example through "programs of activities." These would allow for the addition of further project components to an existing, approved carbon project with a much less costly statistical sampling approach conducted for verifying compliance. Transaction costs should also decrease as local expertise becomes available in developing countries, reducing the need for often expensive international consultants. Moreover, regulatory reforms could rid instruments like the CDM of cumbersome requirements that do not contribute significantly to ensuring climate benefits or broader environmental integrity.

Finally, one should not forget that carbon finance is still a very young and comparatively immature instrument with uncertain and evolving prospects. Although recent market growth rates have been impressive, with both regulatory and voluntary market segments almost doubling in size every year for several years, the overall size of these markets is closely correlated to emission reduction targets adopted by countries and companies. Furthermore, changes to regulations may impact the ability to use international offset projects for compliance or generate credits from certain project types. It would in any case be naïve to expect carbon markets to solve all financing challenges that have plagued conservation and restoration ecology for the last decades.

3.7 Conclusions and outlook

Deforestation, desertification, and other forms of land degradation are among the major environmental challenges the planet faces today. Climate change is directly linked to these in several ways because it contributes and exacerbates ongoing degradation and because it is itself fuelled by emissions from land-use change. There are thus many ways, in principle, to integrate the mitigation of both climate change and land degradation. International carbon markets linked to international climate mitigation efforts have evolved extremely rapidly and now transact many billions of dollars a year. It needs to be kept in mind that carbon markets have only reached a meaningful scale in the last few years and are still very much developing. It would be too much to expect such a new instrument to effectively address several complex global problems at the same time and without contradictions. Nevertheless, their pure financial volume but also the great flexibility they offer through their project-based mechanisms (CDM and voluntary offset projects) offer great potential for combining different environmental agendas. The land-use sector in developing countries includes some of the lowest-cost options for carbon emissions abatement at a very significant scale (Enkvist *et al.*, 2007) (Figure 3.5). Against the backdrop of chronically insufficient and stagnating international donor funding, carbon markets thus hold an immense potential for leveraging funding both for avoiding degradation and for restoration of ecosystems on a global scale.

While carbon credits alone may not be sufficient to cover the costs of forest restoration projects at today's market prices, they can provide a crucial additional income for restoration projects. Using carbon finance effectively can allow restoration activities to be implemented on a larger scale. Although carbon forestry projects value primarily only one ecosystem service, the sequestration of carbon, this does not have to translate into a conflict of goals with restoration aims. Well-designed projects can avoid many of the risks identified in our discussion and focus on integrating non-carbon oriented elements. In many cases, multiple income sources will have to be tapped and there are increasing opportunities to evaluate the multiple benefits of forests. For example, alongside the UNFCCC, various innovative financing mechanisms for other ecosystem services are being explored under the Convention on Biological Diversity (CBD) and the United Nations Convention to Combat Desertification and Land Degradation (UNCCD) (Ebeling and Fehse, 2008). Furthermore, the international community is also exploring potential funding mechanisms for the adaptation of natural ecosystems to expected climate change. Such adaptation funding, together with carbon finance directed at climate change mitigation, may be used to establish the necessary conditions for ecosystems to adapt and for restoring vulnerable systems in a way that makes them more resilient to forthcoming climatic changes and land-use pressures.

Apart from such non-carbon based finance sources, carbon markets themselves are increasingly distinguishing between different types of credits and projects. In particular, projects with clear development and biodiversity co-benefits can often command a price premium, especially on voluntary carbon markets (Ebeling and Fehse, 2008; Hamilton *et al.*, 2008). Whether such "gourmet carbon" will be restricted to niche markets remains to be seen, but forestry projects with an ecologically sound restoration approach are in many ways ideally positioned to monetize on their non-carbon co-benefits and will always have a potential advantage over projects based on single objectives (e.g., energy related projects purely focused on emissions abatement).

It is obvious that carbon sequestration reforestation does not share the multiple goals of ecological restoration and that there are some goals and elements of restoration which carbon forestry will not be able to fulfil optimally. At the same time, a sober analysis should compare the risks and benefits of carbon forestry with a realistic baseline scenario – i.e., what would have happened in the absence of an offset project – instead of an ideal one (Ebeling, 2008). For example, reforestation on degraded lands with non-invasive exotics or a small number of native species is certainly less desirable from a restoration perspective than using a mix of several dozen native tree species. In many cases, however, the realistic alternative is no restoration at all, a much less beneficial outcome by any measure.

Ecologists and restoration practitioners can play an important catalyzing role in realizing synergies through carbon forestry. CDM projects do need to credibly prevent negative environmental and social impacts but they do not as such gain in monetary terms from a focus on valuing and enhancing co-benefits. The emphasis on climate change mitigation may well be appropriate considering that the CDM and voluntary carbon projects are primarily instruments to offset greenhouse gas emissions. However, settling for this outcome could mean that significant opportunities are missed. To fully realize the potential of carbon forestry to contribute to global restoration aims, restoration ecologists need to become more involved in carbon forestry and offer and apply their in-depth knowledge and experience. In many cases it is difficult for carbon project developers to design carbon projects to fit ideally into the local ecological context. They might, however, place great value on such co-benefits in order to ensure good local stakeholder relationships, to sell a good project story to potential carbon buyers, out of a genuine personal conviction, or to design resilient and robust projects from a carbon sequestration perspective. In order to fully realize synergies and avoid conflicts, the goals and design aspects of carbon forestry projects need to be set and mediated by local land-users as well as carbon project developers, restoration ecologists, and funding organizations or carbon buyers.

Carbon markets are still in the process of becoming a more mature and more effective environmental finance instrument. They exist mainly as a result of political agreements, and some uncertainty therefore arises from the limited commitment time-frame of the Kyoto Protocol, which ends in 2012, and its less than global coverage. However, it is virtually certain that climate change will rise in importance as an issue of international concern and that carbon markets will exist beyond 2012. In addition, it is very likely that the land-use and forestry sector will gain much greater prominence in forthcoming climate regimes. Importantly, reforestation could move away from having a single project focus under suggested sectoral mitigation approaches (Boyd *et al.*, 2007). Under such a sectoral approach, any increase in forest cover or carbon stocks in a country could become eligible for receiving carbon credits. The earliest opportunity for realising this may be through the inclusion of "avoided deforestation," or "reduced emissions from deforestation and degradation" (REDD) into a post-Kyoto regime (Ebeling and Yasue, 2008). Although REDD started out as a discussion focused on reducing deforestation, some countries favor the inclusion of carbon benefits through carbon sequestration in a prospective scheme and a recent UNFCCC decision does include this possibility (Ebeling *et al.*, 2008). At the same time, various regulatory carbon markets are taking shape in North America and beyond, and these include forestry much more prominently than the current Kyoto framework.

Forestry, including forest conservation and restoration, is thus firmly anchored in the evolving international climate change agenda. Carbon markets are set to expand alongside growing international efforts to reduce greenhouse gases, and they are bound to focus on the vast mitigation options in the land-use and forestry sector. The main challenge and opportunity for global ecological restoration will thus be to create an effective link with these markets in order to allow for a potentially unprecedented flow of finance into restoration efforts in developing and developed countries.

It is important for conservationists and restoration ecologists to be well aware of the existing challenges of using carbon finance for restoration aims. There are many exaggerated hopes attached to carbon markets, which are partly the result of an imperfect understanding of existing market mechanisms and the surrounding regulatory framework. However, significant opportunities exist to derive multiple restoration benefits from carbon forestry. Ecological restoration is a global challenge and it needs solutions financed on a global scale and implemented locally. Tapping into the largest existing environmental markets can be an important part of this.

References

Aukland, L., Moura Costa, P., Bass, S., Huq, S., Landell-Mills, N., Tipper, R. and Carr, R. (2002). *Laying the Foundations for Clean Development: Preparing the Land Use Sector. A quick guide to the Clean Development Mechanism*. London: IIED- International Institute for Environment and Development Publications.

Aukland, L., Moura Costa, P. and Brown, S. (2003). A conceptual framework and its application for addressing leakage: The case of avoided deforestation. *Climate Policy*, **3**: 123–136.

Bishop, J., Kapila, S., Hicks, F., Mitchell, P. and Vorhies, F. (2008). *Building Biodiversity Business*. London, Gland, Switzerland: Shell International Limited, IUCN- International Union for Conservation of Nature.

Boyd, E., Hultman, N., Roberts, T., Corbera, E. and Ebeling, J. (2007). *The Clean Development Mechanism: Current Status, Perspectives and Future Policy*. Oxford: Tyndall Centre for Climate Change Research, EcoSecurities.

Capoor, K. and Ambrosi, P. (2008). *State and Trends of the Carbon Market 2008*. Washington DC: World Bank.

Chomitz, K. M. (2002). Baseline, leakage and measurement issues: How do forestry and energy projects compare? *Climate Policy*, **2**: 35–49.

Chomitz, K. M. and Lecocq, F. (2003). *Temporary Sequestration Credits: An Instrument for Carbon Bears*. World Bank Policy Research Working Paper No. 3181. Washington DC: World Bank.

Chomitz, K. M., Buys, P., De Luca, G., Thomas, T. S. and Wertz-Kanounnikoff, S. (2007). *At Loggerheads? Agricultural Expansion, Poverty Reduction, And Environment in the Tropical Forests*. World Bank Policy Research Report. Washington DC: World Bank.

Clewell, A., Rieger, J. and Munro, J. (2005). Guidelines for Developing and Managing Ecological Restoration Projects. Tucson, AZ: Society for Ecological Restoration International.

Cooke, J. A. and Johnson, M. S. (2002). Ecological restoration of land with particular reference to the mining of metals and industrial minerals: A review of theory and practice. *Environmental Reviews*, **10**: 41–71.

Cossalter, C. and Pye-Smith, C. (2004). *Fast-Wood Forestry. Myths and Realities. Forest Perspectives No 1*. Bogor, Indonesia: CIFOR-Center for International Forestry Research.

Dessai, S., Schipper, L. F., Corbera, E., Kjellen, B., Gutiérrez, M. and Haxeltine, A. (2005). Challenges and outcomes at the Ninth Session of the Conference of the Parties to the United Nations Framework Convention on Climate Change. *International Environmental Agreements*, **5**: 105–124.

Ebeling, J. (2008). Risks and criticism of forestry-based climate change mitigation and carbon trading. In *Forests, Climate Change and the Carbon Market: Risks and Emerging Opportunities*, ed. C. Streck, R. O'Sullivan and T. Janson-Smith. London, Washington DC: Earthscan, Brookings, pp. 43–58.

Ebeling, J. and Fehse, J. (2008). *Is There a Business Case for High-Biodiversity REDD Projects and Schemes? A Report for the Secretariat of the Convention for Biological Diversity*. Oxford: EcoSecurities.

Ebeling, J., Neeff, T., Henders, S., Moore, C. and Ascui, F. (2008). *REDD Policy Scenarios and Carbon Markets. A Report for the Indonesia Forest Carbon Alliance (IFCA) Commissioned by the Government of Indonesia and the World Bank*. Oxford, Jakarta: EcoSecurities.

Ebeling, J. and Yasue, M. (2008). Generating carbon finance through avoided deforestation and its potential to create climatic, conservation, and human development benefits. *Philosophical Transactions of the Royal Society B*, **363**: 1917–1924.

Enkvist, P. A., Nauclér, T. and Rosander, J. (2007). *A Cost Curve for Greenhouse Gas Reduction. McKinsey Quarterly 2007, Number 1*. Stockholm: McKinsey & Company.

Eswaran, H., Lal, R. and Reich, P. F. (2001). Land degradation: An overview. In *Response to Land Degradation*, ed. E. M. Bridges, I. D. Hannam, L. R. Oldeman, F. W. T. P. D. Vries, S. J. Scherr and S. Sompatpanit. Enfield, NH: Science Publishers Inc., pp. 20–35.

Faeth, P., Cort, C. and Livernash, R. (1994). *Evaluating the carbon sequestration benefits of forestry projects in developing countries*. Washington DC: World Resources Institute.

FAO-Food and Agricultural Organization of the United Nations (2006). *Global Forest Resources Assessment 2005*. Rome: FAO.

Fearnside, P. M. (2001). Saving tropical forests as a global warming countermeasure: An issue that divides the environmental movement. *Ecological Economics*, **39**: 167–184.

Fearnside, P. M., Lashof, D. A. and Moura Costa, P. (2000). Accounting for time in mitigating global warming. *Mitigation and Adaptation Strategies for Global Change*, **5**: 239–270.

Fehse, J. (2008a). Forest carbon and other ecosystem services: Synergies between the Rio conventions. In *Forests, Climate Change and the Carbon Market: Risks and Emerging Opportunities*, ed. C. Streck, R. O'Sullivan and T. Janson-Smith. London, Washington DC: Earthscan, Brookings, 59–70.

Fehse, J. (2008b). Forests and carbon trading. Seeing the wood and the trees. *OECD Observer*, **267**: 41–42.

Feyera, S., Beck, E. and Lüttge, U. (2002). Exotic trees as nurse-trees for the regeneration of natural tropical forests. *Trees – Structure and Function*, **16**: 245–249.

Geldenhuys, C. J. (2004). Concepts and process to control invader plants in and around natural evergreen forest in South Africa. *Weed Technology*, **18**: 1386–1391.

Global Mechanism (2007). *Description of the GM's Strategic Programme for the Compensation for Ecosystem Services*. Rome: Global Mechanism to the UNCCD pages.

Hamilton, K., Sjardin, M., Marcello, T. and Xu, G. (2008). *State of the Voluntary Carbon Markets 2008. Forging a Frontier*. Washington DC, London: The EcoSystem Marktplace, New Carbon Finance.

IPCC-Intergovernmental Panel on Climate Change (2003). *Good Practice Guidance for Land Use, Land-Use Change and Forestry*. Kanagawa, Japan: IPCC.

ITTO-International Tropical Timber Organization (2002). *Guidelines for the Restoration, Management and Rehabilitation of Degraded and Secondary Tropical Forests*. Yokohama, Japan: ITTO.

Jackson, R. B., Jobbágy, E. G., Avissar, R., Baidya Roy, S., Barrett, D., Cook, C. W., Farley, K. A., le Maitre, D. C., McCarl, B. A. and Murria, B. C. (2005). Trading water for carbon with biological carbon sequestration. *Science & Sports*, **310**: 1944–1947.

Lamb, D., Erskine, P. D. and Parrotta, J. A. (2005). Restoration of degraded tropical forest landscapes. *Science*, **310**: 1628–1632.

Landell-Mills, N. and Porras, I. T. (2002). *Silver Bullet or Fools' Gold? A Global Review of Markets for Forest Environmental Services and Their Impact on the Poor*. London: IIED-International Institute for Environment and Development Publications.

Moura Costa, P. and Stuart, M. (1998). Forestry based greenhouse gas mitigation: a short story of market evolution. *Commonwealth Forestry Review*, September 1998.

Moura Costa, P. and Tipmann, R. (2003). Developments under the CDM: Project examples in the Brazilian energy sector. *ReFocus*, Jan–Feb 2003: pp. 42–47.

Moura Costa, P. and Wilson, C. (2000). An equivalence factor between CO_2 avoided emissions and sequestration – description and applications in forestry. *Mitigation and Adaptation Strategies for Global Change*, **5**: 51–60.

Moura Costa, P. H., Stuart, M. D. and Trines, E. (1997). SGS Forestry's carbon offset verification service. In *Greenhouse Gas Mitigation. Technologies for Activities Implemented Jointly. Proceedings of Technologies for AIJ Conference. Vancouver, May 1997*, ed. P. W. F. Edriermer, A. Y. Smith and K. V. Thambimuthu. Oxford: Elsevier, pp. 409–414.

Moura Costa, P., Stuart, M., Pinard, M. and Phillips, G. (2000). Issues related to monitoring, verification and certification of forestry-based carbon offset projects. *Mitigation and Adaptation Strategies for Global Change*, **5**: 39–50.

Neeff, T., Eichler, L., Deecke, I. and Fehse, J. (2007). *Update on Markets for Forestry Offsets*. Oxford: The FORMA project, CATIE.

Neeff, T. and Ebeling, J. (2008). The future of forestry offsets – Will voluntary markets overtake the CDM? In *Greenhouse Gas Market Report 2007 – Building upon a Solid Foundation: The Emergence of a Global Emissions Trading System*, ed. D. Lunsford. Geneva: IETA-International Emissions Trading Association, pp. 132–135.

New Carbon Finance (2009). *Press Release – Carbon Market up 84% in 2008 at $118bn*. London: New Carbon Finance.

Pagiola, S., Agostini, P., Gobbi, J., de Haan, C., Ibrahim, M., Murgueitio, E., Ramirez, E., Rosales, M. and Ruiz, J. P. (2005). Paying for biodiversity conservation services – experience in Colombia, Costa Rica, and Nicaragua. *Mountain Research and Development*, **25**: 206–211.

Pearson, T., Walker, S. and Brown, S. (2006). *Guidebook for the Formulation of Afforestation and Reforestation Projects under the Clean Development Mechanism. ITTO Technical Series 25*. Yokohama, Japan: International Tropical Timber Organization.

Pedroni, L. (2005). Carbon accounting for sinks in the CDM after CoP-9. *Climate Policy*, **5**: 407–418.

Richardson, D. M. (1998). Forestry trees as invasive aliens. *Conservation Biology*, **12**: 18–26.

Robertson, M. M. (2004). The neoliberalization of ecosystem services: Wetland mitigation banking and problems in environmental governance. *Geoforum*, **35**: 361–373.

Rourke, J. T. (2000). *International Politics on the World Stage*. Guilford, CT: Dushkin/ MacGraw-Hill.

Sala, O. E., Chapin III, F. A., Armesto, J. J., Berlow, R., Bloomfield, J., Dirzo, R., Huber-Sanwald, E., Huenneke, L. F., Jackson, R. B., Kinzig, A., Leemans, R., Lodge, D., Mooney, H. A., Oesterheld, M., Poff, N. L., Sykes, M. T., Walker, B. H., Walker, M. and Wall, D. H. (2000). Global biodiversity scenarios for the year 2100. *Science*, **287**: 1770–1774.

Scherr, S., White, A. and Khare, A. (2004). *For Services Rendered. The Current Status and Future Potential of Markets for the Ecosystem Services Provided by Tropical Forests. ITTO Technical Series*. Yokohama, Japan: International Tropical Timber Organization.

Schwarze, R., Niles, J. O. and Olander, J. (2002). *Understanding and Managing Leakage in Forest-Based Greenhouse Gas Mitigation Projects*. Washington DC: The Nature Conservancy.

Stuart, M. D. and Moura Costa, P. (1998). Greenhouse gas mitigation: A review of international policies and initiatives. *Policies that Work for People Series*. London: IIED-International Institute of Environment and Development.

Tomaselli, I. (2006). *Brief Study on Funding and Finance for Forestry and Forest-Based sector*. New York: United Nations Forum on Forests.

UNEP-United Nations Environment Programme (2008). *CDM Pipeline Overview*. Roskilde, Denmark: UNEP Risoe Centre on Energy, Climate Sustainable Development. (www.cd4cdm.org/).

UNFCCC-United Nations Framework Convention on Climate Change (1998). *Kyoto Protocol to the Convention on Climate Change*. Bonn, Germany: UNFCCC.

UNFCCC-United Nations Framework Convention on Climate Change (2001). *The Marrakesh Accords & The Marrakesh Declaration*. Geneva: UNFCCC.

UNFCCC-United Nations Framework Convention on Climate Change (2003). *Report of the Conference of the Parties on its Ninth Session, held at Milan from 1 to 12 December 2003. FCCC/CP/2003/6/Add.2*. Milan: United Nations Framework Convention on Climate Change.

UNFCCC-United Nations Framework Convention on Climate Change (2005). *Caring for Climate. A guide to the Climate Change Convention and the Kyoto Protocol*. Bonn: UNFCCC.

Versfeld, D. B. and van Wilgen, B. W. (1986). Impact of woody aliens on ecosystem properties. In *The ecology and management of biological invasions in Southern Africa*, ed. I. A. W. MacDonald and F. J. Kruger. Cape Town, SA: Oxford University Press, pp. 68–93.

Voluntary Carbon Standard (2008). *Voluntary Carbon Standard 2007.1 – Specification for the Project-Level Quantification, Monitoring and Reporting as Well as Validation and Verification of Greenhouse Gas Emission Reductions or Removals*. London: VCS Association.

World Bank (2006). *BioCarbon Fund Handbook*. Washington DC: World Bank.

4

The value of a restored Earth and its contribution to a sustainable and desirable future

ROBERT COSTANZA

4.1. Introduction

The purpose of the economy *should be* to provide for the sustainable well-being of people. That goal encompasses material well-being, certainly, but also anything else that affects well-being and its sustainability. This seems obvious and non-controversial. The problem comes in determining what things actually affect well-being and in what ways.

There is substantial new research on this "science of happiness" that shows the limits of earning and spending money in supporting well-being. Kasser (2003) points out, for instance, that people who focus on material consumption as a path to happiness are actually less happy and even suffer higher rates of both physical and mental illnesses than those who do not. Material consumption beyond real need is a form of psychological "junk food" that only satisfies for the moment and ultimately leads to depression, Kasser says.

Easterlin (2003) has shown that well-being tends to correlate well with health, level of education, and marital status, and with income only up to a fairly low threshold. He concludes that "people make decisions assuming that more income, comfort, and positional goods will make them happier, failing to recognize that hedonic adaptation and social comparison will come into play, raise their aspirations to about the same extent as their actual gains, and leave them feeling no happier than before. As a result, most individuals spend a disproportionate amount of their lives working to make money, and sacrifice family life and health, domains in which aspirations remain fairly constant as actual circumstances change, and where the attainment of one's goals has a more lasting impact on happiness. Hence, a reallocation of time in favor of family life and health would, on average, increase individual happiness." Layard (2005) echoes many of these ideas and concludes that current economic policies are not improving happiness and that "happiness should become the goal of policy, and the progress of national happiness should be

Ecological Restoration: A Global Challenge, ed. Francisco A. Comin. Published by Cambridge University Press.
© Cambridge University Press 2010.

measured and analyzed as closely as the growth of GNP." Several countries are now interested in alternative measures of progress. For example, the country of Bhutan has recently announced that it will make "Gross National Happiness" its explicit policy goal.

Frank (1999) also concludes that the nation (and the world) would be better off, (that is, overall well-being would be higher), if we actually consumed less and spent more time with family and friends, working for our communities, maintaining our physical and mental health, and enjoying nature.

There is substantial and growing evidence that natural systems contribute greatly to human well-being. Costanza *et al.* (1997) estimated that the annual, non-market value (contribution to human well-being) of the Earth's ecosystem services is substantially larger than global GDP. The UN Millennium Ecosystem Assessment (2005) is a global compendium of ecosystem services and their contributions to human well-being. Vermuri and Costanza (2006) demonstrated a significant relationship between national life satisfaction estimates from surveys and an index of ecosystem services.

So, if we want to assess the "real" economy, all the things that contribute to real, sustainable, human welfare and quality of life, as opposed to only the "market" economy, we have to measure the non-marketed contributions to human well-being from nature, from family, friends and other social relationships at many scales, and from health and education. One convenient way to summarize these contributions is to group them into four basic types of capital that are necessary to support the real, human-welfare producing economy: built capital, human capital, social capital, and natural capital.

The market economy covers mainly built capital (factories, offices, and other built infrastructure and their products) and part of human capital (spending on labor), with some limited spillover into the other two types. Human capital includes the health, knowledge, and all the other attributes of individual humans that allow them to function in a complex society. Social capital includes all the formal and informal networks among people: family, friends, and neighbors, as well as social institutions at all levels, like churches, social clubs, local, state, and national governments, NGOs, international organizations, and the institutions of the market itself. Natural capital includes the world's ecosystems and all the services they provide that support human well-being. Ecosystem services occur at many scales, from climate regulation at the global scale, to flood protection, soil formation, nutrient cycling, recreation, and aesthetic services at the local and regional scales.

4.2 Ecosystem services and natural capital

Ecosystem *functions* refer variously to the habitat, biological, or systems properties or processes of ecosystems. Ecosystem *goods* (e.g., food) and *services* (e.g., waste assimilation) represent the benefits human populations derive, directly or indirectly, from ecosystem functions. A large number of functions and services have been identified (Costanza *et al.*, 1997; Millennium Ecosystem Assessment, 2005; Farber *et al.*, 2006). This analysis uses the seventeen categories listed in Costanza *et al.* (1997). These are shown in Table 4.1. Only renewable ecosystem services, excluding non-renewable fuels and minerals and the atmosphere, are included. Note that there is not necessarily a one-to-one correspondence between ecosystem services and functions. In some cases a single ecosystem service is the product of two or more ecosystem functions whereas in other cases a single ecosystem function contributes to two or more ecosystem services.

The Millennium Ecosystem Assessment (2005) uses the following summary categorization of ecosystem services:

(a) Provisioning services: services from products obtained from ecosystems. These products include food, fuel, fiber, biochemicals, genetic resources and fresh water. Many, but not all, of these services are traded in markets.

(b) Regulating services: services received from regulation of ecosystem processes. This category includes a host of pathways that stem from the functioning of ecosystems and influence people in ways both direct and indirect. These services include flood protection, human disease regulation, water purification, air quality maintenance, pollination, pest control and climate control. These services are generally not marketed, but many have clear value to society.

(c) Cultural services: services that contribute to the cultural, spiritual and aesthetic dimensions of people's well-being. They also contribute to establishing a sense of place.

(d) Supporting services: services that maintain basic ecosystem processes and functions such as soil formation, primary productivity, biogeochemistry and provisioning of habitat. These services affect human well-being indirectly by maintaining processes necessary for provisioning, regulating and cultural services.

It is also important to emphasize the interdependent nature of many ecosystem functions. For example, some of the net primary production in an ecosystem ends up as food, the consumption of which generates respiratory products necessary for primary production.

In general, *capital* is considered to be a stock of materials or information which exists at a point in time. Each form of capital stock generates, either autonomously or in conjunction with services from other capital stocks, a flow of services which may be used to transform materials, or the spatial configuration of materials, to enhance the welfare of humans. The human use of this flow of services may or may not leave the original capital stock intact. Capital stock takes different identifiable forms, most notably in physical forms including natural capital, such as trees, minerals, ecosystems, the atmosphere, etc.; manufactured capital, such as machines and buildings; and the human capital of physical bodies (Costanza and Daly, 1992). In addition, capital stock can take intangible forms, especially as

Table 4.1. *Ecosystem services and functions.*

#	Ecosystem service[*]	Ecosystem functions	Examples
1	Gas regulation	Regulation of atmospheric chemical composition	CO_2/O_2 balance, O_3 for UVB protection, and SO_x levels
2	Climate regulation	Regulation of global temperature, precipitation, and other biologically mediated climatic processes at global or local levels	Greenhouse gas regulation, dimethyl sulfide (DMS) production affecting cloud formation
3	Disturbance regulation	Capacitance, damping, and integrity of ecosystem response to environmental fluctuations	Storm protection, flood control, drought recovery, and other aspects of habitat response to environmental variability mainly controlled by vegetation structure
4	Water regulation	Regulation of hydrological flows	Provisioning of water for agricultural (e.g., irrigation) or industrial (e.g., milling) processes or transportation
5	Water supply	Storage and retention of water	Provisioning of water by watersheds, reservoirs, and aquifers
6	Erosion control and sediment retention	Retention of soil within an ecosystem	Prevention of loss of soil by wind, runoff, or other removal processes, storage of silt in lakes and wetlands
7	Soil formation	Soil formation processes	Weathering of rock and the accumulation of organic material
8	Nutrient cycling	Storage, internal cycling, processing, and acquisition of nutrients	Nitrogen fixation, nitrogen cycle, phosphorus cycle and other elemental or nutrient cycles
9	Waste treatment	Recovery of mobile nutrients and removal or breakdown of excess or xenic nutrients and compounds	Waste treatment, pollution control, detoxification

Table 4.1. (*cont.*)

#	Ecosystem service[*]	Ecosystem functions	Examples
10	Pollination	Movement of floral gametes	Provisioning of pollinators for the reproduction of plant populations
11	Biological control	Trophic-dynamic regulations of populations	Keystone predator control of prey species, reduction of herbivory by top predators.
12	Refugia	Habitat for resident and transient populations	Nurseries, habitats for migratory species, regional habitats for locally harvested species, over wintering grounds.
13	Food production	That portion of gross primary production extractable as food	Production of fish, game, crops, nuts, fruits by hunting, gathering, subsistence farming, or fishing
14	Raw materials	That portion of gross primary production extractable as raw materials	The production of lumber, fuel and fodder
15	Genetic resources	Sources of unique biological materials and products	Medicine, products for materials science, genes for resistance to plant pathogens and crop pests, ornamental species (pets and horticultural varieties of plants)
16	Recreation	Providing opportunities for recreational activities	Eco-tourism, sport fishing, and other outdoor recreational activities
17	Cultural	Providing opportunities for non-commercial uses.	Aesthetic, artistic, educational, spiritual, and/or scientific values of ecosystems

[*] We include ecosystem "goods" along with ecosystem services.
Source: Costanza *et al.* (1997).

information such as that stored in computers and in individual human brains, as well as that stored in the genetic code of individual organisms, and the structure of ecosystems.

Ecosystem services consist of flows of materials, energy, and information from natural capital stock, which combine with manufactured and human capital services to produce human welfare. As noted above, they are the benefits humans derive (directly and indirectly) from functioning ecosystems.

The stock of natural capital has been decreasing rapidly in recent times (Millennium Ecosystem Assessment, 2005). This has affected the flow of ecosystem services. In this chapter the dynamics and value of the world's ecosystem services and natural capital are assessed from 1900 to 2000 and in two hypothetical future scenarios: a "business as usual" (BAU) scenario and a "restored Earth" (RE) scenario. Ecological restoration can thus be expressed in terms of the degree to which we can restore the functions and services of the world's natural capital.

4.3 Ways to restore the Earth's natural capital

There are two basic ways to restore natural capital and the ecosystem services that flow from it: by expanding the land area covered by "natural" ecosystems (those with relatively low levels of human domination), and by changing human practices within all types of ecosystems. For example, we can increase forest area by converting marginal agricultural land to forest, or we can change our practices on existing agricultural land to enhance ecosystem services production. Much of our current agricultural land is marginal for food production and would be more valuable to society used in other ways. For example, Balmford *et al.* (2002) estimated that the net benefits on a global scale of expanding the global nature reserve network to cover 15 percent of the terrestrial biosphere (from about 6 percent currently) and 30 percent of the marine biosphere (from essentially 0 percent currently) would be on the order of US$ 4.4 – US$ 5.2 trillion per year. These net benefits are the difference between the value of the ecosystem services produced by the intact natural ecosystem, minus the value of the most likely human-modified ecosystem. For example, intact mangroves are roughly twice as valuable as shrimp ponds. When this is compared to the roughly US$ 45 billion per year estimated cost of creating and maintaining this global reserve network, one sees that the benefit/cost ratio is about 100:1, an extremely good social investment. Of course, the problem is that many of the ecosystem service benefits of intact natural systems are social, non-marketed benefits, while the benefits of the modified, human-dominated systems are private, marketed benefits. This means that the socially optimal solution will not be achieved unless efforts are made by governments or other groups to protect and restore natural systems, and/or to modify market incentives appropriately.

In this chapter the general scenario used in Balmford *et al.* (2002) will be defined as the "restored Earth" (RE) scenario, and compared with a "business as usual" (BAU) scenario based on current land use and practice trends. The business as usual scenario assumes continued degradation of the Earth's natural capital, both by conversion to urban and agricultural uses and by mismanagement of the remaining natural ecosystems. The RE scenario assumes the Balmford *et al.* (2002) system of terrestrial and marine protected areas (15 percent of the terrestrial biosphere and 30 percent of the marine biosphere). These protected areas are marginal in terms of agricultural potential and establishing the marine areas would actually increase global wild fish production. In addition, the RE scenario assumes changes in management of the remaining agricultural, urban, and heavily managed natural systems to adaptive management regimes that focus on maximizing ecosystem services while also enhancing agricultural productivity and urban quality of life. This scenario is also consistent with the World Resources (2005) recommendations for achieving a restored and sustainable Earth (Table 4.2), with the "Ecotopia" scenario presented in Costanza (2000) and Boumans *et al.* (2002) scenario, and with the Millennium Ecosystem Assessment (2005) "Adapting Mosaic" scenario (see below).

4.4 Estimates of the value of a restored Earth

To estimate the value of a restored Earth, the results of several prior studies were synthesized, including the Global Unified Metamodel of the Biosphere (GUMBO) model (Boumans *et al.* 2002), per unit area estimates of ecosystem services from Costanza *et al.* (1997), and the Millennium Ecosystem Assessment (2005) scenarios. GUMBO was developed to simulate the integrated Earth system and assess the dynamics and values of ecosystem services. It is a "metamodel" in that it represents a synthesis and a simplification of several existing dynamic global models in both the natural and social sciences at an intermediate level of complexity. The current version of the model contains 234 state variables, 930 total variables and 1715 parameters. GUMBO is the first global model to include the feedbacks among human technology, economic production and welfare, and ecosystem goods and services within the dynamic Earth system. GUMBO includes modules to simulate carbon, water, and nutrient fluxes through the *Atmosphere*, *Lithosphere*, *Hydrosphere*, and *Biosphere* of the global system. Social and economic dynamics are simulated within the *Anthroposphere*. GUMBO links these five spheres across eleven biomes, which together encompass the entire surface of the planet. The dynamics of eleven major ecosystem goods and services for each of the biomes are simulated and evaluated. Historical calibrations from 1900 to 2000 for fourteen key variables, for which quantitative time-series data were available, produced an average R^2 of 0.922. A range of future scenarios representing different assumptions about technological change, investment strategies and other factors have been simulated. The relative value of ecosystem services in terms of their contribution to supporting both conventional economic production and human well-being more broadly defined were estimated under each scenario. The GUMBO scenarios are broadly consistent with the Millennium Ecosystem Assessment (2005) scenarios for land-use change and human well-being by the year 2050 (Plate 4.1).

Table 4.2. *World Resources Institute recommendations for a sustainable future.*

Type of action	Recommendations
Manage ecosystems better for higher productivity.	Improve the stewardship of ecosystems by adopting an ecosystem approach to management – recognizing the complexity of ecosystems, and living within their limits. Good stewardship brings higher productivity, which is the foundation of a sustainable income stream.
Get the governance right to insure access to environmental income.	Confer legally recognized resource rights (such as individual or communal, title, or binding co-management agreements). Where possible, decentralize ecosystem management to the local level (community-based, natural resource management), while providing for regional or national coordination of local management plans. Empower the poor through access to information, participation, and justice. Create local institutions that represent their interests and accommodate their special needs.
Commercialize ecosystem goods and services to turn resource rights and good stewardship into income.	Improve the marketing and transport of nature-based goods produced by the poor. Make credit available for ecosystem-based enterprises. Capture greater value from the commodity chain. Partner with the private sector. Take care to keep successful commercial activities sustainable.
Tap new sources of environmental income such as payments for environmental services.	Make the newly developing market of payments for environmental, services more pro-poor by expanding the array of eligible activities and payment schemes. Look upon ecosystem income as a portfolio of many different income sources. Diversify this portfolio to reduce risk, and enhance the bottom line.

Source: World Resources (2005).

In the Millennium Ecosystem Assessment (MEA), the Order from Strength scenario is described as a "regionalized and fragmented world, concerned with security and protection, emphasizing primarily regional markets, paying little attention to public goods, and taking a reactive approach to ecosystem problems." This is the business as usual (BAU) scenario and is consistent with the "Mad Max" scenario used in GUMBO. As shown in Figure 4.1, this scenario implies a large decrease in

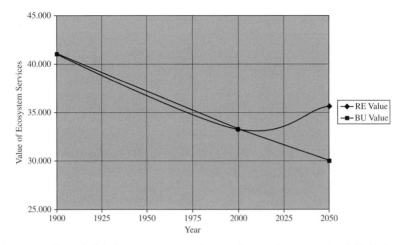

Fig. 4.1. Value of global ecosystem services under Business as Usual (BAU) and restored Earth (RE) scenarios.

all ecosystem services by 2050. Provisioning ecosystem services fall especially in developed countries due to overharvesting. There are also large reductions in regulating and cultural services in both developed and developing countries.

The Adapting Mosaic scenario in the MEA is described as "regional watershed-scale ecosystems are the focus of political and economic activity. Local institutions are strengthened and local ecosystem management strategies are common; societies develop a strongly proactive approach to the management of ecosystems." This is the restored Earth (RE) scenario and is consistent with the Ecotopia scenario in GUMBO. As shown in Figure 4.1, this scenario implies a significant increase in all ecosystem services by 2050. Provisioning services in developed countries increase by a smaller amount than in developing countries due to the fact that these countries were already harvesting near the limit in 2000. These results are broadly consistent with the results from GUMBO.

Table 4.3 is a synthesis of these results broken down by land area for sixteen biomes. The per unit area values of global ecosystem services (from Costanza *et al.*, 1997) are listed for each biome and the resulting total global values for 1900, 2000, and 2050 for the BAU and RE scenarios are estimated. This calculation assumes that the per unit area values for each ecosystem type do not change, and only the areas of each biome change in ways consistent with the scenarios discussed earlier. In fact, we might expect the per unit area values to change as well, but it is hard to guess in which direction and by how much. Per unit area values may increase either due to improved management or due to increased scarcity.

Figure 4.1 plots the estimated total value of ecosystem services from 1900 to 2050 under both scenarios. For the BAU scenario, one can see a roughly 10 percent reduction

Table 4.3. *Summary of average global value of annual ecosystem services for 1900, 2000, and two scenarios for 2050.*

Biome	1900, area (e6 ha)	2000, area (e6 ha)	2050, area (e6 ha)		Total value per ha ($/ha/yr)	1900, total flow value (e9 $/yr)	2000, total flow value (e9 $/yr)	2050, value (e9 $/yr)	
			BAU	RE				BAU	RE
Marine	36302	36302	36302	36302	$577	$20 949	$20 949	$18 934	$20 949
Open ocean	33200	33200	33200	33200	$252	$8 381	$8 381	$8 381	$8 381
Coastal	3102	3102	3102	3102	$4 052	$12 568	$12 568	$10 554	$12 568
Estuaries	180	180	180	180	$22 832	$4 110	$4 110	$4 110	$4 110
Seagrass/algae beds	200	200	100	200	$19 004	$3 801	$3 801	$1 900	$3 801
Coral reefs	62	62	0	62	$6 075	$375	$375	$0	$375
Shelf	2660	2660	2822	2660	$1 610	$4 283	$4 283	$4 543	$4 283
Terrestrial	15323	15323	15323	15323	$804	$20 136	$12 319	$11 040	$14 746
Forest	5793	4855	4723	4955	$969	$5 616	$4 706	$4 578	$4 907
Tropical	2267	1900	1848	2000	$2 007	$4 549	$3 813	$3 709	$4 013
Temperate/boreal	3526	2955	2874	2955	$302	$1 066	$894	$869	$894
Grass/rangelands	3125	3898	4403	3898	$232	$726	$906	$1 023	$906
Wetlands	815	330	241	480	$14 785	$12 054	$4 879	$3 558	$7 097
Tidal Marsh/mangroves	408	165	120	240	$9 990	$4 072	$1 648	$1 202	$2 398
Swamps/floodplains	408	165	120	240	$19 580	$7 982	$3 231	$2 356	$4 699
Lakes/rivers	200	200	200	200	$19 580	$1 700	$1 700	$1 700	$1 700
Desert	1833	1925	1883	1883	$8 498				
Tundra	970	743	620	700					
Ice/rock	1955	1640	892	1367					
Cropland	441	1400	1983	1500	$92	$40	$128	$182	$137
Urban	190	332	379	340					
Total	51624	51625	51625	51625		$41 085	$33 268	$29 974	$35 695

BAU = business as usual; RE = restored Earth.

in the value of ecosystem services from the 2000 conditions. This is surely a very conservative estimate of the losses, and assumes that there are no major collapses of ecosystem functioning, which might occur, for example, as a result of climate change.

The RE scenario, on the other hand, creates a 7 percent recovery overall (19 percent for terrestrial systems) to a state where the value of ecosystem services is 87 percent of the 1900 conditions.

These scenarios are obviously crude estimates of what might be possible and desirable. It is clear, however, that a restored Earth is possible, and it is a much more desirable state than the one we are headed for with the business as usual scenario.

4.5 Paths to a restored Earth

We have determined the desirability of the RE scenario, but how can it be achieved? The first, and probably most important, step is to share this vision widely, and to point out its superiority to the BAU scenario. It is important to emphasize that the RE scenario represents an improvement in the overall sustainable quality of life on Earth. It is not a sacrifice. On the contrary, the BAU scenario is a sacrifice of quality of life in favor of a large amount of unnecessary consumption by a relatively few.

In order to share this vision successfully, it is important to flesh out the RE scenario in quite some detail. Costanza (2000) and the Millennium Ecosystem Assessment (2005) have done this with narrative descriptions of each of their scenarios. A much more elaborate description is probably necessary, however, in order to reach a broader audience and allow them to understand what living in the restored Earth would be like. Novels and films set in this positive possible future would be one way to help bring this vision to life.

Once the vision is widely shared, the paths to achieving it will more easily reveal themselves (Costanza, 2000). Stepping stones along this path include ecological tax reform (Bernow *et al.*, 1998), principles of sustainable governance (Costanza *et al.*, 1998), environmental assurance bonds (Costanza and Cornwell, 1992) and common asset trusts (Barnes, 2006). But interest in and acceptance of these methods will increase dramatically only after a broad consensus on the desirability of the RE vision of the future is achieved.

It is clear that in order to achieve a restored Earth, we must not only preserve remaining natural systems, but also restore degraded systems. The history of socio-ecological systems on Earth is littered with examples of societies that lowered their resilience by degrading their natural systems, often leading to collapse (Diamond, 2005; Costanza *et al.*, 2007). Maintaining resilience and achieving sustainability will require us to back away from the edge of exploiting natural systems to the maximum. It will require the preservation and restoration of natural capital and its precautionary and adaptive management in the face of huge and irreducible uncertainty.

Doing all of this will require us to better integrate our understanding and modelling of humans as creatures embedded in the rest of nature. We must focus our research efforts at least as much on synthesis as we have on analysis and build models that transcend disciplinary boundaries (Costanza, 2003). "Ecosystem services" is an inherently transdisciplinary concept linking human well-being with the rest of nature. Progress in achieving the RE scenario will require much additional research not only on the methods to restore ecological systems, but on the positive impacts of that restoration on human well-being and the sustainability of that well-being.

4.6 Conclusions

Several broad conclusions can be drawn from this sketch:

- The environment is not a luxury good. Ecosystem services contribute to human welfare and survival in innumerable ways, both directly and indirectly, and represent the most of the economic value on the planet, especially for the poor.
- Ecosystem services, and the natural capital stocks that produce them, have been depleted and degraded by human actions to the point that the sustainability of the system is threatened.
- A restored Earth (RE) scenario would increase the value of ecosystem services and the sustainability of life of people on Earth much more than a Business as Usual (BAU) scenario.
- To begin to move to an RE scenario, we must first emphasize and communicate to people the information that this scenario represents an improvement in people's sustainable quality of life, not a sacrifice or step backward.
- A sustainable and desirable future is both possible and practical, but we first have to create and communicate the vision of that world in compelling terms.

References

Balmford, A., Bruner, A., Cooper, P. *et al.* (2002). Economic reasons for conserving wild nature. *Science*, **297**: 950–953.

Barnes, P. (2006). *Capitalism 3.0: A Guide to Reclaiming the Commons*. New York: Berrett-Koehler.

Bernow, S., Costanza, R., Daly, H *et al.* (1998). Ecological tax reform. *BioScience*, **48**: 193–196.

Boumans, R., Costanza, R., Farley, J. *et al.* (2002). Modeling the dynamics of the integrated Earth system and the value of global ecosystem services using the GUMBO model. *Ecological Economics*, **41**: 529–560.

Costanza, R. (2000). Visions of alternative (unpredictable) futures and their use in policy analysis. *Conservation Ecology*, **4** (1): 5. (www.consecol.org/vol4/iss1/art5)

Costanza, R. (2003). A vision of the future of science: Reintegrating the study of humans and the rest of nature. *Futures*, **35**: 651–671.

Costanza, R. and Cornwell, L. (1992). The 4P approach to dealing with scientific uncertainty. *Environment*, **34**:12–20, 42.

Costanza, R. and Daly, H. E. (1992). Natural capital and sustainable development. *Conservation Biology*, **6**: 37–46.

Costanza, R., d'Arge, R., de Groot, R. *et al.* (1997). The value of the world's ecosystem services and natural capital. *Nature*, **387**: 253–260.

Costanza, R., Andrade, F., Antunes, P. *et al.* (1998). Principles for sustainable governance of the oceans. *Science*, **281**: 198–199.

Costanza, R., Graumlich, L. J. and Steffen, W. (eds.) (2007). *Sustainability or Collapse? An Integrated History and Future of People on Earth. Dahlem Workshop Report 96.* Cambridge, MA: MIT Press.

Diamond. J. (2005). *Collapse: How Societies Choose to Fail or Succeed.* New York: Viking.

Easterlin, R. A. (2003). Explaining happiness. *Proceedings of the National Academy of Sciences*, **100**: 11176–11183.

Farber, S., Costanza, R., Childers, D. L. *et al.* (2006). Linking ecology and economics for ecosystem management: a services-based approach with illustrations from LTER sites. *BioScience*, **56**: 117–129.

Frank, R. (1999). *Luxury Fever: Why Money Fails to Satisfy in an Era of Excess.* New York: The Free Press.

Kasser, T. (2003). *The High Price of Materalism.* Cambridge, MA: MIT Press.

Layard, R. (2005). *Happiness: Lessons from a New Science.* New York: Penguin.

Millennium Ecosystem Assessment (2005). *Synthesis.* Washington DC: Island Press.

Vemuri, A. W. and Costanza, R. (2006). The role of human, social, built, and natural capital in explaining life satisfaction at the country level: toward a national well-being index (NWI). *Ecological Economics*, **58**: 119–133

World Resources (2005). *The Wealth of the Poor: Managing Ecosystems to Fight Poverty.* Washington DC: World Resources Institute.

5

Focal restoration

ERIC HIGGS

5.1 Introduction

Since the early days of ecological restoration as an organized practice in the 1980s scholars and practitioners have argued that broader social, cultural, political, economic and ethical concerns affect not only the success of restoration projects but also how we define and conceive of good restoration (Gobster and Hull, 2000; Higgs, 2003; Jordan, 2003). It is indeed the case that in virtually every restoration project success depends on ensuring proper funding, political support, and a supportive social embrace as well as getting the scientific and technical dimensions right. The need for greater engagement seems especially true on an international scale where a multitude of cultural realities enforce slightly different manifestations and norms on restoration practice. With a rapid worldwide expansion of restoration effort in every ecosystem and in every conceivable social setting, there will be a tendency to apply uniform techniques as an efficiency measure, increase the scale of restoration activity, give greater privilege to the role of restoration science, and focus increasingly on the restoration of natural capital. This trend, which I refer to as *technological* restoration, stands in opposition to the kind of restoration that is advocated traditionally and that honors a commitment to *focal* restoration. Focal restoration, in which the practice of restoration enlivens social diversity, participation, economic livelihoods, political integrity, and well-considered values, is an antidote to any version of restoration that neglects a wider, social embrace.

5.2 Effective, efficient and engaging restoration

Parks Canada, the parks management agency in Canada, recently released the first national comprehensive principles and guidelines for ecological restoration of protected natural areas. It rests on the idea that good restoration depends on three qualities. Restoration projects must be *effective*, which means they must meet

Ecological Restoration: A Global Challenge, ed. Francisco A. Comin. Published by Cambridge University Press.
© Cambridge University Press 2010.

clearly defined ecological integrity objectives and rest on a solid understanding of historical structures, patterns and processes. They must also be *efficient* in using materials and labor in order to respect in this case public expenditures and build momentum for other projects. Finally, restoration must *engage* communities through direct involvement in planning, implementation and monitoring of projects as well as building wider social support for restoration. *Engagement* matters a great deal to Parks Canada, an agency that in addition to its mandate to protect ecological integrity also connects "the public, communities and visitors to these special places….[to] facilitate the development of deeper understanding and appreciation of natural systems and the threats they face, and contribute to long-term societal commitment to restoration goals" (Parks Canada Agency, 2007).

The launch of this document at the 2007 Society for Ecological Restoration Conference in San José, California, was cause for celebration. But more than this, it marked a turning point in how we understand restoration. Over the roughly thirty years of professional development of restoration we have witnessed a subtle shift from restoration as a scientific and technical practice to one that incorporates broader economic and social factors. It is not a simple story of change. Rather we see the acknowledgment of social value early in the writings of John Cairns and Tony Bradshaw (among others), who by all accounts were primarily focused on advancing the science of restoration. When Bill Jordan established *Restoration and Management Notes* (now *Ecological Restoration*) in 1981 he focused on the constellation of restoration activity including the stories of patient volunteer restoration projects, trial-and-error informal experiments, the expanding field of restoration ecology, and the creativity embedded in so many projects. Jordan gave voice to those who saw restoration as a way of practicing sustainable living and reconnecting people more profoundly with place. Berger (1979) picked up this theme in his book, *Restoring the Earth*, which chronicles an earlier conference showing the diversity of interest in restoration. Writers such as Mills (1995) and House (1999) gave eloquence to the notion that restoration is as much a cultural as an ecological journey.

The Society for Ecological Restoration (SER) infamously wrestled at defining restoration, producing five definitions in almost as many years. It was not until 2001 that a relatively stable definition formed and took hold: "the process of assisting the recovering ecosystems that have been degraded, damaged or destroyed" (Society for Ecological Restoration International, 2004). While the definition itself belies deeper cultural references, the SER Primer, of which the definition is an integral part, makes a strong gesture: "Some ecosystems, particularly in developing countries, are still managed by traditional, sustainable cultural practices. Reciprocity exists in these cultural ecosystems between cultural activities and ecological processes, such that human actions reinforce ecosystem health and sustainability."

There remain two relatively distinct strands of restoration. One follows the contour of *restoration ecology*, exploring and expanding the scientific reaches of restoration. *Ecological restoration* is broader and incorporates restoration ecology in a larger socially-embedded practice. However, it is now more common to recognize the value that restoration holds for those who become involved and the importance of such care for the integrity of ecosystems. What has happened in the last decade is a gradual shift toward the latter, although the transition is far from complete and perhaps will never be; there is still a distinct tendency to separate more public acts of restoration – community-based projects, volunteer initiatives – from those conducted in the interests of testing ecological and technical ideas. There remains a "two culture" problem (Higgs, 2005) that separates scientific and humanistic understanding, and prevents the former from recognizing how to find an appropriate moral center. Such a moral center is achieved by appeal to social values and close collaboration with multiple disciplines. The fact that there are these two broadly described approaches to restoration does not suggest that one be subsumed by the other. It would be deleterious to the movement as a whole to continue to lionize restoration ecology just as it would be to champion ecological restoration as the exclusive valid approach. The challenge is in figuring out the best way to describe restoration such that the two strands are well combined and mutually respected. To date there have been many examples of such integration in practice but few policy approaches that suggest a systematic approach. Fortunately, the Parks Canada document offers a way forward.

5.3 Cultural variation

The recognition of a cultural setting for ecological restoration opens up the question of how restoration is shaped by cultural variation. Just as approaches to restoration vary from one ecosystem and region to another, so too do they shift in response to local, regional and national values. In the broadest sense we must accommodate the variety and richness not only of the biosphere but also of what anthropologist Wade Davis terms the ethnosphere (Davis, 2007). The fact that traditional peoples around the world have devised myriad sophisticated interrelationships with the land and waters around them must at least give us pause to reflect on the complexities of ecological restoration.

Ecological restoration is dominated by North American scholars and practitioners, with more than 75 percent of SER members from North America. Many of the problems to which restoration is offered as a solution were defined in North America, not the least being the notion that restoration is aimed at returning ecosystems to a pre-settlement landscape, a view that while largely discredited as being culturally and ecologically simplistic still holds some sway. This is not to suggest that restoration in fact had an exclusive origin in North America. Hall (2005) makes this point explicit with his comparative typology of restoration as it developed in the late nineteenth and early twentieth centuries in Europe and North America. He argues that a robust notion of restoration developed in Italy, for example, earlier than it did in the United States. However, the Italian idea of restoration was rooted in landscapes idealized as gardens rather than wildernesses. Much hinges on how people's perceptions differ on what counts as improved or restored. In nineteenth century Italy the ideal landscape was a garden, and restorative efforts were aimed at recovering an untouched landscape or "wasteland" to a gardened condition. In early twentieth century United States, especially through the work and inspiration of Aldo Leopold, the ideal landscape became untouched wilderness, and so restoration was separated from gardening (gardening remained an important practice, but not the subject of restoration). This shift in meaning is significant. It points out that cultural preferences shape what we are interested in restoring and to what ends we aim our efforts.

While the histories of restoration are yet to be written for most regions, we will over time see, I think, a multifaceted and parallel development of restorative practices. It is safe to suggest that most people have adopted restoration as a practice broadly speaking, whether they adhere to the orthodox notion of restoration or another. The point is that the way in which restoration is prominently portrayed as a North American invention detracts from the diversity of restorative approaches taken by people in myriad ecosystems.

It is easy to correct such a dominant theme in restoration history. After all, there is much that can be shared from ecosystem to ecosystem. At the southern tip of Vancouver Island on the west coast of Canada where I live, are the vestiges of Garry oak (*Quercus garryana*) savanna ecosystems. In 2004 at the international SER conference in Victoria, British Columbia, my colleague Nancy Turner brought together specialists in oak savanna restoration from as far away as Illinois to discuss common strategies. Similarly, the approaches developed for restoring native plants in arid lands in Texas are useful in other arid regions worldwide. There is a critical international dimension based on shared understanding of ecosystem characteristics. At best, such convocation brings together the subtle local knowledge of species and their associations to converge on similar challenges in other places. Hence, mangrove restoration specialists share their knowledge carefully when the particularities of local ecosystems are understood. At worst, a tendency to convert ecosystems to biophysical systems and then to ecosystem services leads to a diminishment of complexity.

In the same way that ecosystems vary from place to place, the cultural practices and beliefs that shape the management of these places vary, too. People form distinctive relationships with ecosystems based on cultural values and legacies, economic needs, political relationships and acceptable social practices. It was difficult for me at first to comprehend the depth of the relationship that Songhees First Nation woman Cheryl Bryce had with a small island off the coast of Victoria, British Columbia, and how this shaped her view of ecological restoration (Higgs, 2003; Higgs, 2005). For me, the challenge of restoration involved managing invasive species, returning light fall fires to the meadows species. For her, the restoration constituted the symbolic reconnection with the ecosystems that nourished her ancestors. The act of restoring the island meadows was much more than an ecological one; it embraced cultural recovery, too. Understanding subtle cultural variations depends on the same attentiveness to detail that good ecological observation does. A small farmer with incentives for ecologically-based coffee production in southern Mexico will have different reasons for restoration than an organic farmer in France, who aims to recover sustainable forestry in a small remnant-forested ecosystem. There are few comparative studies that examine such cultural variation and so we are without a comprehensive idea of how motivations for and definitions of restoration vary from one cultural setting to another.

Biological diversity is often used appropriately to justify the importance of an ecosystem restoration project. What of cultural diversity? In the face of rapid climate change, the biological invasions brought about via globalization, and the rush to reconfigure our environmental management efforts in terms of ecosystem services, how best can we honor the values of cultural and biological diversity? The answer lies partly in appreciating the pattern that underlies contemporary change, a pattern that I characterize as technological restoration. Once identified, it becomes easier to imagine and implement reforms that lead us to the effective, efficient and engaging qualities of ecological restoration.

5.4 Technological restoration

Technological restoration is restoration in which efficiency and uniformity are transcendent, and leads both to the projects themselves and to the process of developing those projects being rendered as commodities (Higgs, 2003). An increase in scale of projects, the trend toward more sophisticated techniques and technologies, the development and deployment of national and international standards and, most recently, the rapidly growing interest in the restoration of ecological services all point toward a more technological approach to restoration.

We live in a world saturated with technology. We are enmeshed by the commodities we bring into our lives and our lives are increasingly shaped by technological means of timekeeping and scheduling. To suggest that ecological restoration is technological means that it manifests particular qualities of life in a technological society. We become more used to the conversion of things we value at the core of our lives, many of which are shared commitments to excellence – generosity to family and neighbors, athleticism, creative performance, fine cooking, musicianship, natural history, scholarship – to commodities that can be bought and sold. Take music as an example. The performance of music constitutes a skillful engagement with others that draws on friendship, discipline and exuberance. The conversion of this experience to a downloaded music file that can be listened to independent of original context represents the technological realization of music. There is little question about the utility of the electronic music file: it can be shared readily and modified easily. These are helpful features. The problem comes not in the existence or use of such technologies, but in the fact that they threaten to smother the engagement that comes of musical performance. As the context becomes filled with electronically mediated music distribution, traditional music expression is displaced. That it is not displaced entirely is testament to the strength of things we value in their own right. However, just as the prevalence of electronically mediated music swamps the culture of traditonal music-making, the adoption of technology as a whole tends to cause a gradual deminution of the moral center of our traditional commitments (Borgmann, 1984).

Ecological restoration undergoes this conversion process in part through the natural evolution of practice, which promotes greater professional excellence, improved efficiencies, and the ability of practitoners to undertake larger-scale projects. The bonds that form between people and people and between people and place are loosened when grassroots community projects are gradually replaced by well-designed and funded programs. Specialists are brought in with better equipment, and the benefits of scale begin to operate. All of this is good in at least one sense: more restoration is accomplished with less cost. Those who advocate restoration as a vital strategy have a tough time opposing such logic. However, it is important to examine what is also lost. The intricate social connections that form around restoration projects are diminished. People also become less engaged with the ecosystems they live beside, having less investment in the projects accomplished with outside expertise. People may still appreciate the results of a restoration, but the process in which social and ecological bonds are formed goes missing.

There are three additional forces that amplify technological restoration. First, rapid climate change explodes the tidy notion that restoration is underpinned by historical fidelity and is giving way to non-analog and designer ecosystems. Thus, the act of restoration becomes shaped increasingly by human intentions that mitigate the change and focus ecological programs on adaptation. Second and closely related to climate change, we are more likely to become interested in projects that emphasize ecosystem functions and services (Aronson *et al.*, 2007). There is much to admire about this shift, not the least that it breathes new life into justifications of biological and cultural diversity. The problem is that services, whether directly in the aid of human needs or as proxies for perceived ecological ones, transform our understanding of ecosystems from complex self-regulating entities to those with particular purposes. The technological pattern of conversion is evident, and we are well advised to be cautious around restoring ecological services. Finally, the process of globalization in all its cultural, social, political and economic manifestations is threatening ecological integrity through increasing numbers of aggressive invasive species and also through the homogenization of cultural practices. Together, these three contemporary forces are accelerating and aggravating the rise of technological restoration.

5.5 Focal restoration

The antidote to technological restoration is focal restoration. Some have called this ecocultural restoration, but I prefer a term that goes beyond the combination of ecology and culture; after all it is not simply that we want to intertwine culture and ecology but that we want a way of orienting ourselves with respect to people and ecosystems. Focal restoration is engaging restoration: through the practice of restoration we focus on the things that are of profound importance to us, including skillful practices, community participation and celebration, and abiding connections to places and natural processes. Restoration practice is first and foremost about recovering ecosystems that have been damaged, but it can also recover damaged human relationships. By focusing on things that matter, we are better able to identify the loss of engagement and resist when necessary the patterns of technological restoration. If technological restoration is the conversion of local engagement to a commodity, then focal restoration is the return of that engagement.

In 2006 Jeff Ralph, a graduate student in the University of Victoria's School of Environmental Studies, organized a restoration project as part of his teaching assistant role in our introductory ecological restoration course. Jeff had worked with a community group and city staff for several years to restore the Southwest Woods, a remnant coastal Douglas fir forest in Victoria's Beacon Hill Park. A work party was organized for what turned out to be a spectacular Saturday in October. Sixty students from UVic showed up and cycled through three work sites: ripping English ivy (*Hedera helix*) from an area previously untreated; ripping ivy from a former site of removal; and planting thousands of plants and bulbs. This urban restoration project is challenged by more than invasive species. The Southwest Woods are a relatively safe haven for members of society who need privacy or a place to sleep. In one sense, restoring the ecological functioning of the Southwest Woods stood to compromise the social uses of the site. Through careful and protracted negotiations with city officials, support organizations and individuals, a deeper understanding was reached on what counted as restoration. Many of the typical ecological objectives are being met and at the same time there is sensitivity to a wide variety of social perspectives on the Woods.

More remarkable than the sheer amount of work accomplished in a few hours was the connection that the students made with the place. There was a remarkable energy throughout the day, and several students made a point of mentioning that they had no idea that so much could be accomplished and at the same time how much work remained. Hence, the ecological messages were clear. Another theme emerged: the students felt they achieved a connection to the Southwest Woods that would remain in their memories for a long time, and for some it began a longer term connection to the place. Many wrote about their experience in a later reflection assignment and I was struck by the diversity of their inspiration. More than simply benefiting Beacon Hill Park, this awareness of the ecological and cultural values of restoration resonates in other parts of their lives. This is an instance of focal restoration, and suggests how even brief interventions can result in strong connections to place.

An important implication of focal restoration is that it embraces cultural variety. The decision about why and how to restore resides in the overlapping values of individuals and communities around the world. Focal restoration encourages the celebration of cultural and ecological variety, allowing at the same time circumspection about gigantic schemes for climate change mitigation, control of invasive species and restoring and amplifying ecological services. It is not that we should avoid obligations to cope with the direct and indirect consequences of technological society, but that the technological milieu should not be the one that defines our relationship to communities and ecosystems. Starting from our engaging local relationships we develop a robust, culturally respectful form of restoration.

References

Aronson, J., Milton, S. and Blignaut, J. (2007). *Restoring Natural Capital: Science, Business, and Practice*. Washington DC: Island Press.

Berger, J. (1979). *Restoring the Earth: How Americans are Working to Renew our Damaged Environment*. New York: Alfred A. Knopf.

Borgmann, A. (1984). *Technology and the Character of Contemporary Life*. University of Chicago Press.

Davis, W. (2001) *Light at the Edge of the World: A Journey Through the Realm of Vanishing Cultures*. Vancouver: Douglas & McIntyre.

Gobster, P. and Hull, B. (eds.) (2000). *Restoring Nature: Perspectives from the Social Sciences and Humanities*. Washington DC: Island Press.

Hall, M. (2005). *Earth Repair: A Transatlantic History of Environmental Restoration*. Charlottesville: University of Virginia Press and Center for American Places.

Higgs, E. S. (2003). *Nature by Design: People, Natural Process and Ecological Restoration*. Cambridge MA: MIT Press.

Higgs, E. S. (2005). The two-culture problem: Ecological restoration and the integration of knowledge. *Restoration Ecology*, **13** (1): 159–164.

House F. (1999). *Totem Salmon: Life Lessons from Another Species*. Boston: Beacon Press.

Jordan, W. R. III. (2003). *The Sunflower Forest: Ecological Restoration and the New Communion with Nature.* Berkeley: University of California Press.

Mills, S. (1995). *In Service of the Wild: Restoring and Reinhabiting Damaged Land.* Boston: Beacon Press.

Parks Canada Agency (2007). *Principles and Guidelines for Ecological Restoration.* (www.globalrestorationnetwork.org/).

SER-Society for Ecological Restoration International-Science & Policy Working Group (2004). *The SER International Primer on Ecological Restoration.* Tucson AZ: Society for Ecological Restoration International. (www.ser.org).

6

Ethical dimensions of ecological restoration

REBECCA L. VIDRA AND THEODORE H. SHEAR

6.1 Introduction

Ecological restoration holds great promise for repairing some of the damage humans have caused to ecosystems around the world. It has been hailed as a way of reconnecting people with nature (Shapiro, 1995; Light, 2002; Jordan, 2003), reviving indigenous cultures (Turner *et al.*, 2000), creating art in nature (Turner, 1987), building community (Shapiro, 1995; Jordan, 2003), conserving biodiversity (Wilson, 2002; Gann and Lamb, 2006), enhancing ecological function (National Resources Council, 2001), and as an acid test for theoretical ecology (Bradshaw, 1987). Along with these high and sometimes disparate expectations, it is important to acknowledge the ethical dimensions of restoration (Higgs, 1997).

Through our work as restoration ecologists, we have become interested in exploring ethical concerns with students, researchers, and practitioners. We believe that it is important to increase awareness of the ethical dimensions of restoration brought to light both by those formally trained in philosophy and those who have been directly engaged in the work of restoration. We have been able to involve our students and many others in fruitful and sometimes heated discussions about ethical dilemmas in restoration (Vidra, 2003; Vidra, 2006). Some questions, like those regarding professional ethics, are relatively straightforward to address. We maintain that broader questions of ethics in restoration should continue to be unearthed, refined, and debated by restorationists and philosophers alike.

Discussion and debate surrounding the ethics of restoration began over two decades ago as environmental philosophers questioned the motives behind restoration work. Elliot (1982) labeled restoration as a faking of nature, comparing it to art forgery. Katz (1992a) warned that because restored systems are not natural and are essentially anthropocentric, "nature restoration is a compromise; it should not be a basic policy goal." He further claimed that restoration represented a domination of

Ecological Restoration: A Global Challenge, ed. Francisco A. Comin. Published by Cambridge University Press.
© Cambridge University Press 2010.

man over nature (Katz, 1992b). Ultimately, as restoration projects become more numerous across the globe, we should consider whether restoration success might be used as evidence to support further degradation of pristine ecosystems.

Light and Higgs (1996) have turned the discussion towards the social and political contexts within which restoration occurs. Two of the several philosophers active in the Society for Ecological Restoration International (SER International), they continue to offer provocative insights into the motivations behind restoration work (Higgs, 1997; Light, 2002) and the value, if not the necessity, of including a social component in restoration (Higgs, 2005; Higgs, this volume).

Restoration scientists have also unearthed ethical issues in restoration ecology, noting that balancing human needs and perspectives with the goal of achieving ecological integrity leads to important ethical questions. Cairns (2003) comments on the potential for restoration to encourage further human domination of nature, specifically noting interference with natural recovery processes. Hildebrand *et al.* (2005) offer a pointed critique of the myths of restoration, challenging the central idea that restoration is a successful practice. These scientists question the notion that humans can create natural, functioning systems, although they stop short of advocating abandonment of restoration as a practice.

In a recent study, the majority of professionals belonging to the SER acknowledged that they faced ethical dilemmas in the course of their restoration work (Dickinson *et al.*, 2006) and supported a code of ethics for the society (Carpenter *et al.*, 2006). Several of the professionals commented on particular situations where they were pressured to overlook or augment their findings, sacrifice ecological integrity to obtain funding, and use potentially invasive non-native species for revegetation purposes. It has been our experience that, while restoration practitioners are criticized for being overly prescriptive or allowing too much human interference in nature (e.g., Hildebrand *et al.*, 2005), they face a wide range of ethical dilemmas in their everyday work.

What sustains these discussions of ethics in restoration? We believe that there are two main reasons. First, restoration represents a human intervention in and interaction with nature, regardless of the intentions, that necessitates some control over natural processes. In fact, the very goal of restoration may be to integrate humans with nature, as evidenced by the joint working group of SER and the IUCN Commission on Ecosystem Management, which argues that ecological restoration should both "improve human livelihoods" and "empower local people" (Gann and Lamb, 2006). Second, restoration involves the application of science to practice, two endeavors that operate under a different set of norms and time frames. Indeed, as noted by Light and Higgs (1996), there has been little discussion between philosophers, practitioners and scientists regarding these important issues.

In this chapter, we attempt to outline some of the ethical concerns that we believe are global in nature, either because the answers differ across countries and cultures or because they are questions that all of us face, regardless of our nationality. We have chosen not to focus on professional ethics in restoration as they have been dealt with elsewhere (Carpenter *et al.*, 2006; Dickinson *et al.*, 2006) and may be dependent upon contexts that shift from country to country. Our goal is to condense the rich literature of restoration ethics into a few salient issues which will inspire future discussion by all restorationists, regardless of their training in ethics or geographical location.

6.2 Definitions and expectations of ecological restoration

In 2006 Gale Norton, Secretary of the United States Department of Interior, proudly proclaimed that 700,000 acres of wetlands had been gained in the United States between 1998 and 2004. Norton used this increase to show her Administration's support for wetland restoration projects. However, the definition of wetlands was expanded to include man-made storm retention ponds created on golf courses, residential developments, and along roadsides. Few restorationists would agree that these isolated ponds represent functioning wetland communities, yet this pronouncement attracted a great deal of media attention. In this case, restoration was used as a buzzword to garner public support rather than an accurate description of the projects.

It is incumbent upon the ethical restorationist to advocate for an honest definition of ecological restoration. We have on countless occasions asked students, practitioners and fellow researchers to define restoration. Only in rare cases can they articulate a comprehensive, precise definition. Surprisingly, many cannot even make a start.

As the field of restoration has grown, many different definitions of ecological restoration have been employed. The Society for Ecological Restoration International alone has gone through several different definitions before arriving at its current one: "Ecological restoration is the process of assisting the recovery of an ecosystem that has been degraded, damaged, or destroyed" (www.ser.org). The current joint working group of SER and the IUCN Commission on Ecosystem Management has also adopted this definition while developing a global call for restoration (Gann and Lamb, 2006). However, the Society of Wetland Scientists (www.sws.org) defines wetland restoration more specifically as "actions taken (…) that result in the reestablishment of ecological processes, functions, and biotic/abiotic linkages that lead to a persistent, resilient system integrated within its landscape." The restoration of the human–environment relationship, urged by Allison (2004) and others, is not addressed by these definitions.

Different definitions belie different motivations behind restoration projects, leading to different expectations of success. In the United States, a great deal of restoration work has been driven by mitigation efforts which carefully define success using measurable characteristics, such as number of trees per acre in a wetland restoration site. Clewell and Aronson (2006) recently referred to this as a "technocratic rationale" for restoration and argue that this motivation does not create "a strong bond between nature and culture."

Restoration may also be undertaken specifically to reconnect people with nature under the "idealistic rationale" (Clewell and Aronson, 2006). Ordinary citizens may engage in restoration work to support the "land ethic," to assuage their guilt regarding human's influence on the earth, or to build community. Restoration has been hailed as a way for people to reconnect with nature and to their own role in ecosystems, serving as a focus for social movements (Shapiro, 1995; Jordan, 2003). These types of restoration projects may have very different objectives, with success measured in terms of human rather than ecological benefit.

Conserving biodiversity through restoration is yet another distinct motivation, termed the biotic rationale by Clewell and Aronson (2006). Whether enhancing species richness or improving populations of specific species is the goal of these types of projects, measures of success might differ from those projects undertaken in the mitigation or social context. Conserving or enhancing biodiversity may be a much narrower target than achieving ecological integrity.

Clearly, the motivations behind a particular project will determine the types of restoration attempted and the ways success will be measured. While there is some agreement on what restoration is, this agreement has come at a cost: restoration is now defined so broadly that it's being abused as a buzzword and hailed as the solution to the world's problems.

6.3 Restoration as a process: an art, practice or science?

Successful restoration is not only dependent on the expectations of the final product. The process of restoration contributes another set of success criteria. A subtle distinction has been made between the practice of ecological restoration and the science of restoration ecology (Higgs, 1994; Van Andel and Aronson, 2006). Others have claimed that restoration is an art involving an aesthetic component (Turner, 1987). Is the actual process of doing restoration an art, a practice, or a science? When those involved in restoration view the process differently, ethical issues are bound to arise.

In 2003, we partnered with the North Carolina Museum of Art in Raleigh, North Carolina to design an Art and Environment Park on approximately one hundred acres of open space adjacent to the museum buildings (Plate 6.1). The museum staff envisioned integrating sculptures into restored landscapes as variable as Piedmont prairie, longleaf pine savanna, upland hardwood forest and riparian corridors. They spoke of the visual impact of the space, with beautiful transitions between tiny reproductions of these ecosystems. We struggled with the challenges of creating an ecosystem in a place that it was unlikely to have ever occupied naturally, with integrating human features into these ecosystems, and with balancing the whims of nature with the talents and whims of artists. The initial space, occupied by over-grazed pasture, heavily invaded forest, and a degraded stream channel, posed several logistic challenges. As ecologists, we developed a plan for restoring the existing ecosystems given the predictable onslaught of continued human interfer-ence (e.g., exotic species introductions, heavy recreational use). We were more concerned with creating a sustainable, native ecosystem while the museum regarded the artistic aesthetics of the creation more highly. We have enjoyed watching the installation of the restoration project but both the process and the endpoint were not entirely consistent with our vision.

Considering restoration as a practice is useful in some contexts, as in a practice of connecting people to nature. However, the practice of restoration can lead to a prescriptive approach. For example, people practice medicine partially by applying accepted knowledge and techniques to particular situations.

In our experience, stream restoration in the eastern United States is one such example. Based on a geomorphic strategy (Rosgen, 1994), reference sites are chosen (in our region, usually a single stream segment) and new stream channels are carefully constructed to meet a set of engineering principles, ostensibly in reference to pristine sites. This type of restoration is often done in a mitigation context, to restore streams that have been degraded by construction or to move streams away from construction sites. This prescriptive approach is one of the myths of restoration that Hildebrand *et al.* (2005) call on restorationists to reexamine in their quest for more successful projects.

There are many situations where a prescriptive approach is useless, as in the large-scale, multi-billion dollar Everglades restoration project. Prescriptions for restoring water flow or eradicating exotic species will likely not include plans for addressing political concerns in southern Florida or for balancing agricultural needs with those of ecosystems (Grunwald, 2006). Yet, in other cases, such as the design of carbon mitigation projects in developing countries, a prescriptive approach is desirable. Our continuing ethical challenge, therefore, is to practice restoration with flexibility while embracing diverse approaches and outcomes.

The science of restoration ecology has greatly expanded over the past few decades, as evidenced by the increasing number of articles in academic journals (Ormerod, 2003). Several books have recently been published that describe scien-tific advances in restoration, including the Island Press/SER series. Approaching the process of restoration scientifically is perhaps what is meant by using restoration as an acid test for theoretical ecology. We examine the challenges of doing so in the next section.

6.4 Restoration as the acid test: the next big lie?

Using restoration to test the basic tenets of ecology is a worthwhile goal. Certainly we need to integrate our knowledge of biological and physical processes to create successful restoration projects in terms of ecological integrity. Ecologists have viewed restoration as an opportunity to create large-scale experiments for this acid test, yet the challenges of doing so are often insurmountable.

Ecological experiments require careful consideration of controls, replications and treatments. Ideally, we would be able to take a set of degraded sites, leave some alone as control sites, and apply the same treatment(s) to the remaining sites, with enough replicates to achieve statistically significant results on realistic scales (Michener, 1997). The reality is that individual projects often have different goals and differing levels of funding and thus may have different desired endpoints. Projects also need to address site-specific constraints such as topographic relief, soils, adjacent inputs and contamination. While one project may restore an existing ecosystem, another may create an ecosystem in a place where it never previously existed. These are just some of the constraints to conducting traditional experiments in the field. The ethics of leaving control sites alone to further degrade is also an important challenge. This is particularly important when dealing with invasive exotic species eradication, where a site left unrestored could potentially become further degraded and even lead to degradation of adjacent sites. Yet, this challenge is not unique to ecological restoration; for example, in medical testing, it is common to use placebos in the course of clinical trials.

In terms of testing the basic tenets of ecology, restoration ecologists have done a better job of applying certain ecological paradigms to their work. For example, the promise of assembly rules has been recently evaluated in a recent volume (Temperton *et al.*, 2004). Our own experience with creating realistic experiments in restoration has been mixed. For example, we recently conducted a small-scale experiment to observe the effect of removing the dense exotic understory of a heavily invaded deciduous forest (Plate 6.2) (Vidra *et al.*, 2007). We found some interesting results, particularly that continuing removal of exotic seedlings is a necessary part of the management plan because of continued propagule pressure. We also found that the initial recovery of native vegetation was dominated by a few weedy species, suggesting that additional plantings may be necessary. In spite of space (the forest covered only 45 acres) and time constraints (the forest was heavily disturbed and the plots destroyed three years after the experiment was put in place), this study offered the opportunity for establishing an experimental framework to test restoration possibilities (Vidra *et al.*, 2007).

It is seldom possible, regardless of the scientific discipline, to publish "negative results." It is even more difficult to publish case studies of individual restoration projects. Case histories can be published in technical journals, trade journals, and posted on Internet sites. In their *Guidelines for Developing and Managing Ecological Restoration Projects* (Clewell *et al.*, 2005), SER International strongly encourages publishing case studies of restoration projects. Yet, case studies are difficult to publish in top-tier journals.

Restoration is being touted as the acid test and hailed as a great opportunity for conducting experiments. Yet, in a recent survey of over fifty wetland restoration projects attempted as mitigation for road construction impacts, several projects ostensibly included research components but were not designed to allow for mean-ingful analysis (Brinson and Rheinhardt, 2002). Restoration projects have been thought of as research efforts across the globe but without appropriate experimental design, adequate control of covariables, etc., it is difficult at best to extract mean-ingful, publishable lessons from these efforts.

In reality, the constraints of individual projects and sites limit their value as experimental sites and the results will therefore almost always be mixed. We need to think more carefully about how to present these types of results and put an appropriate amount of pressure on journals to publish meaningful, if not statistically significant, results. How we as restoration ecologists translate these results into recommendations for the practice of restoration will continue to be both a logistical and ethical challenge.

6.4.1 Participation and decision-making in restoration

Ecological restoration is often practiced principally as a technical affair, with technologists combining plants (and sometimes animals), soil, water, chemicals, and energy in (hopefully) exacting combinations to revitalize ailing ecosystems. This technocratic approach has left many unsatisfied and searching for a more inclusive and expansive approach. Ecological restoration is often held up as a new paradigm for reconnecting humans to nature, usually the urban dwellers isolated from the farms and forests that provide the products that feed and shelter them.

Jordan (2003) labels this as the new communion with nature, with restoration ritualized to become an essential element of the human community. Every week-end in cities throughout Europe and North America, we know of citizens that volunteer their labor to local restoration projects. The Earth Restoration

Service (www.earthrestorationservice.org) aims to support a global network of environmental restora-tion projects by matching volunteers with projects that need their skills. The Global Restoration Network of SER International provides a large database of collected practical experiences and fosters the creative exchange of expertise among practi-tioners and researchers worldwide (www.globalrestorationnetwork.org).

Higgs (1997) argues that good restoration must include historical, social, cultural, political, aesthetic and moral components. He finds a democratic, participatory potential within the practice of ecological restoration, which may or may not be realized depending upon the politics surrounding a particular restoration project. The restoration ultimately fails if it does not fulfill its democratic potential. Jensen (2001) reminds us that democratic and participatory are not equal, with democracy requiring equal say in decision-making, not just participation in implementation. He counters that good restoration need not be democratic, and that there are dangers to be had in fully democratizing any land management activity.

The distinction between democratic and participatory restorations will become increasingly important, as more groups define ecological restoration as restoration of a cultural landscape. This concept is easy to accept, as human manipulation of the environment on a global scale (for example global climate change) leaves no square meter of the Earth's surface unaffected by human activity. Of particular interest is the case of the indigenous peoples seeking restoration of traditional land-based cultural practices as a vital means of continuing their traditional cultures. Often these cultural practices require tribal control over ancestral lands and resources. Many indigenous people still have some land that they control, but most new lands that could be made available for cultural restoration through ecological restoration are publicly or privately owned by others.

Martinez (2000), organizer of SER's Indigenous Peoples Restoration Network and an influential advocate for return of land management to indigenous peoples, professes the powerful principle that land is not fully restored until the historic native people have been returned to it. This idea has gained some acceptance in Canada and the USA, where many public lands are now being managed coopera-tively with "First Nations".

Defining ecological restoration as restoration of indigenous peoples and their land management practices can lead to conflict over who can participate in the restoration, and how. If authentic restoration is fundamentally dependent on inclu-sion of indigenous peoples, then the remainder, often the bulk of local populations, is at risk of being relegated to either diminished or subservient roles. If they are seeking Jordan's (2003) communion with nature, they may not experience it in those roles.

The engagement of indigenous peoples in restoration is predicated upon the application of traditional ecological knowledge (knowledge of the ecology and management of ecosystems resulting from generations of trials and practical experiences). Few proponents of the application of traditional ecological knowledge have differentiated between the management practices of indigenous peoples and those management practices as applied by those peoples (Berkes *et al.*, 2000; Ford and Martinez, 2000; Kimmerer, 2000). If the first is the aim, then any restorationist regardless of origin can be trained in traditional ecological knowledge. Conflicts that might arise would likely be between applications of culturally based and scientifically based techniques. In the model of restoration described by Martinez (2000), the indigenous people would always hold trump cards of historical authenticity and relevance. Of concern to us is who ultimately gets to participate, and how.

For example, consider that Australians and North Americans of European descent might accept that good or complete restoration is the domain of the indigenous peoples of their respective continents. They might derive sufficient joy from observing restoration as it is performed by the indigenous people, or even more simply from the knowledge that it is occurring. This would be the result of an acceptance of indigenous people as components of a restored "wildness." The remainder of the population, as interlopers in the restoration, would be left to define themselves in some way apart from nature.

This conflict must be resolved for restoration to advance, particularly as local populations in most parts of the world are expected to increase for many decades. Sauer (2003) suggests that anyone can become "indigenous" by behaving "as-if-indigenous." For her, groups organized for conservation and restoration are analogous to landscape-dependent tribes. In this manner, every human has institutions through which to commune with nature in an authentic way.

6.4.2 *Acknowledging the ethical dimension*

Recent writings on ethics in restoration, this chapter included, tend to list potential ethical issues (Cairns, 2003; Hildebrand *et al.*, 2005; Vidra, 2004). While illuminating the particular issues that may arise in restoration is important background work, restorationists need to acknowledge, and perhaps even embrace, the ethical dimensions of ecological restoration. If we accept that restoration will involve a necessary amount of human intervention with all the potential pitfalls that brings, we need to be ready for the ethical concerns that will inevitably arise. Some of these questions span different cultures, some are better addressed in local contexts (Table 6.1). It is our hope that those involved in all aspects of restoration join the conversation about ethics in restoration.

Table 6.1. *Ethical dimensions of ecological restoration and the scale of operation or concern.*

Ethical dimension	Scale of concern
Restoration as human intervention in nature	Global
Definitions of restoration	Local to global
Realistic expectations of restoration	Local to global
Definition of success criteria	
Process of restoration	Global
Experimental ethics	National to global
Publishing restoration science	Global
Democratic approaches to restoration	Local to global
Participation as decision-makers	Local to global

References

Allison, S. K. (2004). What do we mean when we talk about ecological restoration? *Ecological Restoration*, **22** (4): 281–286.

Berkes, F., Colding, J. and Folke, C. (2000). Rediscovery of traditional ecological knowledge as adaptive management. *Ecological Applications*, **10** (5): 1251–1262.

Bradshaw, A. D. (1987). Restoration: The acid test for ecology. In Jordan, W. R. III., Gilpin, M. E. and Aber, J. D. (eds.). *Restoration Ecology: A Synthetic Approach to Ecological Research*. Cambridge University Press.

Brinson, M. M. and Rheinhardt, R. D. (2002). *An Evaluation of the North Carolina Department of Transportation Wetland Mitigation Sites: Selected Case Studies. Phase 2 Report*. Raleigh: Institute of Transportation Research and Education, North Carolina State University.

Cairns, J. Jr. (2003). Ethical issues in ecological restoration. *Ethics in Science and Environmental Politics*, **2003**: 50–61.

Carpenter, A., Finley, E., Gao, Y., Lin, C. *et al.* (2006). Developing a code of ethics for ecological restorationists. *Ecological Restoration*, **24** (2): 105–108.

Clewell, A. F., Reiger, J. and Munro, J. (2005). *Guidelines for Developing and Managing Ecological Restoration Projects*, 2nd ed. Tucson AZ: Society for Ecological Restoration International.

Clewell, A. F. and Aronson, J. (2006). Motivations for the restoration of ecosystems. *Conservation Biology*, **20** (2): 420–428.

Dickinson, W., Ferreyra, J., Imbesi, K. L. *et al.* (2006). The ethical challenges faced by ecological restorationists. *Ecological Restoration*, **24** (2): 102–105.

Elliot, R. (1982). Faking nature. *Inquiry*, **25**: 81–93.

Ford, J. and Martinez, D. (2000). Traditional ecological knowledge, ecosystem science, and environmental management. *Ecological Applications*, **10** (5): 1249–1250.

Gann, G. D. and Lamb, D. (eds.) (2006). *Ecological Restoration: A Means of Conserving Biodiversity and Sustaining Livelihoods* (version 1.1). Tucson AZ and Gland Switzerland: Society for Ecological Restoration International.

Grunwald, M. (2006). *The Swamp: The Everglades, Florida and the Politics of Paradise*. New York: Simon & Schuster.

Higgs, E. S. (1994). Expanding the scope of restoration ecology. *Restoration Ecology*, **2**: 137–146.

Higgs, E. S. (1997). What is good ecological restoration? *Conservation Biology*, **11**: 338–348.

Higgs, E. S. (2005). The two-culture problem: Ecological restoration and the integration of knowledge. *Restoration Ecology*, **13**: 159–164.

Hildebrand, R. H., Watts, A. C. and Randle, A. M. (2005). *The myths of restoration ecology. Ecology and Society*, **10** (1): 19. (www.ecologyandsociety.org/vol10/iss1/art19/).

Jensen, J. (2001). The ambiguous promise of democratic restoration. Abstract in *Restoration Acros Borders*: Niagara Falls: The Society for Ecological Restoration's 13th Annual Conference.

Jordan, W. R. III. (2003). *The Sunflower Forest: Ecological Restoration and the New Communion with Nature*. Berkeley: University of California Press.

Katz, E. (1992a). The big lie: Human restoration of nature. *Research in Philosophy and Technology*, **12**: 231–242.

Katz, E. (1992b). The call of the wild. *Environmental Ethics*, **14**: 265–273.

Kimmerer, R. W. (2000). Native knowledge for native ecosystems. *Journal of Forestry*, **98**: 4–9.

Light, A. (2002). Restoring ecological citizenship. In Minteer, B. and Taylor, B. P. (eds.). *Democracy and the Claims of Nature*. Lanham MD: Rowman and Littlefield.

Light, A. and Higgs, E. S. (1996). The politics of ecological restoration. *Environmental Ethics*, **18**: 227–247.

Martinez, D. (2000). Many indigenous worlds or the indigenous world? A reply to my "indigenous" critics. *Environmental Ethics*, **22**: 291–310.

Michener, W. K. (1997). Quantitatively evaluating restoration experiments: Research design, statistical analysis, and data management considerations. *Restoration Ecology*, **5** (4): 324–337.

National Resources Council (2001). *Compensating for Wetland Losses under the Clean Water Act*. Washington DC: National Academy Press.

Ormerod, S. J. (2003). Restoration in applied ecology: Editor's introduction. *Journal of Applied Ecology*, **40**: 44–50.

Rosgen, D. L. (1994). A classification of natural rivers. *Catena*, **22**: 169–199.

Sauer, L. (2003). All restoration is indigenous. Abstract in *Assembling the Pieces–Restoration design and landscape ecology: Society for Ecological Restoration International's 17th Annual International Conference*. Tucson AZ: Society for Ecological Restoration.

Shapiro, E. (1995). Restoring habitats communites and souls. In *Ecopsychology: Restoring the Earth, Healing the Mind*, ed. A. D. Kanner, T. Roszak and M. E. Gomes. New York: Sierra Club Books, pp. 224–239.

Temperton, V. M., Hobbs, R. J., Nuttle, T. and Halle, S. (2004). *Assembly Rules and Restoration Ecology*. New York: Island Press.

Turner, F. (1987). The self-effacing art: Restoration as imitation of nature. In *Restoration Ecology: A Synthetic Approach to Ecological Research*, ed. W. R. Jordan III, M. E. Gilpin and J. D. Aber. Cambridge University Press, 47–50.

Turner, N. J., Ignace, M. B. and Ignace, R. (2000). Traditional ecological knowledge and wisdom of aboriginal peoples in British Columbia. *Ecological Applications*, **10**: 1275–1287.

Van Andel J. and Aronson, J. (2006). *Restoration Ecology: The New Frontier*. Malden: Blackwell Science Ltd.

Vidra, R. L. (2003). What are your ethical challenges? *Ecological Restoration*, **21** (2): 120–121.

Vidra, R. L. (2004). *Implications of Exotic Species Research for Urban Forest Restoration. PhD Dissertation*. Raleigh, NC: Department of Forestry, North Carolina State University.

Vidra, R. L. (2006). Studying the ethics of ecological restoration: An introduction. *Ecological Restoration*, **24** (2): 100–101.

Vidra, R. L., Shear, T. H. and Stucky, J. M. (2007). Effects of exotic species removal in an exotic-rich urban forest. *Journal of the Torrey Botanical Society*, **134**: 410–419.

Wilson, E. O. (2002). *The Future of Life*. New York: Knopf.

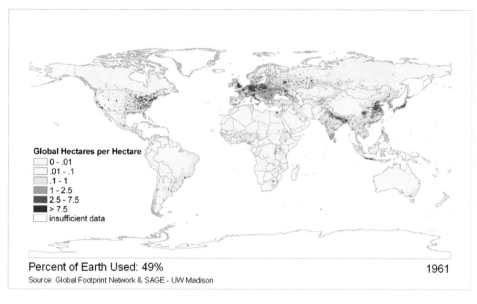

Global Hectares per Hectare
- 0 - .01
- .01 - .1
- .1 - 1
- 1 - 2.5
- 2.5 - 7.5
- > 7.5
- insufficient data

Percent of Earth Used: 49%   1961

Source: Global Footprint Network & SAGE - UW Madison

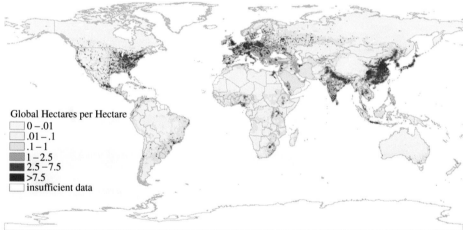

Global Hectares per Hectare
- 0 – .01
- .01 – .1
- .1 – 1
- 1 – 2.5
- 2.5 – 7.5
- >7.5
- insufficient data

Percent of Earth Used:121%   2001

Source: Global Footprint Network & SAGE - UW Madison

Plate 1.1. Recent change in the ecological footprint between 1961 and 2001. *Source:* Global Footprint Network and SAGE, University of Wisconsin-Madison.

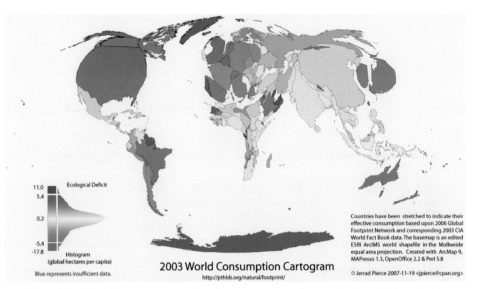

Plate 1.2. World map representing the areas of each country in relation to its relative consumption. *Source:* J. Pierce (http://pthbb.org/natural/footprint).

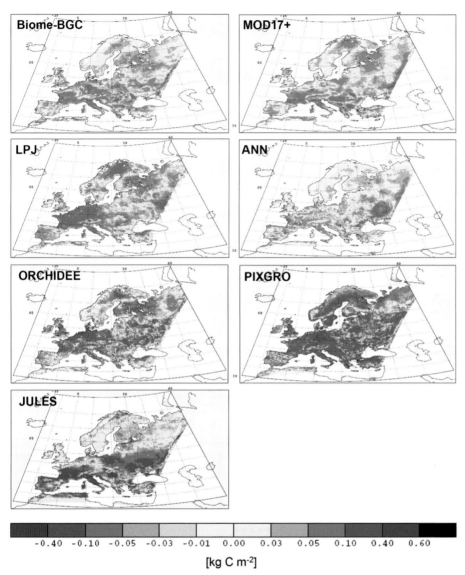

Plate 2.1. Anomaly in net ecosystem production in 2003 compared to the average in 1998–2002 using different bottom-up models. Red areas show reduction in NEP. Analysis and figure from Vetter *et al.*, 2007.

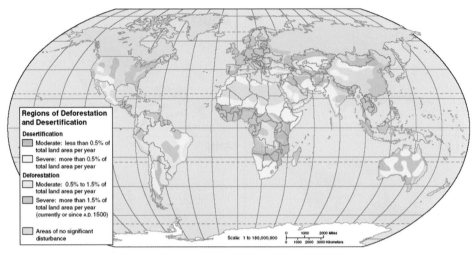

Plate 3.1. Regions of deforestation and desertification (Source: Rourke, 2000).

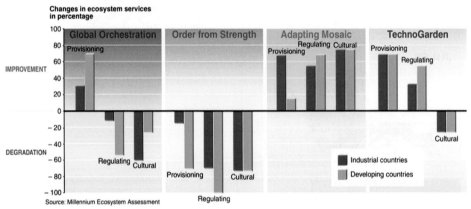

Plate 4.1. Changes in ecosystem services as estimated in the Millennium Ecosystem Assessment (2005) under four future scenarios. These scenarios correspond broadly to the scenarios used in the GUMBO model, and in this study, as follows: Global Orchestration = Big Government, Order from Strength = Mad Max = business as usual (BAU), Adapting Mosaic = Ecotopia = restored Earth (RE), Techno Garden = Star Trek. One can clearly see that the effects on ecosystem services are worst in the Order from Strength (BAU) scenario and best in the Adapting Mosaic (RE) scenario.

Plate 6.1. The North Carolina Museum of Art in Raleigh, NC, USA is integrating art with restoration by showcasing outdoor sculptures, like this bamboo fence, within small restored ecosystems like the Piedmont prairie, seen here on the left-hand side.

Plate 6.2. Volunteers remove vegetation from an experimental plot. All vegetation was removed and exotic species kept from reinvading for a two-year period in order to measure recruitment and recovery of native vegetation.

Plate 7.1. A 60 year old plantation of *Flindersia brayleyana* in north eastern Australia. This was originally established as a monoculture but has been colonised by a variety of native species from an adjoining natural rainforest. The site now has considerable species diversity (Photograph J. Firn).

Plate 7.2. Small farm woodlot in northern Vietnam established by farmer to complement agricultural cropland. The hill in the background has been reforested with fast growing eucalyptus by a state-owned forestry company.

Plate 7.3. Reforested landscape in northern Vietnam. The valley floors have been converted to rice paddies while the hill areas have been reforested. The resulting landscape is a mosaic of agriculture and plantation forests.

Plate 7.4. Deforested landscape in northern Laos showing forest fragments scattered among areas of agricultural land. The forest areas include regrowth forest as well as intact forest. Possible interventions might include protecting the remaining forest areas as well as reforesting some of the lands between the forest patches.

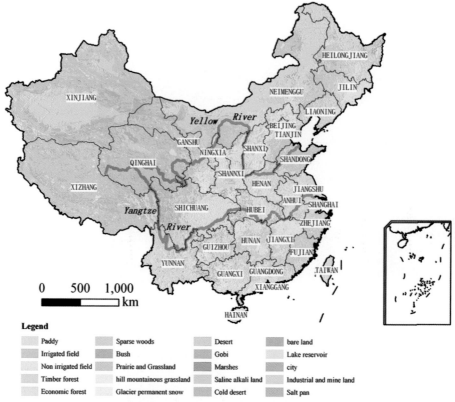

Plate 8.1. The land use or cover and provinces in China.

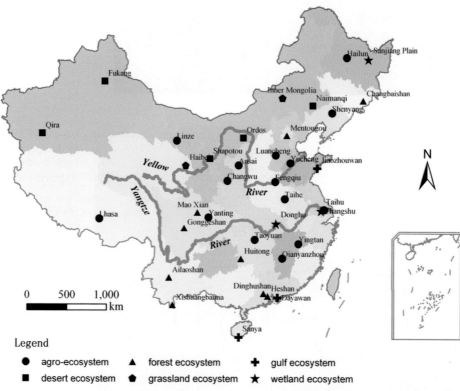

Plate 8.2. The field research stations in the Chinese Ecological Research Network.

Plate 8.3. The fire-damaged forest stand in Yunnan Province, southwestern China (Photo by Y. Lu).

Plate 8.4. The severe gully erosion landscape on the Loess Plateau (Photo by W. Zhao).

Plate 8.5. Qinghai-Tibet railway at an altitude of over 4,000 meters above sea level (Photo by X. Qi).

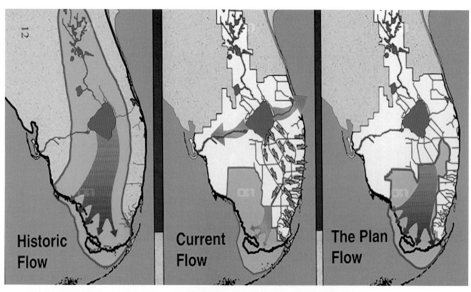

Plate 9.1. Historic flow, current conditions, and planned hydrologic restoration of the Florida Everglades, south Florida, USA (Illustration courtesy of US Army Corps of Engineers).

Plate 9.2. State of Ohio website announcement for farmers to apply for funds to set aside wetlands, riparian forests, and other filter strips in the Scioto River Basin in south-central Ohio. Funds were provided by the US Department of Agriculture CREP (Conservation Reserve Enhancement).

Plate 9.3. The Mesopotamian Marshlands of Iraq. Top, a typical scene before marshlands were destroyed; bottom, marshlands after drainage in the early 2000s.

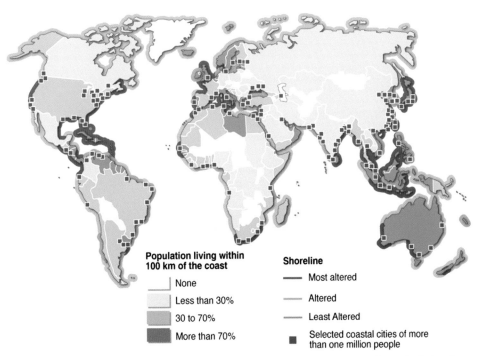

Plate 10.1. World distribution of disturbed coasts and percentage of people living in the coastal zone by country. After 2007 *UNEP/GRID-Arendal Maps and Graphics Library.* Cartographer/Designer *Philippe Rekacewicz, Hugo Ahlenius,* Retrieved 17:25, April 2008 from (http://maps.grida.no/go/graphic/coastal-population-and-altered-land-cover-in-coastal-zones-100-km-of-coastline).

Plate 10.2. Contrast of coastal types (cliff in the foreground, beach in the background) in the coastal zone just south of Barcelona. A human-modified system (a marina) can be observed between both sites.

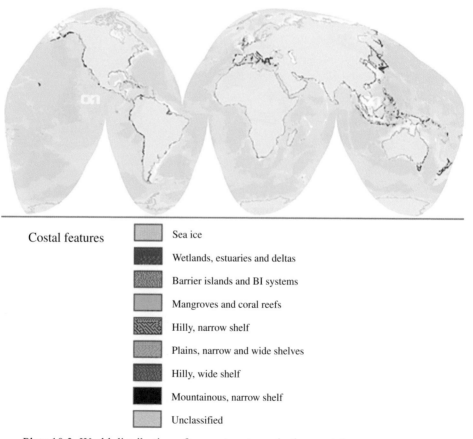

Costal features

- Sea ice
- Wetlands, estuaries and deltas
- Barrier islands and BI systems
- Mangroves and coral reefs
- Hilly, narrow shelf
- Plains, narrow and wide shelves
- Hilly, wide shelf
- Mountainous, narrow shelf
- Unclassified

Plate 10.3. World distribution of ecosystem types in the coastal zone.

Plate 10.4. Planting mangrove trees (top) and recovering water flows (bottom) should be complementary actions for the restoration of degraded mangrove ecosystems in the Yucatan coast (southeastern Mexico).

Plate 11.1. Major existing patterns of the Greater Barcelona Region. On the lower right is the Mediterranean Sea while the large black area is the Barcelona metropolitan area including city; black: built area. Gray = natural vegetation; white (on land) = agriculture; black line = stream/river network and urban region boundary (roughly around periphery). Mapped area from left to right approx. 135 km. See Forman (2004). Produced by X. Mayor; courtesy of Barcelona Regional.

Plate 11.2. Diagrammatic mapped summary of many planning solutions for the Greater Barcelona region. Dotted line = approximate boundary of urban region; black area = large natural area or emerald; straight strip = overland corridor or ribbon (various types); alternating black-and-white strip = stream corridor connecting emeralds; C = coastal beach, wetland, and woodland habitat restoration; area with thin dashes = large agricultural area; area with thick black dashes = agriculture-nature park; GH = area of greenhouses; strip with white or black-and-white segments = targeted stream or river commonly with sewage pollution; N = targeted water body with nitrogen pollution; I = targeted river with industrial pollution; W = watershed protection for reservoir; G = area targeted for growth; X = limited growth, no growth, or building-removal area; large dotted area = green net area; P = Great Park; H = heavy industry center; T = truck transport center; o = underpass or overpass for walkers and wildlife to cross highway; area around river system at eastern end = ecotourism area. See Forman (2004).

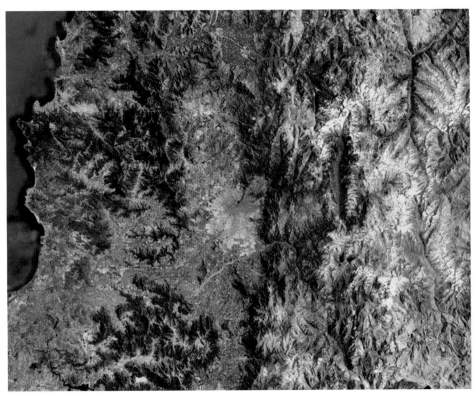

Plate 11.3. Landsat satellite image of Santiago, Chile metropolitan area in urban region between mountains and sea. Gray area near center = built metropolitan area; north/south mountains with white ice/snow areas on right = the Andes; dark area on far left = Pacific Ocean. Metropolitan area is approx. 25 km diameter.

Plate 13.1. Constructing ways of communication across barriers for human cooperation and peace between Israeli and Palestinian people in the Alexander River area. Top: the lower Alexander River before restoration (1996) (left), and after restoration (2006) (center); sewage from the rehabilitated treatment ponds of Tul Karem crosses under the security fence, into Israel, (right). Bottom: the Nablus Stream Emergency Sewage Project, the security wall and Tul Karem in the background (left), Israelis, Palestinians and Germans meet and talk "only sewage" (2003) (right).

Plate 13.2. Establishing agreements for lake restoration as a basis for development in Burkina Faso. Top: Lake Bam drying out during the dry season (June 2006) (left), Lake Bam flooding during the wet season (September 2006) (right). Bottom: Lake Bam is the source of life for the whole region (left), signing a letter of intent by Burkina Faso Ministry of Agriculture, Australian IRF and Israeli ARRA (2004) (right).

Part II

Towards the practice of ecological restoration
on a global scale

7

Undertaking forest restoration on a landscape scale in the humid tropics: matching theory and practice in developed and developing countries

DAVID LAMB

7.1 Introduction

In recent years large areas of tropical forest have been cleared to create more agricultural land. It is becoming evident that this clearing process has often gone too far since many of these lands have subsequently been abandoned and now lie in a degraded state (ITTO, 2002; Schroth *et al.*, 2004; Palm *et al.*, 2005; Rudel *et al.*, 2005). Fire in these areas may burn nearby intact forest and erosion in them often affects adjoining agricultural lands. Many are occupied by exotic weeds. One obvious solution is to reforest these areas. This assumes the drivers of the original degradation process can be controlled and further disturbances can be prevented. Such reforestation would almost certainly assist in promoting the sustainability and future productivity of the nearby agricultural lands and would help counter the current global trend towards increasing biodiversity loss.

The key question is: how might this reforestation be done on a scale that matches the magnitude of the problem? That is, given the current mix of silvicultural knowledge, policies and technologies, what restoration on a landscape scale? One obvious solution is to simply facilitate natural succession development. This may occur if further disturbances cease and plants and wildlife are able to recolonize the degraded areas from any remnants of natural forest that are still present. Such recovery occurred once farming stopped in Puerto Rico (Aide *et al.*, 2000). Similarly in Hong Kong, significant forest recovery occurred once a simple monoculture was present that facilitated further successional development (Nicholson, 1996; Zhuang and Corlett, 1997). But it is often difficult to exclude recurrent disturbances such as fire or prevent the harvesting of timber or other resources that some local people may depend on. In addition, many degraded landscapes are now without significant patches of residual forest. This may mean that the recolonization of these areas is likely to be non-existent or slow especially if they are now occupied by grasses. Under these circumstances various forms of active intervention will be needed.

Ecological Restoration: A Global Challenge, ed. Francisco A. Comin. Published by Cambridge University Press. © Cambridge University Press 2010.

Artificially restoring the original forest ecosystems is difficult. Current ecological theory provides only limited guidance about how to re-assemble forest ecosystems (Temperton *et al.*, 2004). Although there has been considerable work carried out in recent years into how decreasing or increasing biological diversity affects ecosystem functioning, the overwhelming impression from this literature is that we are still unable to make many predictions about the consequences of the loss of species from an ecosystem (Hooper *et al.*, 2005). Consequently, we do not know which species or what number of species might be especially critical for the restoration process. Nor do we know if the sequence in which species are added to a site matters (Booth and Swanton, 2002). However, several useful generalizations have emerged from this work. One is that the consequences of species loss from a natural ecosystem are often species- or context-specific. This means that the loss of some species may have rather more consequences than the loss of others (Doak and Marvier, 2003; Simberloff, 2003). Another is that the taxonomic identity of the species lost is usually less important than the type of species. That is, the ecological functions a species is able to perform are more important than its taxonomic identity. This suggests that when restoring a degraded ecosystem, certain combinations of species may be able to improve eco-system functioning rather more than random assemblages.

Most of these studies have been carried out using comparatively short-lived field trials, often involving just grasses, and it is not clear how applicable they are to multi-species communities of longer-lived trees. Nor is it clear, in the case of restoration, just how to identify which combination of trees might form complementary mixtures and which might be unstable. Some studies have used mathematical models involving long successional time periods or multiple generations of particular species to explore the functional role of diversity over time (e.g., Naem, 2003; Meurk and Hall, 2006). These models usually assume successional development takes place in an undisturbed environment. But the real world is more complex than this and, once disturbed by human activities, most ecosystems are rarely able to escape further disturbances. Under these circumstances, and in the absence of better information, the process of reforestation has become driven by pragmatic choices and socioeconomic imperatives more than by ecological theory.

7.2 Reforestation at a particular site

Most reforestation in the tropics is currently carried out using fast-growing, exotic species such as eucalyptus or acacias (Cossalter and Pye-Smith, 2003). The silvicultural techniques to do this are reasonably well known (Evans, 1992) and some large plantations have been established by industrial timber users such as pulpwood companies. The purpose of these plantations is to generate an economic product rather than to restore the original forest and most of these industrial plantations are monocultures grown on short rotations of perhaps ten years. The advantage of this type of reforestation is that it yields a rapid financial return and so can be profitably replicated over large areas. It may also restore some of the original ecological processes and functions. More commonly, however, those operating in these plantations are different to those operating in the original forest (Cossalter and Pye-Smith, 2003). For example, water use may be substantially greater in plantations of fast growing species grown on short rotations than in natural forests (Bruijnzeel, 2004) and possibly soil loss also, depending on understory conditions. Not all monocultures contain fast growing exotic species and some using higher value species are managed for sawlogs on rotations of twenty to forty years. Over time these plantations may acquire considerable biodiversity if natural forest is nearby (Keenan *et al.*, 1997) (Plate 7.1).

A second approach might be to seek to fully restore the original forest ecosystems. This involves re-establishing the original biodiversity and species communities of the site and is often referred to as forest restoration (Lamb, 2001; Lamb and Gilmour, 2003). This is difficult to achieve for both technical and financial reasons. Nonetheless, some encouraging results have been achieved using species-rich plantings (Parrotta *et al.*, 1997; Tucker and Murphy, 1997; Elliott *et al.*, 2006). If the plantings are near natural forest then seed-dispersing wildlife can accelerate recolonization by additional plant species and facilitate the development of wildlife habitats (Keenan *et al.*, 1997). The chief disadvantage of this approach is that it requires considerable knowledge about the autecology of a large number of species and is very expensive to implement. This means it is difficult to be of use on a large scale except, perhaps, in some special circumstances (Elliott *et al.*, 2006).

There is a third approach which is mid-way between these two and combines the advantages of simple commercial plantations and the benefits from forest restoration. This can be referred to as rehabilitation and it involves establishing plantations containing mixtures of species, including commercially valuable species (Lamb and Gilmour, 2003). In this case some, but not all, of the original tree biodiversity is established, thereby enhancing the functional recovery of the site. The smaller number of species means it is more feasible to implement. At the same time, the use of some commercial species means there is a greater incentive to reforest larger areas than might be the case with a restoration planting. Multi-species plantations are obviously more complex to manage and their design raises a series of difficult questions; for example, just which species to use and how many species, or functional types of species, are needed to restore particular ecological processes (Schwarz *et al.*, 2000; Swift *et al.*, 2004; Hooper *et al.*, 2005). However, there are some promising results (Piotto *et al.*, 2004; Montagnini and Jordan, 2005; Petit and Montagnini, 2006; Erskine *et al.*, 2006).

In theory, each of these approaches might be scaled up and applied over large areas. In practice, however, any kind of reforestation is difficult to achieve even on a small area. This may be due to a lack of technical knowledge or because of biophysical, socioeconomic or institutional reasons (Table 7.1). Some of these knowledge and biophysical problems can be overcome by using well-known exotic species that are able to tolerate infertile sites and are commercially useful. But these species do not suit all circumstances (Evans, 1992; Cossalter and Pye-Smith, 2003). In addition there are also socioeconomic and institutional problems which are often even more difficult to resolve. These problems are especially difficult in degraded tropical landscapes because large numbers of the people living in these areas are often poor. For these communities reforestation can be much less attractive than other land uses such as agriculture or even agroforestry because the initial costs are much higher and the benefits are too slow in coming. Any reforestation that aims to cover large areas must, therefore, somehow balance the need to restore biodiversity and ecological functioning on degraded lands with the need to simultaneously improve human livelihoods.

Table 7.1. *Constraints commonly inhibiting farmers or households from undertaking reforestation on degraded lands.*

Type of constraint	Example
Biophysical	Land simply too degraded (soil fertility, pollution, weeds and pests)
	Fires too frequent
	A lack of seed/seedlings of desired tree species
Lack of technical knowledge	Appropriate species to plant
	The seed and seedling biology of these species
	Plantation designs and/or community assembly rules
	Stand management methods
Socio-economic	Insufficient landholdings (all land needed for household food production)
	Opportunity costs of reforestation too high
	Initial costs of reforestation too high
	Finance for investment in reforestation unavailable
	Uncertain markets and prices for future goods or services provided by plantations
	Attitudes towards nature
Institutional	Uncertain land tenurial arrangements
	Bureaucratic arrangements
	Inappropriate governmental reforestation policies (on incentives, loans or subsidies)
	Constraints on harvesting or marketing of products produced

7.3 Reforestation of degraded sites in the wet tropics of Australia and Vietnam

Notwithstanding these problems, there has been increased interest by farmers throughout the tropics in undertaking reforestation on part of their land. This is the case most especially when they have secure tenure of their land and trees (or at least secure and long-term access rights) as well as having the rights to harvest these trees (Chambers and Leach, 1989; Dove, 1992; Gilmour and Nurse, 1995; MARD, 2001; Tran, 2006). There has been a variety of reasons for this increased interest, including a rise in the market for forest goods such as fire-wood or timbers (often caused by a decline in the supply of these from natural forests) and the provision of incentive payments or subsidies from the government to improve the supply of various ecological services from these deforested landscapes.

Farmers in both Australia and Vietnam have taken part in this reforestation effort, although for different reasons. In Australia they were motivated by the possibility of growing high-value rainforest trees for commercial benefit and also to restore biodiversity to deforested land. In Vietnam the primary reason was to improve household income. The different approaches used are summarized in Table 7.2. Interestingly, many of the reforestation problems being encountered by the two sets of farmers are often quite similar, despite differences in the environmental history of the two countries, the economic circumstances of the farmers and in the types of reforestation being attempted.

7.3.1 Australia

Most of the tropical rainforests on good soils and gentle topography in the wet tropics of north eastern Australia have been cleared for agriculture and, until recently, few farmers were interested in reforestation. Any timber market would have been dominated by trees being harvested from natural rainforests still remaining in the region. Most of these remaining forests are found on hill areas and were reserved as state forests although smaller, privately owned fragments also remained in the agricultural areas. Some plantations (of the exotic *Pinus caribaea* and the native *Araucaria cunninghamii*) were established by the government forestry agency but few farmers were interested doing likewise.

However, interest in reforestation was kindled following the cessation of logging in the natural rainforests in 1988. After this time, most residual rainforests were placed into the protected area network. Several government compensation programs and a general concern about conservation then prompted a number of landowners to reforest small areas of their farms.

Table 7.2. *Types of reforestation carried out in wet tropics of Australia and Vietnam.*

Type of reforestation	Tropical Australia	Vietnam
Plantation monocultures	Plantations grown on long rotation (>30 year) for sawlogs; largely exotic pines though some with the native *Araucaria*; often large areas (>1000 ha.).	Plantations grown on short rotations (<10 year); mostly exotic Eucalyptus or Acacia though some with high-value native species. Most plantations in small individual patches.
Multi-species plantations	Widespread but mostly small scale; use native high-value species; use of enhanced species richness primarily to foster biodiversity conservation.	Widespread and always small scale; include exotics and native high-value species; use of enhanced species richness primarily to hasten cash flows and diversify income sources.
Restoration plantings	Mostly small scale plantings of native species undertaken by non government organizations	Rarely undertaken
Encourage natural regrowth	Scattered areas only; often on margins of undisturbed natural forest; many have significant levels of biodiversity; many enriched by seed-dispersing wildlife.	Common but rarely associated with undisturbed forest; considerable variation in composition and biodiversity content; probably only limited enrichment by seed-dispersing wildlife but some silvicultural enrichment with commercially useful species.

Several types of reforestation were used. Some farmers were primarily interested in achieving commercial outcomes from their plantings, while others were interested in plantings that sought to achieve some financial benefits while also generating a biodiversity gain. Thus plantings varied from simple monocultures to mixed species plantings although almost all used high-value native rainforest tree species (in contrast to the government's use of *Pinus caribaea* in plantations). Most of the tree species used in the commercial plantings needed rotations of around thirty to forty years. Some non-commercial restoration plantings to restore biodiversity were also carried out though most of these were funded by government programs or carried out by non-government-organizations, rather than by farmers (Erskine *et al.*, 2005).

The key problems facing those carrying out these several programs fell into two groups. One involved a lack of ecological and silvicultural knowledge about the species being used (what species to use of the several hundred available? what were the site preferences of these species? how to design multi-species commercial plantations or, in the case of restoration plantings, what methods should be used to re-assemble forests resembling the original ecosystems?). An especially difficult issue for landowners interested in achieving both production and conservation aims was in deciding how to develop plantations that achieved both goals. However, through trial and error, progress was made in resolving many of these problems and ultimately some 3200 hectares of land were replanted over a ten-to-fifteen year period (Vize *et al.*, 2005). These plantings were almost entirely made up of native species and most were planted in mixtures. The modal number of species used in the commercial plantation mixtures at any particular site was eight to ten species and over eighty species per site were included in the biodiversity restoration plantings. This meant that biodiversity at a site level had been significantly enhanced. Even monoculture plantations were able to acquire additional species brought in by seed-dispersing birds from remnants of natural forests scattered across the landscape (Keenan *et al.*, 1997). However most of the plantings were small (less than five hectares) meaning that the consequence of the plantings on a landscape scale was much less significant.

The second problem concerned the economics of reforestation. It had been expected that many of these new plantations might be used to replace the high-value logs harvested from natural forests before the logging ban had come into place. Government support had assumed the planting program would become self sustaining as trees were harvested from the new plantations. Unfortunately, the high-value timber market virtually disappeared once the supply from the natural forests declined and logs were only available episodically. This occurred while the new plantations were still young. The collapse in the market meant growers had no idea of the financial returns they might achieve through reforestation. Because of this, the rate of planting declined substantially once government support ceased in 2000 (Erskine *et al.*, 2005).

7.3.2 *Vietnam*

Very little undisturbed forests still remain in Vietnam and only small remnants are found outside the steeper mountain areas. Some extensive regrowth areas are present although most of these are also in hillier country. Most agricultural landscapes have only a small number of forest patches remaining and there is little wildlife still present in these agricultural landscapes that is able to disperse seeds from or between these residual forests. The government of Vietnam is currently seeking to substantially increase the country's forest cover (Gilmour *et al.*, 2000; MARD, 2001; Ohlsson *et al.*, 2005). The aim is to protect the environment, improve livelihoods and increase the wood supply for industry and households by reforesting five million hectares over a ten-year period. The plan involves creating two million hectares of protection and special-use forests (one

million hectares from natural regeneration and one million hectares from planting) and three million hectares of production forests (reforestation involving two
million hectares of timber trees and one million hectares of woody cash crops
such as fruit trees, rubber etc.). In the past the government tried to do this via its
own forestry agencies but small farmers will have an increasingly prominent
role in future. This follows the significant increase in agricultural productivity
that came after the de-collectivization of agriculture in the late 1980s. Rural
households are being given legal tenure over land but most farms are small and
the areas in these that are potentially available for reforestation are usually less
than five hectares (Plate 7.2).

The dilemma facing farmers who must decide whether or not to plant trees is in
knowing what the opportunity costs and the financial returns might be from doing
so. Many rural households are only now emerging from poverty and rarely have
savings they can invest in long-term farm activities such as forestry. Tree planting
must improve livelihoods if it is to be adopted on any scale. Tree farming is also a
new land use for most (although many already practice agroforestry). This means
that, like the Australian farmers, they must also grapple with the dilemma of which
species to use and how to establish and manage any new plantation.

In the early 1990s most tree planting was done using fast growing exotic species
from genera such as *Eucalyptus* and *Acacia*. This was because these species
tolerated the degraded site conditions that are common in much of Vietnam and
came as a silvicultural "package" with readily available seed, known planting
methods etc. They were also grown on short rotations (less than ten years) meaning
there was a comparatively speedy cash return. In more recent years there has been a
greater emphasis on also using native species because they have higher timber
values and can be used in a more diverse range of markets. Some have been grown
in monocultures and others in simple mixtures of a few species (McNamara *et al.*,
2006). However, whatever the biodiversity advantages these native species may
offer they must still improve livelihoods if they are to be widely used and the key
disadvantage of many of these species is that they are slower growing than the exotic
species. Silvicultural methods needed to use these species are still being resolved
but current plantation designs involve mixtures of fast (exotic) and slow growing
(native) species as well as multi-purpose species able to provide fruit or other foods
as well as timber. These designs generate early cash flows before the final clear
felling (e.g., by allowing for the removal of the exotics after, say, ten years and
leaving the natives to grow for another twenty years). They also generate a variety of
products for different markets. However, the decision about whether to use them or
not will ultimately be made by individual farmers based on their assessment of how
the planting of these species will improve their financial circumstances and not
about any biodiversity gains they might generate.

Several comparisons can be drawn from these two sets of experiences:

- Many land owners in both countries are potentially interested in reforestation but for different reasons. Both sets of landowners require some financial return from investing in reforestation if they are to replant large areas, although many Australian landowners are also interested in the possibility of reforestation to generate some conservation benefits.
- Despite this interest, reforestation is a comparatively risky land use in both countries because of opportunity costs, establishment costs, the length of timber rotations and the uncertainty over future prices.
- Multi-species plantations can be attractive to farmers in both countries but for different reasons. In Vietnam these are seen as a means of hastening cash flows and diversifying income sources. In Australia they are attractive because of the perceived conservation benefits these provide. In both cases there are still considerable difficulties in designing plantations that will fulfill these objectives and it is often easier to use monoculture.
- The uncertain timber market and the decline in government reforestation incentives led to a decline in the rate of reforestation using high-value rainforest tree species in Australia (notwithstanding the interest in creating conservation benefits). In Vietnam the market for timber products is still high and the availability of some government incentives means there is still considerable interest in reforestation using native species.
- Significant biodiversity gains can be achieved through managing regrowth forests but there is a difference in the two countries in the capacity of such forests to recover unaided. There is evidence that biodiversity recovery can be rapid in Australian regrowth forests, most especially in those where the intensity of disturbance was low and where natural forest is nearby. The recovery of biodiversity is more difficult in Vietnam because of the reduced populations of former seed-dispersing wildlife and the less frequent presence in the landscape of natural forest patches.

The two sets of experiences suggest that some forms of reforestation on degraded lands are possible but that extensive restoration is unlikely to ever be feasible simply because no farmer is likely to contemplate reforesting all their land. The two case studies also suggest that, of the farmers who undertake reforestation, many are interested in the possibility of using mixed species plantings (i.e., rehabilitation, as defined earlier). This highlights the need for ecologists to understand how to assemble multi-species communities and to identify the particular species that generate the greatest functional benefits. But even if this information were available, different farmers would make still different choices about what species to use and what type of plantations to establish, and many would still use monocultures, because their economic circumstances are different.

Moreover, there is also a scale problem. The reforestation projects carried out by these individual farmers obviously form the starting point of any program to overcome degradation. However, most of these are too small to be able to influence key ecological processes, such as the maintenance of viable populations of species or hydrological cycles that necessarily operate on a landscape scale. In both cases, trees were largely planted where these Australian and Vietnamese farmers thought their individual opportunity costs were lowest and the sizes of the areas planted were those they could afford. No particular attention was paid to where their neighbors were planting. Nor were trees planted to improve the connectivity between residual forest fragments or to protect riparian areas, although some more targeted planting, including corridors, was done in some of the Australian operations (Lamb and Erskine, 2008). This poses a major dilemma: how to integrate the reforestation activities of the many individual landowners who may be present in a particular area in order to generate the functional benefits that only operate at these larger scales.

7.4 Forest restoration across landscapes

If the ecological benefits from reforestation are to be maximized, a way must be found to integrate the many separate decisions made by individual landowners present in a particular catchment or landscape. Maginnis *et al.* (2005) refer to this as forest landscape restoration, by which they mean a process that aims to regain ecological integrity and enhance human well-being in deforested or degraded land-scapes. In this ideal process each farmer would know how their reforestation efforts complement and build upon the activities of others. However, three questions need to be addressed if forest degradation is to be overcome at this scale:

- how much reforestation is needed in a particular landscape to achieve the desired functional outcomes?
- where should these new forests be located in order to most efficiently enhance the conservation of biodiversity and restore ecological processes?
- what type of reforestation should be used at particular localities?

Before answering these questions it is necessary to understand the condition of the landscape that is to be reforested and the circumstances of the human populations present in that landscape.

7.4.1 The ecological mosaic

Landscapes are usually made up of a mosaic of different vegetation types and land uses (Gilmour, 2005a). They may contain agricultural areas, remnant natural forest, patches of forest regrowth, degraded former agricultural sites and urban areas. These spatial patterns may reflect differences in soil fertility or topography. Successional changes may be underway in some of these land units. For example, forest regrowth may have commenced on old abandoned farmland. In the Australian situation described earlier there were significant patches of natural forest still present scattered through the largely agricultural landscape. The more extensive areas on steeper sites form part of the protected area network while many other, smaller remnants remain present on individual farmers' land. In Vietnam there are fewer remnants of undisturbed forest in the agricultural landscapes but many big areas of abandoned grasslands and shrublands and other highly degraded sites on hillslopes. Many valley floors have been converted into paddy rice areas. Reforestation can only occur on hill areas (Plate 7.3).

It is rarely possible that all deforested and unused areas can be restored at once and, more commonly, priorities must be set. The areas of each of these landscape units as well as their spatial distribution will determine just how a forest landscape restoration program is tackled. The key ecological principles of landscape ecology upon which such a program must be based are reasonably well known (e.g., Harrison and Fahrig, 1995; Bennett, 2003; Turner *et al.*, 2001). Five particularly important principles are:

- It is preferable to utilize existing natural regrowth forests wherever possible.
- Large forest patches are more useful as conservation reserves than small patches.
- Long-lived forest patches are preferable to short-lived patches (i.e., plantations grown on long rotations are preferable to plantations grown on short rotations).
- Connectivity between existing forest fragments should be enhanced to facilitate species movement across landscapes (by bridging plantations that form corridors or by establishing many small plantations or "stepping stones" between the remaining forest areas).
- Plantings along riparian strips can act as filters to limit erosion and sedimentation into streams.

There are a number of alternative forms of intervention that might be made within the landscape mosaic using these principles (Table 7.3). These interventions may involve restoration plantings, rehabilitation plantings or even industrial plantations using monocultures of exotic species.

7.4.2 The socioeconomic mosaic

Of course, landscapes are not just bio-physical mosaics. Rather, they also contain mosaics of socioeconomic situations and circumstances. Some areas of the landscape will be densely populated and others will not. Some will be occupied by tenured land owners while others will be occupied by farmers who do not have a formal title to the land they use. Some farms will be large while other land-users may have small holdings. These farmers will also differ in terms of household incomes and the degrees to which they are likely to want to engage in "risky" new land-use activities, such as reforestation. But these various stakeholders are not the only ones with a legitimate interest in the land-use decisions being made in the area. Others might include hydro-electric authorities, protected-area managers, conservation groups, etc. Together, these various groups form

Table 7.3. *Alternative points of intervention or priorities for restoration in "degraded" landscapes.*

Types of reforestation interventions	Reason
Protect areas of degraded or secondary forest and facilitate natural regrowth (perhaps by enrichment planting).	Many degraded forests are capable of recovering much of their original biodiversity if protected and if residual natural forest is nearby. The cost of doing this is likely to be less than replanting.
Enlarge small residual forest fragments by planting species-rich forest around their margins (i.e. restoration plantings).	To diminish the role of the "edge effect" and increase the effective inner core habitat area.
Create buffer zones around residual natural forest fragments using commercial timber plantations (i.e. monocultures or rehabilitation plantings).	To provide buffers against fires, weeds etc. and to reduce "edge effects".
Create habitat corridors between residual forest areas (restoration or rehabilitation plantings).	To facilitate the movement and genetic interchange of poorly dispersed species.
Establish new forest patches in the agricultural matrix between residual forest areas.	To act as "stepping stones" that facilitate the movement and genetic interchange of more easily dispersed species.
Create protective zones along riverine areas.	To act as filters limiting the movement of soils into streams.
Reforest eroded areas on hill slopes.	To stabilize hill slopes and limit erosion.
Reforest recharge areas in landscapes prone to salinity.	To restore key hydrological processes.

a complex mix of people who differ in objectives, wealth, political power and influence (Gilmour, 2005a). There may be disputes over resource ownership between some of these various stakeholders.

In the Australian situation most farmers have freehold land and can choose how it is managed. Their choice will vary with the quality of land and few are likely to want to reforest what they consider to be good agricultural land. On the other hand, declines in the prices of some agricultural crops in recent years (e.g., tobacco, sugar) mean that some farmers might be open to trying new activities if they perceive these to be profitable. However, there are also many external stakeholders in this region who are interested in watershed management and regional conservation issues and may have a quite different set of priorities. These stakeholders are often politically influential and can influence government policies that may impinge on reforestation decision-making by landowners.

In Vietnam many farmers have recently acquired long-term leases over their land and also have the right to choose how it is managed. Their choice is likely to depend on the amount of land they have and on their present household income. Households with only small areas of land and limited incomes are unlikely to be able to afford to grow trees even if they were interested. There are also some farmers who currently have *de facto* rather than *de jure* rights which mean that they, too, may be reluctant to invest in a long-term activity like tree growing. Perhaps the largest external stakeholder in Vietnam is the national government which has a policy of rapidly increasing forest cover to 43 percent of the nation's land area (MARD, 2001). It has developed a number of policies including the offering of financial incentives that are aimed at persuading farmers to undertake more reforestation to allow the government to meet this goal. The extent to which these incentives influence individual farmers' decisions obviously depends on their circumstances and their perceptions of market opportunities.

Both the Australian and Vietnamese landscapes (in the widest sense of the word) are, therefore, hugely complex. Random or opportunistic interventions in these ecological and socioeconomic mosaics are unlikely to be useful. Rather, there is a need to establish an approach or mechanism by which interventions might be carried out in a more strategic fashion. To be successful, this should improve the ecological integrity of the landscape as well as enhance the well-being of all the stakeholder groups present.

7.5 Implementing forest landscape restoration

Two things are needed to implement forest landscape restoration. One is a list of priorities showing where in the landscape the process should start. Success often fosters further success, so it is important that the first interventions generate rapid and obvious benefits. The second is an agreement from the stakeholders on whose lands this initial reforestation take place that these interventions are acceptable. There are several steps by which these two objectives might be achieved and these are outlined in Table 7.4 and discussed further in ITTO/IUCN (2005).

This approach outlined below is similar in many respects to participatory land-use planning projects such as those outlined by Jackson and Ingles (1998), ITTO (2002) and Suraswadi *et al.* (2005). It differs from the methodology used to undertake restoration at particular sites outlined by SER International (Clewell *et al.*, 2005) because there is no reference area against which comparisons can be made and because the trade-offs between stakeholders mean the overall objectives may be less precise.

7.5.1 Understanding the landscape mosaic

The starting point is to acquire an understanding of the spatial mosaic of both the biophysical and socioeconomic elements in the landscape. In the case of the

Table 7.4. *Stages in the development of the process by which forests might be re-established on a landscape scale.*

Step	Element	Types of issues or questions to be resolved
1	Understand the landscape mosaic	What are the biophysical components of the landscape mosaic: what are the spatial patterns of forest and agricultural lands; what are the biodiversity patterns and trends; where are the environmental problems such as erosion? What of the socioeconomic mosaic: where do people live; what are the land uses generating economic wealth; what are the economic and social trends?
2	Identify the stakeholders involved	Who are the resident and non-resident stakeholders? Are they in the public or private sectors? Are they legal landowners or de facto owners? Are they wealthy or poor?
3	Develop a planning process	Decide how much land to reforest. Where to reforest? What sort of reforestation to undertake?
4	Negotiate priorities and trade-offs between stakeholders	Process must take into account the variety of viewpoints likely to be held by stakeholders and that these views may change over time since both ecological and socioeconomic systems are dynamic.
5	Implement and monitor	Are new regulations needed to prevent further degradation? Is there a need for compensation payments to certain stakeholders? Might incentive payments be useful? Are there markets for goods and services generated by the new forests? How to practice adaptive management?

biophysical elements, the key issues are to determine what vegetation types and land-use activities are currently present and just exactly where these are located. In particular, this means knowing how much undisturbed natural forest or regrowth forest remains and how much other land might be most easily reforested (e.g., less fertile agricultural land or unused, degraded land). Another key issue is the extent of degradation that has occurred: what biota are present? How many species are threatened and where are all these species located? Likewise, where are the environmental problems such as erosion points? Similar questions apply in the case of the socioeconomic elements: where are the human populations found? What are their various land use activities and how have their spatial patterns changed over time? What are the economic drivers that have caused changes in land use activities? The nature of these bio-physical and socioeconomic mosaics will indicate the environmental problems to be addressed and the types of interventions that might be possible. The areas most readily available for reforestation will probably be those that are only lightly used or farmland that has been abandoned (Plate 7.4).

7.5.2 Identifying stakeholders

The second step is to identify the stakeholders (both internal and external) with legitimate concerns related to the landscape. Internal stakeholders are the people now living in the area, such as farmers and local townspeople. External stakeholders are those living outside the landscape, such as farmers living at lower elevations, downstream water users, fishermen, protected-area managers etc. These various stakeholders will differ in the amounts of land they own or manage, in whether they have tenure, in the degree to which their livelihoods depend on land-use activities in the particular landscape, in their wealth and in their political power. All will have some interest in the nature and location of any reforestation that might take place (although the degree of interest will not necessarily be identical). Note that farmers without tenure are unlikely to be interested in any form of reforestation involving lengthy time periods.

7.5.3 Planning reforestation

The third step requires the development of reforestation plans. These plans will necessarily address the three key questions outlined earlier (i.e., how much reforestation, where to reforest and what kind of reforestation.) A useful approach is to develop alternative options or scenarios which can then be debated and evaluated by stakeholders.

How much reforestation

There has been considerable research by landscape ecologists concerning the first of these questions. For example, there is little doubt that the threat of species loss increases the more a forest is fragmented and cover is lost (Fahrig, 2003). But there is some uncertainty over the idea of a critical threshold level of cover below which species are lost. Andren (1994) suggests that many birds and mammal species are lost once deforestation causes habitat cover to fall below 10 to 30 percent of the original area. Tischendorf (2001) refers to values of 30 to 50 percent of the landscape. However, Lindenmayer and Luck (2005) review some of these issues and note there must be uncertainty about the precise nature of community level thresholds since multiple interacting processes may operate at different scales and affect diverse species in different ways. Indeed, they suggest that populations of many species probably begin to decline *above* a nominal threshold level of cover at which a species is lost from a region. Nor is it clear that thresholds work in reverse after reforestation – will species richness necessarily recover immediately a supposed landscape threshold has been exceeded?

Irrespective of whether or not thresholds can be specified, large-scale reforestation in extensively deforested landscapes may be difficult to achieve in practice, especially if the former forest land is now being used productively for, say, some agricultural purpose. Nor is it likely that many rural communities will be concerned with reforestation if significant amounts of residual forest remain. On the other hand, many communities might be willing to reforest a landscape with large areas of unused or low quality agricultural land, especially if much of this is steep and seen to be eroding. Furthermore, the reforestation option will probably become even more attractive if there is some kind of incentive payment or subsidy to cover the initial costs.

But the issue of future benefits or, conversely, opportunity costs will always be critical – how might these particular landowners directly benefit from reforestation and what opportunities might these landowners forgo by planting trees? In the Australian situation, many landowners in the wet tropics were willing to reforest small areas of their farms with native tree species as part of the regional reforestation scheme. But the high costs of reforestation and the uncertainty about future timber markets meant few were willing to commit more than a few hectares to tree growing. The net result was that 3,200 hectares of plantations were established but on most farms the reforested areas were less than five hectares in size (Vize *et al.*, 2005). These plantations represent a trivial increase in the overall forest cover in the region.

The Vietnamese situation is significantly different. Since the early 1990s the land allocated to farmers often includes land designated as "forest" as well as agricultural land. The "forest" land may have been nominated to become protection forest or production forest, and each category may include land already containing forest or land still requiring reforestation. This means the amount of reforestation that is possible will depend on the effectiveness of the original land allocation process. However, in many cases there are discrepancies between official government land-use plans and land-use activities that are actually occurring on the ground. For example, Ohlsson *et al.* (2005) reported that some areas supposedly available for reforestation are, in fact, used for food production. Tran (2006) reports similar findings. This means it may be difficult to reach any particular reforestation target (or threshold) by planning within the reforestation sector alone, at least in the immediate future, and that a more effective short-term approach might be to simply create enabling conditions that make reforestation a more attractive land-use activity to landholders. These might involve providing things like silvicultural knowledge, market information, credit for reforestation etc.

Where to reforest

The second question concerns the location of these plantings. Much research to date has concerned the process of forest fragmentation and the implications this might have for species conservation (e.g., Kareiva and Wennergren, 1995; Bowne and Bowers, 2004; Neel *et al.*, 2004). If the area reforested is likely to be limited, at least in the short term, then the spatial patterns of reforestation and degree of connectedness between forest areas will have important consequences for the rates of various ecological processes (Tischendorf, 2001). Some of the options have been described earlier and are outlined in Table 7.3.

All of these forms of reforestation may generate ecological benefits but it is not possible to specify which of these options should take priority. This will depend on the spatial patterns of existing forest within the existing landscape mosaic and the characteristics of any threatened species (e.g., their vagility) or the severity of erosion at particular localities etc. But there is a trade-off to be made between public benefit and private cost. Any reforestation will depend on the impediments to reforestation described above being overcome and compensation may need to be paid to a landowner whose land is reforested because it occupies a particularly critical spatial location.

What type of reforestation

The third question concerns what sort of reforestation takes place at each of these sites. Various alternatives were outlined in Table 7.2. From a biodiversity conservation or watershed protection point of view it would be preferable for all reforestation to be carried out using species-rich "restoration" plantings. These are likely to recreate most quickly the habitat conditions necessary for the conservation of threatened species. However, they are unlikely to be established in many developing countries without financial assistance (but see Elliott *et al.*, 2006) and will normally only be carried out over comparatively small areas because of the costs of doing so. Simple monocultures with fast-growing exotic species are often an attractive option for large industrial growers. These industries usually prefer large contiguous plantations on flat land close to road networks. Small plantations scattered across a number of carefully chosen locations within a landscape are likely to be much less attractive. Despite this, industrial enterprises sometimes find it useful to form agreements with small growers to obtain timber resources. The attractiveness of the arrangement to these growers obviously depends on the price offered by the industrial user and this will usually decrease with increasing distance from the mill.

The third alternative includes monocultures of high-value native timber species or various forms of multi-species rehabilitation plantings. These latter may be agroforestry plantings or timber plantations. Though these may be more technically difficult to establish and manage they offer some advantages. Although the site (or alpha) diversity in such plantations might be modest, the overall landscape (or gamma) diversity could be high. In the case of the northern Australian example, the alpha diversity in multi-species plantations was around 8 to 10 species at each site but the gamma diversity in plantations across the region was 175 species. A similar pattern has evolved in parts of northern Vietnam. Most private plantations have less than eight tree species but surveys in seventy-seven plantations across the region found sixty-four species were present, as reported in ICARD, 2003. Under these circumstances significant populations of differ-ent tree species can develop across the landscape. In the wet tropics of Australia these types of plantations have been found not to be as effective in creating wildlife habitats as restoration plantings but they are satisfactory for some species (Kanowski *et al.*, 2005).

7.5.4 Negotiating priorities and trade-offs

Different stakeholders commonly have different views on how resources might be used and just what type of reforestation should take place (Suraswadi *et al.*, 2005). Those with an interest in conservation are more inclined to favor restoration plantings using complex planting designs, while most farmers usually prefer agro-forestry systems or tree plantations that generate an early cash flow. Under these circumstances, it is often useful to develop a series of alternative options or scenarios to explore the various ways in which a forest landscape restoration program might take place. These scenarios differ in terms of the amount of refor-estation to be done, the locations of this reforestation within the area and the types of reforestation that should be undertaken at particular places in the landscape. Maps, three-dimensional models or computer models can show the spatial patterns and parts of the landscape involved. These scenarios then provide the basis for discus-sion among stakeholders over the priorities to be adopted and the trade-offs needing to be made. It is commonly assumed that landscape patterns affect the rate of ecological processes. Hence, one way of assessing the overall worth of the various reforestation interventions planned in a particular scenario might be to use a set of attributes or indices that describe the landscape pattern emerging from the implementation of that option. The best scenario then becomes that with the highest "score."

The success of such an approach obviously depends on the indices used. Some possible indices are suggested in Table 7.5, which also shows some possible socio-economic indices. Some of these attributes are metrics commonly used by landscape ecologists to assess change (e.g., forest area: edge ratio, degree of connectivity between remnant patches), while others are more especially relevant to restoration (e.g., number of untreated erosion points, proportion of farmers engaging in refor-estation). It is not difficult to think of others that might be used. For example, it could be useful to distinguish between the number of farmers planting trees voluntarily and the number needing some kind of subsidy or incentive before engaging in reforestation. There is no single set of attributes and the nature of the ecological and socioeconomic situation will determine those that might be useful.

Some of the ways in which these negotiations and trade-offs might be undertaken are described by Kusumanto (2005). Since the final choice may benefit some stakeholders at the expense of others it may be necessary to develop some kind of

Table 7.5. *Examples of possible indices to use in evaluating alternative reforestation scenarios.*

Ecological indices	Socio-economic indices
Total area of forest (undisturbed natural forest, regrowth forest and plantations).	Number or proportion of landowners with tree plantations.
Total area of "high quality" (multi-species forest) forest.	Number or proportion of landowners with monoculture plantations of exotic tree species.
Proportion of landscape covered by forest.	
Number of forest patches >50 ha.	Number of farmers able to harvest forest products for household or on-farm use.
Area of "core" forest (not affected by edge effect).	Number or proportion of landowners receiving financial benefit from harvesting trees (currently or within next 15 years etc).
Forest area: edge ratio.	
Degree of connectivity between forest remnants.	Costs of incentives or subsidies to enable reforestation.
Modal distance between forest remnants.	
Proportion of riparian area protected.	Net present commercial value of plantations and forests.
Proportion of land >25° slope deforested.	
Number of erosion points remaining untreated.	Total payments received from recipients of ecological services.
	Opportunity costs of reforestation.

incentive or compensation program that allows for some sharing of costs and benefits (Jackson and Ingles, 1998).

7.5.5 *Implementing and monitoring*

It is rarely possible to make confident predictions about precisely how ecological services such as water yields, sedimentation or biodiversity will be altered by the change in the nature of the landscape mosaic caused by forest landscape restoration (although it is assumed the changes will be beneficial). Current research into how the populations of particular species are affected by the distribution of fragments of natural forest shows how difficult this is (e.g., Wiegand *et al.*, 2005). It is obviously more difficult when the forests present include patches of natural forest, multi-species plantations and monocultures of exotic tree species that are harvested at varying intervals. Swift *et al.* (2004) also point out the likelihood that the relationship between biodiversity and functional outcome varies with scale. It is likely to be even more difficult to predict the ways human livelihoods will be affected by the new reforestation patterns. Under these circumstances, it is necessary that some form of monitoring and adaptive management is undertaken, so that changes in circumstances or unexpected developments can be accommodated. This is discussed further by Gilmour (2005b) and Gasana (2005). As our predictive understanding increases it may be possible to refine and improve the indices described in Table 7.5, and use them to evaluate future scenarios.

7.6 Discussion and conclusions

Deforestation is often caused by powerful macroeconomic forces and any attempts to reforest degraded landscapes must recognize this rather than assume the problem is simply a technical matter. This is especially the case when attempting to restore forest cover over large areas and not just on particular farms or in local areas. Forest landscape restoration assumes that a coordinated approach across a landscape will generate greater benefits than leaving any reforestation program to the individual decisions of many separate landowners who are acting independently. The key problem, therefore, lies in finding ways of making the reforestation that should be undertaken an attractive land-use option for the individual farmers involved, especially for those whose families may have been involved in the original deforestation (Jackson and Ingles, 1998).

The two case studies differ in this respect. The Australian farmers were amenable to some reforestation using complex plantation designs but were unwilling to invest too much of their land or resources in the absence of a market for forest products or without a subsidy from the public for the ecological services these plantations might produce. On the other hand, the Vietnamese farmers appeared more willing to carry out reforestation because their timber markets seemed more encouraging. However, their interest in biodiversity was largely linked to its capacity to improve their economic circumstances and provide some form of "insurance." Some form of forest landscape restoration will be possible in both cases but each will take some time to achieve and each will probably use a different pathway.

This raises the issue of who promotes the whole process of forest landscape restoration and takes on the coordinating role to achieve the changes. One possibility is that the central planners in a government agency do it. However, it is probably useful to involve those most adversely affected by the current (degraded) state in the planning process. Gilmour (1990) describes the situation in Nepal where villages over a large area responded to a shortage of forest products by spontaneously developing indigenous silvicultural systems to manage nearby forests even though these were still formally under government control. This suggests that planning might be done by a cooperative group or coalition formed from the stakeholders present (some of whom might be government bodies). Such a group would require a significant amount of coordination if a large area is to be reforested, but this would be a logical approach for stakeholders negatively affected by the degradation process. A third possibility, somewhere between these two, is a special regional organization, perhaps led by a government agency, but including representatives of different stakeholder groups. Gottfried *et al.* (1996) discuss these possibilities and suggest that the best outcomes are more likely to occur in small-scale organizations than in large ones. Further, they argue that planning groups who are close to local decision-makers (e.g., farmers) are likely to be more sensitive to unfavourable outcomes. Such groups are able to react quickly to events and will be more effective in achieving landscape-scale goals. These views point to the second and third options being more attractive than the first.

The ultimate purpose of forest landscape restoration is to convert a currently degraded landscape into one that is both ecologically and socioeconomically resilient, that is, one that can accommodate future changes and still maintain its overall structure and function by being able to adapt and re-organize itself (Folke *et al.*, 2003). The ecological component of this resilience depends on there being a diversity of functional types within the biota present and these being connected by mobile links, such as seed-dispersing wildlife moving between core habitat areas within the landscape mosaic. This provides what Folke *et al.* (2003) refer to as ecological memory and allows for an increased buffering capacity. Hence, the more forest patches and the more structurally complex these are, the greater this capacity. It should be recalled that the Vietnamese field situation described earlier was one where there were rather fewer wildlife species able to disperse seed than in the Australian field situation. To that extent, it may be harder to build ecological resilience in the Vietnamese case than the Australian case unless more active planting is carried out. The existing levels of gamma diversity noted earlier give grounds for some optimism.

The socioeconomic component of resilience depends on an improvement and diversification of income sources for the landowners present. But it also depends on the capacity of the various stakeholders to influence ecological conditions, learn from this experience and share knowledge (Walker *et al.*, 2004). This requires that there are innovators and experimenters, entrepreneurs and imple-menters as well as knowledge carriers and networkers (Folke *et al.*, 2003). In practice, this means that forest landscape restoration is more likely to be successful if the various participants can look beyond the reforestation occurring at a particular site and instead think of what is happening across the landscape. The idea of an ecological threshold area for reforestation was discussed earlier. A more crucial threshold may be when the numbers of landowners or farmers involved in reforestation or participating in reforestation learning networks reaches some critical level (a resilience threshold?). At this point, the whole process becomes self-organized and self-sustaining because the key stakeholders believe it is worthwhile. After this time, increased reforestation, increased landscape hetero-geneity, improved ecological functioning and improvement to livelihoods from reforestation become almost inevitable.

References

Aide, T. M., Zimmerman, J. K., Pascarella, J. B., Rivera, L. and Marcano-Vega, H. (2000). Forest regeneration in a chronosequence of tropical abandoned pastures: implications for restoration ecology. *Restoration Ecology*, **8** (4): 328–338.

Andren, H. (1994). Effects of habitat fragmentation on birds and mammals in landscapes with different proportions of suitable habitat: A review. *Oikos*, **71**: 355–366.

Bennett, A. F. (2003). Linkages in the landscape: The role of corridors and connectivity in wildlife conservation. Gland, Switzerland and Cambridge: IUCN-International Union for Conservation of Nature.

Booth, B. D., and Swanton, C. J. (2002). Assembly theory applied to weed communities. *Weed Science*, **50**: 2–13.

Bowne, D. R., and Bowers, M. A. (2004). Interpatch movement in spatially structured populations: A literature review. *Landscape Ecology*, **19**: 1–20.

Bruijnzeel, L. A. (2004). Hydrological consequences of tropical forests: Not seeing the soil for the trees. *Agriculture, Ecosystems and Environment*, **104**: 185–228.

Chambers, R. and Leach, M. (1989). Trees as savings and security for the rural poor. *World Development*, **17** (3): 329–342.

Clewell, A., Rieger, J. and Munro, J. (2005). *Guidelines for Developing and Managing Ecological Restoration Projects*. SER International-Society for Ecological Restoration International (www.ser.org).

Cossalter, C. and Pye-Smith, C. (2003). *Fast-Wood Forestry: Myths and Realities*. Jakarta: Center for International Forest Research.

Doak, D. and Marvier, M. (2003). Predicting the effects of species loss on community stability. In *The Importance of Species: Perspectives on Expendability and Triage*, ed. P. Karieva, P. and S. Levin. Princeton: Princeton University Press, pp. 140–160.

Dove, M. (1992). Foresters' beliefs about farmers: A priority for social science research in social forestry. *Agroforestry Systems*, **17**: 13–41.

Elliott, S., Blakesly, D., Maxwell, J. F., Doust, S., Suwannaratana, S. (eds.) (2006). *How to Plant a Forest: The Principles and Practice of Restoring Tropical Forests*. Chiang Mai: The Forest Restoration Research Unit, Chiang Mai University, Thailand.

Erskine, P., Lamb, D. and Bristow, M. (2005). *Reforestation in the Tropics and Sub-Tropics of Australia using Rainforest Trees*. RIRDC Publication No. 05/087. Canberra: Rural Industries Research and Development Corporation. (https://rirdc.infoservices.com.au/items/05–087).

Erskine, P., Lamb, D. and Bristow, M. (2006). Tree species diversity and ecosystem function: Can tropical multi-species plantations generate greater productivity. *Forest Ecology and Management*, **233**: 205–210.

Evans, J. (1992). *Plantation Forestry in the Tropics: Tree Planting for Industrial, Social, Environmental and Agroforestry Purposes*. Oxford: Clarendon Press.

Fahrig, L. (2003). Effects of habitat fragmentation on biodiversity. *Annual Review of Ecology and Systematics*, **34**: 487–515.

Folke, C., Colding, F. and Berkes, F. (2003). Synthesis: Building resilience and adaptive capacity in social-ecological systems. In *Navigating Social-Ecological Systems: Building Resilience for Complexity and Change*, ed. F. Berkes, J. Colding and C. Folke. Cambridge University Press, pp. 352–384.

Gasana, J. (2005). Monitoring and evaluating site-level impacts. In *Restoring Forest landscapes: An Introduction to the Art and Science of Forest Landscape Restoration*. ITTO Technical Series No. 23. Yokohama: International Tropical Timber Organization, 125–134.

Gilmour, D. A. (1990). Resource availability and indigenous forest management systems in Nepal. *Society and Natural Resources*, **3**: 145–158.

Gilmour, D. A. (2005a). Understanding the landscape mosaic. In *Restoring Forest landscapes: An Introduction to the Art and Science of Forest Landscape Restoration*. ITTO Technical Series No. 23. Yokohama: International Tropical Timber Organization, pp. 43–51.

Gilmour, D. A. (2005b). Applying an adaptive management approach in FLR. In *Restoring Forest landscapes: An Introduction to the Art and Science of Forest Landscape Restoration*. ITTO Technical Series No 23. Yokohama: International Tropical Timber Organization, pp. 35–42.

Gilmour, D. A., Nguyen, V. S. and Tsechalicha, X. (2000). *Rehabilitation of Degraded Forest Ecosystems in Cambodia, Lao PDR and Vietnam*. Cambridge: IUCN-International Union for Conservation of Nature.

Gilmour, D. A., and Nurse, M. C. (1995). Farmer initiatives in increasing tree cover in central Nepal. In *Farm Forestry in South Asia*, ed. N. C. Saxena and V. Ballabh. New Delhi, Thousand Oaks and London: Sage Publications, pp. 87–103.

Gottfried, R., Wear, D. and Lee, D. (1996). Institutional solutions to market failure on the landscape scale. *Ecological Economics*, **18**: 133–140.

Harrison, S. and Fahrig, L. (1995). Landscape pattern and population conservation. In *Mosaic Landscape and Ecological Processes*, ed. L. Hansson, L. Fahrig and G. Merriam. London: Chapman and Hall, pp. 293–308.

Hooper, D. U., Chapin, F. S., Ewel, J. J., Hector, A., Inchausti, P., Lavorel, S. *et al.* (2005). Effects of biodiversity on ecosystem functioning: A concensus of current knowledge. *Ecological Monographs*, **75**: 3–35.

ICARD-Information Center for Agriculture and Rural Development in Vietnam (2003). *Forest development and the planting of native trees in three mountainous provinces*. Ha Noi, Vietnam. Information Center for Agriculture and Rural Development (unpublished report).

ITTO-International Tropical Timbers Organization (2002). *ITTO Guidelines for the Restoration, Management and Rehabilitation of Degraded and Secondary Tropical Forests*. ITTO Policy Development Series No. 13. Yokohama: International Tropical Timber Organization.

ITTO/IUCN-International Tropical Timber Organization/International Union for Conservation of Nature (2005). *Restoring Forest Landscapes: An Introduction to the Art and Science of Forest Landscape Restoration*. ITTO Technical Series No. 23. Yokohama: International Tropical Timber Organization.

Jackson, W. J., and Ingles, A. W. (1998). *Participatory Techniques for Community Forestry: a Field Manual*. Gland, Switzerland and Cambridge: International Union for Conservation of Nature and World Wide Fund for Nature.

Kanowski, J., Catterall, C., Proctor, H., Reis, T., Tucker, N. and Wardell-Johnson, G. (2005). Biodiversity values of timber plantations and restoration plantings for rainforest fauna in tropical and sub tropical Australia. In *Reforestation in the Tropics and Sub-tropics of Australia using Rainforest Trees*. RIRDC Publication No 05/087, ed. P. Erskine, D. Lamb and M. Bristow. Canberra: Rural Industries Research and Development Corporation, pp. 183–205.

Kareiva, P. and Wennergren, U. (1995). Connecting landscape patterns to ecosystem and population processes. *Nature*, **373**: 299–302.

Keenan, R., Lamb, D., Woldring, O., Irvine, T. and Jensen, R. (1997). Restoration of plant biodiversity beneath tropical tree plantations in Northern Australia. *Forest Ecology and Management*, **99**: 117–131.

Kusumanto, T. (2005). Applying a stakeholder approach in FLR. In *Restoring Forest Landscapes: An Introduction to the Art and Science of Forest Landscape Restoration*. ITTO Technical Series No 23. Yokohama: International Tropical Timber Organization, pp. 61–70.

Lamb, D. (2001). Reforestation. In *Encyclopedia of Biodiversity*, ed. S. Levin. San Diego: Academic Press, pp. 97–108.

Lamb, D. and Gilmour, D. (2003). *Rehabilitation and Restoration of Degraded Forests*. Gland, Switzerland and Cambridge: International Union for Conservation of Nature and World Wide Fund for Nature.

Lamb, D. and Erskine, P. (2008). Forest restoration at a landscape scale. In *Living in a Dynamic Tropical Forest Landscape*, ed. N. Stork and S. Turton. Oxford: Wiley-Blackwell, pp. 510–525.

Lindenmayer, D. B., and Luck, G. (2005). Synthesis: thresholds in conservation and management. *Biological Conservation*, **124**: 351–354.

McNamara, S., Tinh, V. T., Erskine, P., Lamb, D., Yates, and D., Brown, S. (2006). Rehabilitating degraded forest land in central Vietnam with mixed native species plantings. *Forest Ecology and Management*, **233**: 358–365.

Maginnis, S., Rietbergen-McCracken, J. and Jackson, W. (2005). Introduction. In *Restoring Forest Landscapes: An introduction to the Art and Science of Forest Landscape Restoration*. ITTO Technical Series No. 23. Yokohama: International Tropical Timber Organization, pp. 11–13.

MARD-Ministry of Agriculture and Rural Development (2001). *Five Million Hectare Reforestation Partnership: Synthesis Report*. Hanoi: Ministry of Agriculture and Rural Development.

Montagnini, F. and Jordan, C. (2005). *Tropical Forest Ecology: The Basis for Conservation and Management*. Berlin: Springer.

Meurk, C. D., and Hall, G. (2006). Options for enhancing forest biodiversity across New Zealand's managed landscapes based on ecosystem modeling and spatial design. *New Zealand Journal of Ecology*, **30**: 131–146.

Naem, S. (2003). Models of ecosystem reliability and their implications for the question of expendability. In *The Importance of Species: Perspectives on Expendability and Triage*, ed. P. Karieva and S. Levin. Princeton University Press, 107–139.

Neel, M. C., McGarigal, K. and Cushman, S. A. (2004). Behaviour of class level landscape metrics across gradients of class aggregation and area. *Landscape Ecology*, **19**: 435–455.

Nicholson, B. (1996). Tai Po Kau Nature Reserve, New Territories, Hong Kong: a reafforestation history. *Asian Journal of Environmental Management*, **4**: 103–119.

Ohlsson, B., Sandewall, M., Sandewall, R. K. and Phon, N. H. (2005). Government plans and farmer intentions: A study on forest land use planning in Vietnam. *Ambio*, **34**: 248–255.

Palm, C., Vosti, S. A., Sanchez, P. A. and Eriksen, P. J. (eds) (2005). *Slash and Burn: The Search for Alternatives*. New York: Columbia University Press.

Parrotta, J., Knowles, O. H. and Wunderle, J. (1997). Development of floristic diversity in 10-year-old restoration forests on a bauxite mined site in Amazonia. *Forest Ecology and Management*, **99**: 21–42.

Petit, B. and Montagnini, F. (2006). Growth in pure and mixed plantations of tree species used in reforesting rural areas of the humid region of Costa Rica, Central America. *Forest Ecology and Management*, **233**: 338–343.

Piotto, D., Viquez, E., Montagnini, F. and Kanninen, M. (2004). Pure and mixed forest plantations with native species of the dry tropics of Costa Rica: A comparison of growth and productivity. *Forest Ecology and Management*, **190**: 359–372.

Rudel, T., Coomes, O. T., Moran, E., Achard, F., Angelson, A., Xu, J. and Lambin, E. (2005). Forest transitions: Towards a global understanding of land use change. *Global Environmental Change*, **15**: 23–31.

Schwartz, M. L., Brigham, C. A., Hoeksema, J. D., Lyons, K. G., Mills, M. H. and van Mantgem, P. J. (2000). Linking biodiversity to ecosystem function: Implications for conservation biology. *Oecologia*, **122**: 297–305.

Schroth, G., Fonseca, G., Harvey, C., Gascaon, C., Vasconcelas, H. and Izac, A. M. (eds.) (2004). *Agroforestry and Biodiversity Conservation in Tropical Landscapes*. Washington: Island Press.

Simberloff, D. (2003). Community and ecosystem impacts of single species extinctions. In *The Importance of Species: Perspectives on Expendability and Triage*, ed. P. Karieva and S. Levin. Princeton: Princeton University Press, pp. 221–233.

Suraswadi, P., Thomas, D. E., Pragtong, K., Preechapanya, P. and Weyerhauser, H. (2005). Northern Thailand: Changing smallholder landuse patterns. In *Slash and Burn: the Search for Alternatives*, ed. C. Palm, S. A. Vost, P. A. Sanchez and P. J. Eriksen. New York: Columbia University Press, 355–384.

Swift, M., Izac, A. M. and van Noordwijk, M. (2004). Biodiversity and ecosystem services in agricultural landscapes–are we asking the right questions? *Agriculture, Ecosystems and Environment*, **104**: 113–134.

Temperton, V. M., Hobbs, R., Nuttle, T. and Halle, S. (eds.) (2004). *Assembly Rules and Restoration Ecology*. Washington DC: Island Press.

Tischendorf, L. (2001). Can landscape indices predict ecological processes consistently? *Landscape Ecology*, **16**: 235–254.

Tran, B. D. (2006). *Potential to integrate high-value native tree species into the upland farming systems of Hoa Binh Province, Vietnam*. Unpublished M. Agric Thesis, School of Agriculture Food and Wine, University of Adelaide.

Tucker, N. and Murphy, T. (1997). The effects of ecological rehabilitation on vegetation recruitment: Some observations from the Wet Tropics of North Queensland. *Forest Ecology and Management*, **99**: 133–152.

Turner, M., Gardner, R. H. and O'Neil, R. V. (2001). *Landscape Ecology in Theory and Practice: Pattern and Process*. New York: Springer.

Vize, S., Killin, D. and Sexton, G. (2005). The community rainforest reforestation program and other farm forestry programs based around the utilization of rainforest and tropical species. In *Reforestation in the Tropics and Sub-tropics of Australia: Using Rainforest Trees*, ed. P. Erskine, D. Lamb and M. Bristow. RIRDC Publication No 05/087. Canberra: Rural Industries Research and Development Corporation, pp. 7–22.

Walker, B., Holling, C. S., Carpenter, S. R. and Kinzig, A. (2004). Resilience, adaptability and transformability in socio-ecological systems. *Ecology and Society*, **9** (2) 5. (www.ecologyandsociety.org/vol9/iss2/art5).

Wiegand, T., Revilla, E. and Moloney, K. (2005). Effects of habitat loss and fragmentation on population dynamics. *Conservation Biology*, **19**: 108–121.

Zhuang, X. Y., and Corlett, R. T. (1997). Forest and forest succession in Hong Kong, China. *Journal of Tropical Ecology*, **14**: 857–866.

8

Land degradation and ecological restoration in China

BOJIE FU, DONG NIU, YIHE LU, GUOHUA LIU AND
WENWU ZHAO

8.1 Introduction

China is a vast country with a diverse physical environment (Plate 8.1). It has a long geological history and most of its lands had been formed as early as the end of the Mesozoic era. These unusual characteristics provide favorable conditions for the survival and development of a large number of plants, animals and vegetation types (W. H. Li, 2004). However, China is also a large country with a long history of agricultural development and livestock breeding (China National Committee for the Implementation of the UN Convention to Combat Desertification, 1992). It has experienced fast population growth since the 1960s resulting in a population of about 1.3 billion in 2005. However, the most suitable land resources for human survival only account for less than one-third of the total land area in China. More than 80 percent of China's population is supported by less than one-third of the land resources (Gao *et al.*, 1999). Therefore, disparities between human development needs and regional ecological and environmental health have been hardly avoidable to date.

The People's Republic of China was founded on the basis of a fragile economy resulting from years of war. The survival of the big and increasing population in China was one of the main objectives of the country before 1976. Since 1978, China has been engaged in a massive transformation of its economy and society including administrative restructuring (Ma, 2005), gradually dismantling the command economy, opening up to the outside world, allowing markets to function in increasing portions of the economy, and reducing benefits provided by collective and governmental organizations (Banister and Zhang, 2004). The gross domestic product (GDP) in China multiplied about eleven times over the period 1978–2004, resulting in an annual growth rate of about 9.4 percent, while the economic value of the consumption of rural population, urban population, and government multiplied five times, eleven times, and ten times in the same period, respectively (Ma and Ning,

Ecological Restoration: A Global Challenge, ed. Francisco A. Comin. Published by Cambridge University Press.
© Cambridge University Press 2010.

2006). However, the rapid growth of the population, economic production and consumption have brought great pressure on natural resources and the environment in China. The intensity of human activities multiplied seventy-seven times over the period 1952–2000 (W. G. Zhang, 2005), and this caused an equivalent widening of the gaps between resource demand and supply. At the same time, China and the rest of the world have become closely connected in the process of globalization. The rest of the world can exacerbate resource and environmental problems in China through imports and exports (Liu and Diamond, 2005). The per capita ecological footprint in China increased 77.2 percent between 1978 and 2003 (Chen *et al.*, 2005). Therefore, China's large-scale and fast-growing economy on a resource-limited territory combined with its big population in a globalizing world leads to various kinds of land degradation problems.

There are numerous reports on land degradation, such as desertification and its assessment (Zha and Gao, 1997; Yang *et al.*, 2005), degradation of forest ecosystems (Z. S. He, 2003; W. H. Li, 2004), ecosystem degradation in wetlands (P. Z. Xu and Qing, 2002; Liu and Pen, 2003; Li and Zhou, 2005), grassland degradation (Wang and Li, 1999; Z. Q. Xu *et al.*, 2005), and the degradation of croplands (Tan *et al.*, 2005). It is rational to consider these land degradation problems in the light of the interactions between global and regional factors driving environmental change (Ge *et al.*, 2005). Land degradation has gradually become a constraint for sustainable socioeconomic development and regional ecological security. Consequently, the future development of China is facing critical challenges from resource and environment issues such as climate change, farmland and food supplies, water resources, energy, mineral resources, environmental quality, and the potential risk of some large-scale projects.

Facing increasing trends of ecosystem degradation caused by increasing population pressure and economic development, the Chinese government has launched some large-scale ecological engineerings to recover the degraded ecosystems. However, caution needs to be taken in implementing such projects. For example, although desertification can be caused by both natural and socioeconomic factors, mismanagement is responsible for most current desertification (Zhu and Liu, 1988; Mitchell *et al.*, 1998). In particular, major corrective ecological engineering projects are really very complex, and inappropriate strategies and methodologies may produce unexpected results (Mitchell *et al.*, 1996). Therefore, a synthetic scheme or regime for restoring different degraded ecosystems with different degraded characteristics is urgently needed.

The recent decades saw many studies on sustainable ecosystem restoration and vegetation rehabilitation both in China and abroad, e.g., forest ecosystem restoration (Harrington, 1999; Moore *et al.*, 1999; Stephenson, 1999; X. W. Li *et al.*, 2001; Allen *et al.*, 2002; Hasenauer, 2002; Brown *et al.*, 2004; Gundale *et al.*, 2005), ecosystem management (Pavlikakis and Tsihrintzis, 2000), rehabilitation of mangrove ecosystems (Field, 1998; Ellison, 2000a, 2000b; Lin *et al.*, 2005), control measures for degradation of soil and water ecological systems (Tong *et al.*, 2003), grassland restoration (Martin *et al.*, 2005), land-use effects on ecosystem restoration (Kettle *et al.*, 2000; Foster *et al.*, 2003), evaluation of ecosystem restoration (Fulé *et al.*, 2001; 2002), ecosystem restoration methodologies (Swetnam *et al.*, 1999; Covington *et al.*, 2000), and ecoregions and ecosystem management in China (B. J. Fu *et al.*, 2004b).

The current status of ecological restoration research in China has been briefly described in articles on the first and second Chinese symposia on restoration ecology, and on the historical development of ecological restoration research since the 1950s (X. Zhao, 2001; Ren and Peng, 2003). In the current situation of diverse types of degraded ecosystems, including forests, grasslands, wetlands, croplands and river basins, the past and current research findings are not sufficient to provide reasonable bases for sustainable ecosystem restoration. There is little attention paid to integrative analysis concerning multiple-ecosystem degradation and restoration countermeasures in China. The purpose of this chapter is to promote a broad, flexible and comprehensive perspective on ecological restoration of degraded lands based on an analysis of the general status of land degradation and ecological restoration researches and practices in China.

8.2 Land degradation and related impacts

Land degradation can be categorized into various types, such as soil erosion, desertification, salinization or alkalization, land impoverishment and land destruction, according to their origins and characteristics (H. Liu, 1995). The three principal profiles of land degradation are: (1) widespread in development with complicated types and distinctive areal differentiations; (2) fast in development with certain periodicity; and (3) high in intensity with serious ecological and social consequences.

8.2.1 Degradation of different ecosystem types

Forest ecosystems

Rapid population growth, coupled with the development of agriculture and urban construction, as well as improper forest management, has brought about an acceleration in rate of the degradation of forest resources in China (W. H. Li, 2004). Forest ecosystem degradation can be defined by changes in the forest structure, ecological function, stand quality, symbiosis and the soil seed bank. Degraded forest ecosystems are diverse in ecological types and widespread geographically from the

Table 8.1 *Characteristics of forest ecosystems in China.*

	1973–1976	1977–1981	1984–1988	1989–1993	1994–1998
Area of forest (10,000 ha)	12186	11010	12465	13370	15894
Volume stock of forest (10^8 m^3)	86.56	90.28	80.91	90.87	100.86
Volume stock of unit (m^3/ha)	90.1	94.4	75.4	83.7	78.1
Young forest (10^4 ha)	3843	3345	3958	4133	
Middle-aged forest (10^4 ha)	2627	3474	4171	4716	
Mature forest (10^4 ha)	3133	2744	2091	2011	

Source: The forest resource statistic data, State Forestry Administration, P. R. China.

northeast to the southwest, spanning over a series of bio-climatic zones (Liu *et al.*, 2000), as shown in Plate 8.1.

There is now more forested land in China and the area covered continues to go up year by year, whereas in previous decades it was decreasing. However, the area covered with high-density forests has decreased, as has the unit stock volume; the quality of forest resources is low; the forest age structure is unbalanced, with a higher proportion of young and middle-aged components, but a low proportion of mature ones (Table 8.1); the proportion of secondary forest with low biodiversity is high, and the fragmentation of forest ecosystems has clearly increased. Generally, the quality of forest ecosystems has decreased in the last decade (State Environmental Protection Administration of China, 2004).

Grassland ecosystems

Grassland ecosystems occupy about 41.7 percent of the total area of China and rank as the largest terrestrial landscape type playing a pertinent role for ecosystem conservation and socioeconomic development (Ministry of Agriculture, 2003a). The grassland ecosystem types are diverse and the top three types of grassland (Yun, 2002) are alpine cold meadow (16.22 percent), warm desert grassland (11.47 percent) and warm grassland (10.46 percent). Nevertheless, the grassland area average by population in China is only half of the world average (Han and Gao, 2005). Furthermore, mismanagement has been extensive for nearly all kinds of grassland in China: the utility rate of grassland is only 80 percent, and the proportion of artificial grassland is low; livestock husbandry is nearly completely dependant on nature and, therefore, high quality grassland areas have decreased significantly; at the same time, the grasslands have degraded seriously leading to a continual decrease of their livestock-carrying capacity; the area affected by insects and mice has increased significantly, with small fluctuations, from 4,300 km^2 in 1991 to 8,000 km^2 in 2002; while productivity has decreased to a level of only about 30 to 50 percent of that in the 1960s (Wang and Li, 1999). There are 1 million km^2 grassland, about 70.7 percent of the total grassland

Table 8.2. *Area (km²) of wetland in Dongtin and Hanjiang in the lower and middle reaches of Yangtse River during different periods.*

Years	1930s	1950s	1970s	1980s	1990s
Dongtin	4206	4009	2508	2147	1503
Hanjiang	8330	5960	2373	2983	2608

Source: Zhang, M. X. *et al.* (2001).

area, suffering from severe degradation; grassland areas are suffering the most from desertification (Wang, 2005).

Wetland ecosystems

There are 659,400 km² of wetlands in China, 10 percent of the total wetlands in the world, with a surface of 259,400 km² for natural wetlands, and 400,000 km² for artificial ones (M. X. Zhang *et al.*, 2001; H. Y. Liu, 2005). The areas of swamps, lakes, coastal intertidal zones, and shallow seawater beaches account for about 46.14 percent, 35.08 percent, 8.37 percent and 10.41 percent of the total wetlands area, respectively. These wetland ecosystems provide habitats for a large number of flora and fauna species. However, the wetland area has decreased along the coast. Most of the wetland in Sanjiang plain was converted into farmland at an annual rate of about 3.79 percent from 1949 to 2000, resulting in a significant decrease of the wetland area from 534.5 km² at the end of the 1940s to 83.5 km² in the early twenty-first century. The areas of Lake Dongtin and Lake Hanjiang, in the lower and middle reaches of the Yangtze River, decreased by 64.3 percent and 68.7 percent in the period from 1930s to 1990s (M. X. Zhang *et al.*, 2001), respectively (Table 8.2). The area of mangroves along the southern Chinese coast decreased substantially from 42,000 hectares to 14,600 hectares between the 1950s and the late 1990s. The functions of the wetlands have been inevitably impaired with the significant decrease of the area. The degraded wetland functions may affect vegetation cover, water quality and retention capacity, biodiversity, food web complexity, productivity, and the connectivity with other water bodies (Qiu and Wu, 1996; He, 2004; Ren *et al.*, 2004; Lin *et al.*, 2005).

Farmland ecosystems

Generally, the total farmland area decreases significantly with rapid industrialization and urbanization. However, the dynamics of farmland areas are quite distinctive, with an increase from 786,000 hectares in 1949 to 5,240,000 hectares in 2000. This is mainly because of large-scale land reclamation for farmland. However, the total

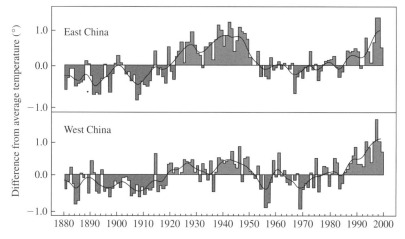

Fig. 8.1. The trends of temperature difference from average temperature during the past 120 years. After Qin *et al.* (2005).

amount of farmland surface and the amount per capita has decreased continuously in the last decade, resulting in an average of 0.1 hectares of farmland per capita. Of the farmland lost during the period between 1988 and 2000, 56 percent was converted to urban and suburban built-up areas, 21 percent was converted to forest or grass-lands, and 16 percent was converted to bodies of water (State Environmental Protection Administration of China, 2004). Along with the decrease in the farmland area, farmland quality has experienced some degradation due to mismanagement and insufficient agriculture inputs. Alkalized/salinized farmlands have increased significantly and their productivity has decreased accordingly. Large areas of farm-land have been irrigated by polluted water, resulting in soil pollution that has caused human health problems.

8.2.2 *The main ecological problems forming the background of land degradation*

The trend of aridification and climate warming

The trend to aridification has continued unchanged while the frozen zone and glaciers, desert and other ecosystems have been changing with the variation of environment since the late Pleistocene. Even in relatively warm and humid periods, the general trend of aridification was invariable. The climate has become warm and dry in the last 200 years (Figure 8.1). It became clearly warmer during the 1920s to the 1940s, and after 1980 (Figures 8.1 and 8.2) (Ding and Dai, 1994; Qin *et al.*, 2005). The winters have been warm for the last sixteen years in China (Figure 8.2).

Climate warming has been observed in most of China with a large regional variation. The mean annual temperature has increased by 0.5°C north of latitude

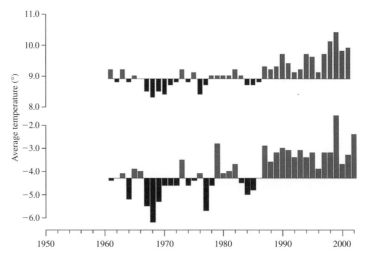

Fig. 8.2. The temperature change during a recent fifty-year time period in China. After Qin *et al.*, 2005.

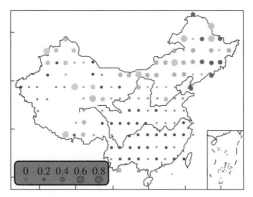

Fig. 8.3. The rates of temperature increase in regions of China over the last fifty years. After Xu *et al.*, 2003.

35°N. It has increased by 1.0°C in the areas of the middle and lower reaches of the Yellow River, the north part of the North China Plain, and the north of Northeast China (Figure 8.3) (Y. Xu *et al.*, 2003). Climate warming in the northern part of China has occurred largely in winter, while the southern part of China generally has experienced summer climate warming (Yue *et al.*, 2001). Other signals of climatic warming include glacier shrinkage of 3,900 km^2 in the past four decades, the snowline uplifting by between 30 and 60 meters, the reduction of frozen areas by 12 to 13 percent along the Tibet Road since the 1960s, and the reduction of seasonal maximum thickness of the frozen layer by about 0.6 to 1.0 meters in recent decades.

Table 8.3 *Land area (10 000 km²) affected by water erosion of different intensities in China in the year 2000.*

Erosion categories	Slight	Medium	Serious	Very serious	Extremely serious
Area	83.05	55.49	17.83	5.99	2.51

Source: Xu, F. *et al.* (2003).

Table 8.4 *Land area (10 000 km²) affected by water erosion in different regions in 2000 in China.*

Regions	Northwest and Tibet	Southwest	Northeast	North China	East China	Central China	South China
Area	63.99	41.82	15.41	15.24	13.14	13.13	2.16

Source: Xu, F. *et al.* (2003).

Water and wind erosion

The total area suffering from wind and water erosion is more than 3.56 million km², of which 1.91 million km² are affected by wind erosion. Water erosion mainly takes place in the Loess plateau and the southwest hilly area. The water-eroded area was 1.65 million km² at the end of 1990s, and has decreased by 0.14 million km² compared to that at the end of 1980s. The water-eroded area is larger in northeast China and Tibet, and the intensity and distribution pattern there is different (Tables 8.3 and 8.4) (Y. Xu *et al.*, 2003). The wind erosion has been occurring mainly in northern China and affects sandy land, desert, and grassland with sandy soils.

Desertification

China is one of the countries facing a serious problem of desertification. In the past forty years, desertification rates have reached between 0.81 and 1.64 percent annually in northern China (Lin and Tang, 2002). The total area of the affected land is approximately 2,636 million km², covering 27.46 percent of the total territory of China. The slight, medium, serious, and extremely serious desertified lands account for 23.9 percent, 37.4 percent, 16.5 percent and 22.2 percent of the total, respectively. The desertified land is mainly located in the western part of China, with the most serious desertification in the autonomous regions of Xinjiang and Inner Mongolia.

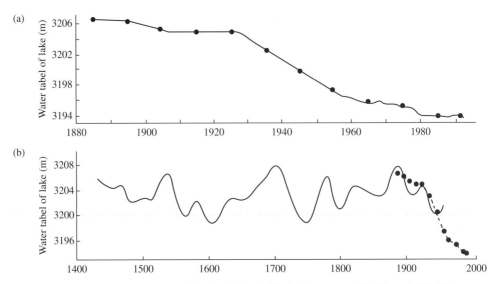

Fig. 8.4. Change in the water level of Qinghai lake in recent 100 (a) and 600 (b) years. After Feng *et al.*, 2000.

The desertification process has experienced a continual acceleration (T. Wang *et al.*, 2004). The desertification rate in the 1990s was double that of 1970s. The annual expansions of desertified land were, respectively, 1,560 km^2, 2,100 km^2, 2,460 km^2, 3,600 km^2 in the periods between 1950s to mid 1970s, mid 1970s to mid 1980s, and the early 1990s to late 1990s.

Shortage of water resources

The per capita water resources in China are much lower than the world average and China is ranked as one of the most water-impoverished countries in the world (State Environmental Protection Administration of China, 2004). Furthermore, the water resources of many large rivers in Asia, such as the Yellow River, the Yangtze River and others, are unevenly distributed with 83.7 percent of water in the southwest and 16.3 percent in the northwest. This pattern makes the water-shortage problems more serious. At the same time, water consumption has also increased, while water quality has declined recently. For example, water consumption from Tarim River increased from 5 billion m^3 in the 1950s to 15.3 billion m^3 in 1998, and the mineral content of this water increased by between 4.83 and 5.24 percent annually. Another example is that the water level of Qinghai lake has fallen significantly (Figure 8.4) with about 670 km^2 of the water surface area being lost (Feng *et al.*, 2000). Furthermore, during the 1990s over 30 percent industrial waste water and 90 percent domestic waste water were discharged without proper treatment, leading to another kind of water shortage because of poor water quality (Min and Cheng, 2002).

8.2.3 Causes of land degradation

Disadvantaged natural conditions

China has unique natural conditions in the world: mountains, hills, and highlands cover 65 percent of the total territory; 33 percent is arid and desert regions; 70 percent is strongly affected by the East Asia monsoon and 35 percent by soil erosion; 55 percent is unsuitable for human production activities, and 30 percent of farmland is unfavorable for crop growth. The Qinghai-Tibet plateau, the Roof of the World, accounts for about 17 percent of the total territory. The average altitude of China is 1.73 times higher than the world average. These characteristics, together with mismanagement, make the land resources of China easily susceptible to degradation. The unfavorable natural conditions are the controlling factors for land degradation over a long timescale. However, human factors operate at a much shorter timescale than natural conditions. For example, the time scale for natural factors contributing to desertification is 1000 to 10,000 years, whereas that for human influence is only about 10 to 100 years (Lin and Tang, 2002).

Population and economic development pressure

The population has already reached over 1.3 billion in China and has been increasing continuously at a speed of about 8.8 percent annually in the last 350 years. Certainly, vast resources are required to feed so huge a population, and this itself induces land degradation, especially as China is a developing country. The pressure of the huge population is the most siginificant cause of land degradation in China.

The population make huge demands on resources of all kinds in this country which is developing faster than any other. Intensive economic and social activities have brought about ecosystem changes, e.g. forest and grassland have been con-verted into cropland, and land has been over-grazed. The economic development factor interacting with the population pressure makes this situation more serious.

The vigorous urbanization process in China is the most direct result of the combined pressures of an increasing population and economic development . One of the important measures of urbanization is the proportion of the population living in cities and towns. Following the founding of the People's Republic of China, the urbanization measured by the population indicator had increased to about 41.8 percent by 2004 (Hu, 2006), 3.9 times the 1949 level, even though some retrogress had been observed in the period between 1960 and 1977 (Figure 8.5). The growing population was concentrated in cities and towns. Accordingly, many kinds of human processes were intensified in urban areas including natural resource demand and consumption, product manufacturing, and pollutant discharge. Another spatial explicit measure of urbanization is the

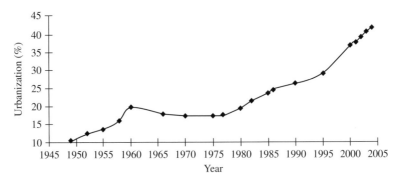

Fig. 8.5. The trend of urbanization in China.

expansion of urban built-up areas. The urban land of China was reported to have increased by 817,000 hectares during 1990–2000, and about 79 percent of this expansion was previously cultivated land (Liu *et al.*, 2005b). Both population migration to urban areas and urban land expansion will continue under the pressure of economic development (Jing, 2006) and institutional, as well as political factors (Z. Zhao, 2006).

Human activity

China has a huge population and a very long history of agricultural practices destroying forests and other types of vegetation cover to increase the land under cultivation. The impact of human activity has become more significant in the last 300 years and especially in recent years. For example, 38.17 billion tons (about 29.4 tons per capita) of soil and rock were removed in China, which is 1.84 times the world average (Zhou and Niu, 2000). Soil and rock removed in arable farming, animal husbandry, forestry, mining, quarries, infrastructure development and urban construction were 22.60, 1.65, 1.20, 4.80, 1.45, 5.36 and 2.11 billion tons respectively. Undoubtedly, such activities affect ecosystems and environment significantly.

Global climate change

The simulation results obtained by the IAP/LASG Global Ocean-Atmosphere-Land System (GOALS) model, the RIEMS GOALs model and the CSIRO MK2 IEM model show that the climate in China might warm markedly in the future; however, no steady trend of precipitation change in the next fifty years was found. Despite different patterns of warming in various parts of China, the forecast for temperature and precipitation dynamics show similar trends to those of the last fifty years. The structure and functions of ecosystems will change as a result of climate change.

8.2.4 The impacts of land degradation

Biodiversity loss

China, one of the main points of origin for seed plants on Earth, inherited the flora elements of the Northern continent in the Tertiary period, Tethys Sea, and the ancient southern continent of Gondwanaland. Next only to Brazil and Colombia, China has more than 30,000 species of seed plants. However, it is said that about 15–20 percent of the total number of species will disappear in this century unless there is an improvement of ecosystem management (S. X. Xu *et al.*, 2002). Moreover, more and more natural vegetation has been replaced by crops and plantations, with much less biodiversity and an unstable community structure caused by agricultural and industrial development. Land degradation has caused significant biodiversity problems. The reported number of endangered vertebrate species in the late 1980s was 398, accounting for about 7.7 percent of the total vertebrate species ever recorded in China (L. Z. Chen, 1993). In the last 50 years, 200 plant species became extinct and the endangered higher plant species reached 15–20 percent of the total number of species (L. Z. Chen, 1994).

Frequently occurring disasters

Frequent sand and dust storms hit both sandy areas and the surroundings of arable lands. Sand and dust storms caused severe adverse environmental impacts in extensive areas in North China, including the two major metropolises of Beijing and Tianjin in 2000. Furthermore, there is an increasing occurrence of debris flows, landslides, rainstorms and large floods destroying numerous houses and croplands in China at present. On a national scale, the disasters have north to south zonal and east to west azonal distributions (Q. H. Gao, 2003). The absolute direct losses caused by disasters in eastern and southern China are much heavier than those of the western and northern parts. However, the relative direct losses are higher in central and southern China and decrease gradually to the east, west and north. The rate of occurrence of natural disasters is increasing notably, but in a fluctuating manner.

Gross domestic production (GDP) loss

Land degradation causes a vast loss in GDP every year in China. Desertification alone is capable of causing loss of property amounting to 54 billion Chinese Yuan. Annually, about 13 million hectares of croplands are threatened by wind and dust storms resulting in a decrease in cereal yield. It is estimated that nearly 100 million hectares of rangeland and steppe have been impacted by severe wind erosion incidents. Over a thousand water conservation facilities have been destroyed by sand movements. Similarly, a total of 800 kilometers of railways and several thousand kilometers of highway in desert regions are impacted by sand accumulations and wind hazards. Water erosion is also serious. It is estimated that about 430,000 km^2 of the Loess Plateau is affected by soil loss and water erosion (China National Committee for the Implementation of the UN Convention to Combat Desertification, 1992). Generally, land degradation exacerbates various kinds of disasters that result in economic loss amounting to between 5 and 13 percent of the total GDP (State Environmental Protection Administration of China, 2004). Furthermore, land degradation leads to poverty in less developed regions, forming a vicious cycle of degradation and poverty (State Environmental Protection Administration of China, 2003).

8.3 Ecological restoration

8.3.1 Ecosystem monitoring for ecological restoration

To achieve effective ecosystem conservation and ecological restoration, it is necessary to improve understanding of the structure and function of ecosystems. The Chinese Ecosystem Research Network (CERN), supervised by the Chinese Academy of Sciences (CAS), was founded in 1988 to meet this need. CERN joined the International Long Term Ecological Research Network (ILTER) in 1993 and is also a member of the Global Terrestrial Observing System (GTOS). The mission and objectives of CERN focus on (1) ecological monitoring to continuously measure and record the dynamics in ecosystem structure, processes, and functions, (2) ecological research to understand ecosystem dynamics and the underlying mechanisms in response to environmental changes and human perturbation, and (3) ecological applications to explore and demonstrate ecological techniques and options to restore, enhance, and sustain ecosystem services.

CERN consists of thirty six field stations distributed in different ecosystems including thirteen agro-ecosystem research stations, nine forest ecosystem research stations, two grassland ecosystem research stations, six desert ecosystem research stations, three wetland ecosystem research stations, and three gulf ecosystem research stations (Plate 8.2). There is also one synthesis center and five subcenters in CERN. The synthesis center was created to integrate all kinds of data and to ensure data sharing. The subcenters are in charge of the standardization of the field investigation activities on water, soil, air, biology, and aquatic ecology. The agro-ecosystem research stations focus on the water cycle in the crop–soil–atmosphere system, crop productivity, nutrient cycling, soil, crop, and sustainable agro-ecosystem management. The forest ecosystem research stations focus on the structure, function, biodiversity, productivity, material cycling, carbon sequestration and overall forest dynamics as well as sustainable forest management. The research stations for grassland ecosystems deal with biodiversity, productivity, and material cycling, grazing management, and grassland degradation and restoration. Research stations for lake ecosystems focus on aquatic ecology and biodiversity, biological productivity and food web structure, nutrient cycles, lake eutrophication, water pollution control and aquatic ecosystem management. Research stations for desert ecosystems focus on desertification process control, water conservation, oasis ecology, and restoration ecology. Research stations for arid ecosystems focus on arid agriculture, soil and water conservation, ecological land development, restoration and management.

Besides CERN, there are also other ecosystem monitoring facilities, such as the China Forest Ecosystem Research Network (CFERN) and the China Network of Soil Conservation Monitoring (CNSCN). Websites for CERN and CFERN (http://www.cern.ac.cn and http://www.cfern.org/) have been set up (B. Wang *et al.*, 2004). The web-based data sharing system is also in use not only for CERN data but also for other information sources, such as grassland data (Yuan *et al.*, 2004).

8.3.2 Ecosystem assessment for ecological restoration and sustainable management

In order to manage an ecosystem efficiently, it is necessary to assess ecological risk on multiple scales. B. J. Fu *et al.* (2004b) have given a detailed risk assessment of soil erosion, desertification and acid rain, and put forward working schemes for future research on a national scale. On a sub-national scale, the conditions and trends of major ecosystem types and their related ecosystem services and human well-being in western China have been comprehensively assessed (J. Y. Liu *et al.*, 2005a). However, further efforts are needed, such as ecological risk forecasting and detailed information about the local environment on a comparatively large scale for regional decision-making.

The division of a country into regions for the purposes of ecological restoration is known as ecological regionalization. It is a foundation for ecological assessment, rational management and sustainable utilization of ecosystems and natural resources. It can provide a scientific basis for constructing healthy ecological environments and making policies for environmental management (Fu *et al.*, 2004a). Using a synthetic analysis of the characteristics of the ecology and environments of China, B. J. Fu *et al.* (2004a) have discussed the principles of ecological regionalization and proposed indices and nomenclature for ecological regionalization in China. According to their results, there are three domains, thirteen ecoregions and fifty-seven ecodistricts (B. J. Fu *et al.*, 2004b) in China (Figure 8.6). Their scheme can be used as a framework for ecosystem assessment, restoration, and

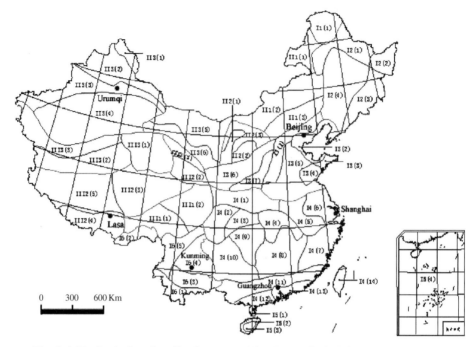

Fig. 8.6. Ecological regionalization map. After Fu *et al.*, 2004b.

sustainable management. Their results help to identify appropriate measures for the management of forest, grassland, agriculture and wetland ecosystems.

8.3.3 *Ecological restoration on various scales and at various geographical locations*

Chinese people have rich experience and traditional knowledge to fight the adverse effects of ecosystem degradation. As early as the 1950s, the government of China organized scientific surveys and studies on the lands in degradation and have given priority to desertification control in severely affected regions. Since the 1970s, China has initiated and implemented successively such major ecological restoration projects as the Three North Regions Shelterbelt Construction Project, the Coastal Shelterbelt Project, the National Action Program to Combat Desertification, the Plain Farmland Protective Shelterbelt Project, and the Shelterbelt Construction Project along the middle and upper reaches of the Yellow River. All these projects have had successful results, with many convincing models emerging (China National Committee for the Implementation of the UN Convention to Combat Desertification, 1992). More recently, large-scale ecological restoration and conservation projects have been initiated, including natural forest protection projects, grain for green projects, reforestation projects, lake restoration projects, soil and water conservation projects, and desertification control projects. And at the same time, 2194 protected areas of various kinds were established by the end of 2004, accounting for about 14.8 percent of the Chinese territory (J. W. Li *et al.*, 2006). Accordingly, great progress has been made in the field of ecological restoration and conservation (State Environmental Protection Administration of China, 2004). Plates 8.3 and 8.4 illustrate the types of land degradation that are being tackled.

Forest ecosystem restoration

For the ecological restoration of forest ecosystems, the Chinese government formally approved and started to implement six state key forestry development programs (W. H. Li, 2004). Conservation of natural forest and construction of nature reserves are the most effective approaches in maintaining and strengthening the multi-functional nature of forests. Afforestation, in combination with watershed management, provides a solution for restoring the ecological function of forests. Therefore, the Natural Forest Protection Program (NFPP) and the Grain for Green Program (GGP) have been given much importance (S. G. Zhang *et al.*, 2002; Y. C. Lu and Zhang, 2003; Z. Feng *et al.*, 2005; Z. Xu *et al.*, 2005).

The GGP has four main stages (S. Li, 2003) including mobilization (1949–1998), small-scale experiment and demonstration (1999–2001), large-scale implementation (2002–2010), and sustainable management (2011–2020). In the second stage, about 1.16 million hectares of sloped cropland and another 1 million hectares of waste land were reforested. According to the GGP blueprint, 2.94 million hectares of sloped cropland and 3.47 million hectares of waste land will be reforested by the end of the third stage. The last two stages require a budget of about 472.1 billion Chinese Yuan. The suitability of various regions for GGP was also evaluated and they were partitioned into various types in a hierarchical manner (S. Li and Zhai, 2004).

The NFPP was first initiated in Sichuan province in 1998 and approved for wide implementation by the state council in 2000 (Y. C. Lu and Zhang, 2003). The primary objectives of NFPP are to: (1) restore natural forests in ecologically sensitive areas, (2) protect existing natural forests from excessive cutting, and (3) maintain a multiple-use policy in natural forests (W. H. Li, 2004). The NFPP covers seventeen provinces and autonomous regions in the upper reaches of the Yellow River and the Yangtze River and the main natural forest regions of Hainan Island, Northeast China, and the Inner Mongolia and Xinjiang autonomous regions. Overall, about 106.4 billion Chinese Yuan will have been invested in the NFPP by 2010 (http://www.tianbao.net).

Grassland ecosystem restoration

The ecological restoration of grassland in China is being undertaken through the establishment of three systems, five priority regions, and eight projects in the period 2001–2010 (Ministry of Agriculture, 2003b). The three systems are the breeding system of excellent pasture species, the ecological monitoring and early warning system, and the scientific support system. The five priority regions include the Northern China arid and semi-arid grassland region, the Qinghai-Tibet plateau cold grassland region, the grassland region of the middle and upper reaches of the Yellow River and the upper reaches of the Yangtze River, the Northeast and North China semi-humid and humid grassland region, and the southern China hilly grassland region. There are 3.41 million km^2 of usable grasslands in the above five priority regions. Enclosure grassland conservation, rotational grazing, grassland upgrading by seeding and fertilizing, water-saving irrigation, air seeding, grass planting, grassland disaster control, and the establishment of grassland nature reserves are the eight ecological conservation and rehabilitation projects that will cover about 0.9 million km^2.

Once the severe ecological degradation is under control, another ten years' (2011–2020) effort is needed to improve the ecological health of the grassland ecosystems in China (Z. L. Wang, 2005). These efforts include rehabilitating over 60 percent of the grasslands suitable for rehabilitation, and creating 0.8 million km^2 of grassland plantations and upgraded grasslands. Then further actions are needed including the creation of conservation and rehabilitation or recreational areas, scientific research and capacity building, and grassland industrial development and ecological regulation.

Over 7 billion Chinese Yuan were invested in grassland ecosystem management from 2000 to 2004. After five years (2001–2005) of conservation and rehabilitation, the vegetation cover of grasslands had clearly improved (Y. Li, 2005). It was also found that community-based grassland management was advantageous (Banks *et al.*, 2003).

Wetland ecosystem restoration

Many ecological engineering approaches have been used at various wetland sites to tackle ecological degradation problems (S. Zhao, 2005). These approaches include mangrove revegetation (Liao *et al.*, 2005), wetland park construction (Lei, 2005), wetland treatment (Deng *et al.*, 2005), wetland construction for agricultural production (J. C. Shi, 2004), landscape design (Liu and Guo, 2003; J. Zhang *et al.*, 2005), population relocation (Xiong *et al.*, 2004), submerged plant harvesting and garden style ecological management of reed (Shang *et al.*, 2003), and collective technologies for coastal ecological restoration (H. L. Li *et al.*, 2003). The ecosystem functions and environmental quality of the wetlands in local sites have been improved significantly (Luo *et al.*, 2003; Jiang *et al.*, 2005).

As a contracting country to the Convention on Wetlands since 1992, China has made notable progress on the conservation and restoration of wetland resources (Huang, 2004). 353 wetland nature reserves had been established by the end of 2002, covering about 0.4 million km^2. Within these reserves, 0.16 million km^2 of natural wetland and 33 national key waterfowl species have been protected and are in good condition. Thirty wetland sites have been recognized as internationally important wetlands, with a total area of 34,600 km^2. According to the Chinese wetland conservation project plan, the following projects will have been accomplished by the end of 2030: ecological rehabilitation of 14,000 km^2 of wetland, 53 national level demonstration areas of wetland conservation and smart utilization, 713 wetland nature reserves, and 80 internationally important wetland sites. These will put 90 percent of the natural wetlands of China in an effective conservation state that guarantees the functioning and ecosystem services of the wetlands.

Ecological rehabilitation of wasted mining land

Land reclamation of mine sites in China began in the late 1970s but has been practiced more widely since the promulgation of the Regulations of Land Reclamation in 1988 (M. S. Li, 2006). The degraded wasteland in need of ecological rehabilitation typically comprises stripped areas (59 percent), open-pit mines (20 percent), tailings dams (13 percent), waste tips (5 percent), and land affected by mining subsidence (3 percent) (Miao and Marrs, 2000). The progress of restoration was slow at the early stages but has accelerated recently. By 1994, almost 400,000 hectares had been rehabilitated throughout China, which is 13 percent of the total area of degraded land (3 million hectares). The rehabilitation approaches include the improvement of sediments or soils, revegetation and phytoremediation (Z. H. Ye *et al.*, 2000; Wong and Luo, 2003; F. Q. Chen *et al.*, 2004; Wang and Li, 2005).

The extent and quality of rehabilitation varies significantly (M. S. Li, 2006). In terms of mine type, coal, bauxite and ferrous metal-mined lands have a higher restoration rate (>10 percent) while it is lower (<5 percent) for most non-ferrous metal-mined lands. Comparing regions, the mine site reclamation in Inner Mongolia (12.4 percent) and the middle and lower reaches of the Yangtze River Basin (15.7 percent) are the best, whereas reclamation on the Qinghai-Tibet Plateau is the poorest (<1 percent). Large-scale state-owned mines are generally in a better state than the small-scale privately or collectively owned mines.

Ecological agriculture as a tool for restoration of degraded land

Agricultural activities in China have a history of 7000 or 8000 years. Many traditional agricultural practices are in accord with the principles of modern ecology.

Therefore they are usually treated as the basic reference and source of elements for the design and development of ecological agriculture (Ellis and Wang, 1997; W. H. Li, 2003). Ecological agriculture is a workable alternative that has the potential to mitigate the negative impacts of modern conventional agriculture and, at the same time, to overcome the limitations of traditional agriculture in meeting the needs of China's growing population (T. Shi, 2002). Ecological agriculture represents a site-specific manifestation of the application of ecological economics, among others, to agricultural practices. Preliminary results from ecological agricultural projects include: higher productivity and commodity output with less cost; higher stability of the farm system during natural disasters; more concern with the environment and ecosystems; harmony among farmers and agricultural administrators; and improved rural landscapes. It can promote economy within the different branches of agriculture, maximizing productivity and minimizing the negative environmental impacts that lead to land degradation.

The ecological agricultural approaches can not only be used to solve the problems of resource shortage and environmental pollution of agricultural landscapes (Wen and Pimentel, 1992; Korn, 1996; Song *et al.*,1997; Q. Guo, 1999; P. Shi and Li, 1999), but they can also be used in the ecological restoration of the lands degraded by mining activities (Miao and Marrs, 2000), severe soil erosion (G. Liu, 1999), desertification (G. M Wang *et al.*, 2005), water resources depletion (Geng *et al.*, 2001; Chang *et al.*, 2005), and the restoration of waste brownfield sites (X. Li *et al.*, 2004). These approaches have proven effective in addressing land degradation problems (X. J. Ye *et al.*, 2002).

8.3.4 Some sector initiatives for ecological land rehabilitation

There are other on-going projects for ecological restoration that may produce great achievements in the future.

Chinese Academy of Sciences' Action Plan for the Development of Western China

This plan concentrates on the assessment of environmental factors and their impacts on the main regional ecological process, the actual modes and key technologies for practical ecological rehabilitation or reclamation of different types. Major research and demonstrations have been conducted on five sites including Tarim River, Otindag sandy land, Heihe River, Minjiang River and the Loess plateau. A typical succession of vegetation and soil after 40 years' restoration in the Shapotou desert station of the Chinese Ecosystem Research Network (CERN) showed that moving dunes could turn into natural grasses, shrubs and lower plants (Figure 8.7) (W. F. Xiao *et al.*, 2003).

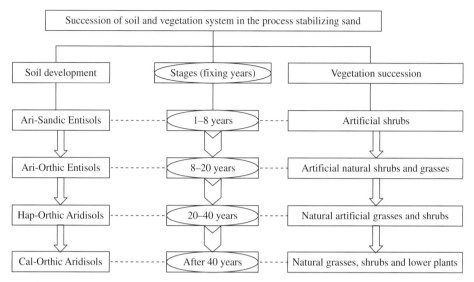

Fig. 8.7. Succession of vegetation and soil over forty years' restoration in Shapotou station. After Xiao *et al*, 2003.

Restoration in the Hexi Corridor

The inland river basin of the Hexi Corridor is a complex ecosystem consisting of ice and snow belts, frozen soil zones, oasis and desert. The water coming from the mountains plays a key role in the ecosystem conservation and development of the oasis and desert. The effective way to restore degraded land in this area is rational management and suitable techniques of water-efficient farming, including constructing irrigation channels, reducing irrigation areas, rationalizing species composition and the grid size of shelter forest, adopting advanced irrigation management, and the integration of all the above measures. Through effective management in this area, the water consumption decreased by 20–25 percent, while production increased by 42 percent.

Restoration of degraded grasslands in the Xilingol Steppe

Overgrazing is the main reason for grassland degradation, and the following techniques are effective for restoration: fencing of degraded grasslands, rotational grazing and delayed grazing. Using these techniques degraded grasslands in the Xilingol Steppe were restored in just one year. Long-term observation of aboveground biomass and climate has shown that precipitation from January to July is the primary climatic factor causing fluctuation in community production. Ecosystem stability increases progressively along the hierarchy of organization levels, i.e. community stability arises from compensatory effects among the major components at species and functional type levels (Bai *et al*., 2004).

Restoration of degraded grasslands in the Otindag sandy land

Wind erosion and desertification are the main problems of land degradation in the Otindag sandy land. The main techniques applied in this area included the fencing of degraded grassland, using checkerboards of biological material in seriously degraded areas and then conducting air-seeding of native species. The degraded ecosystem was significantly restored in five years.

Ecosystem restoration on the Loess Plateau

Population and economic development pressure are the main reasons for recent land degradation on the Loess plateau. Therefore an optimal eco-productive paradigm on the Loess plateau is needed to rehabilitate the degraded ecosystem and to feed local people. It should focus on ecologically and economically sound land use and management for regional sustainable development. Measures such as greenhouse agriculture, apple plantations, and livestock production may be helpful in promoting local economic development.

If there is no human disturbance on the degraded land on the Loess plateau, the vegetation could rehabilitate naturally. Long-term monitoring has shown that degradation has not destroyed the whole functioning of the ecosystem or the soil seed bank, so the potential exists for natural restoration of vegetation. The degraded landscape patches on the Loess plateau should be restored in twenty-five years.

Sustainable vegetation restoration – "three-circle" pattern in the Ordos Plateau

The landscape on the Ordos Plateau can be classified into three types (or terraces): hard hill, soft hill and low land. Based on this landscape classification, a "three-circle" pattern to combat desertification has been proposed. The first circle is a highly efficient agricultural and pastoral area; the second circle is runoff garden; and the third circle is the fenced natural and artificial vegetation restoration area. The ratio for the three circles is 1:3:6 (X. S. Zhang, 1994). Thus with high input and output agriculture in the first circle and some shrub species with high economic value in the second circle, high quality living conditions can be provided for local people. After the cessation of over-disturbance by human activity, the severely desertified land in the third circle can be subsequently restored (Zheng and Zhang, 1998).

The actions of the State Environmental Protection Administration of China (SEPAC) and local governments

Six key tasks pertinent to ecological conservation and rehabilitation were put forward in the eleventh five-year (2006–2010) plan of SEPAC (Xie, 2005) at national level. The first task is to facilitate the optimization of industrial development and resources utilization based on the regionalization of ecological functions. This task guarantees the large-scale context for ecological restoration. The second task is to advance the construction and management of protected areas with special attention to management capacity. The third task is to promote the harmonious development of regional socio-economy and the environment through a series of ecological demonstration projects. This task forms the direct base for various ecological restoration projects. The fourth task is to strengthen rural environmental protection. This requires many ecological restoration actions in rural and rural–urban transition areas. The fifth task is to focus on the control of ecological degradation and destruction with the enhancement of ecological conservation stewardship on resource exploitation. The sixth task is to strengthen biodiversity conservation and biosecurity management in accordance with the line of duty specified in the Convention on Biological Diversity. In order for these tasks to succeed, legislation on the environment needs to be improved, strengthened, and executed (Du, 2005; Y. F. Li, 2005; W. G. Wang, 2005). SEPAC also provides many relevant recommendations on policy concerning ecological conservation, development and rehabilitation, including natural resources conservation, human-dominated landscape management, revegetation, desertification control, biodiversity conservation, and eco-tourism development (State Environmental Protection Administration of China, 2003, and 2004). Under the guidance of the central government and SEPAC, different levels of local government have been planning and implementing their ecological demonstration or eco-development projects on various scales from eco-village to eco-township, eco-city, and eco-province (Bian, 2003; R. S. Wang, 2005; W. G. Zhang, 2005).

8.3.5 *The main challenges for degraded land restoration*

China is a developing country with a large human population and diversified environmental conditions. The drive for socioeconomic development is very strong, and at the same time, environmental quality, resource usability and ecological security are also important concerns for the success of regional sustainable development.

Therefore, it is crucial to harmonize the relationships between human population growth, regional economic development, environmental conservation, and ecological restoration (Fu and Lu, 2006). There are many real-world problems to face, such as unfavorable or harsh biophysical conditions, insufficient financial support, huge population pressure, and poverty, all of which hinder the development of ecological restoration.

China is currently in a state of rapid socioeconomic development. Many new problems concerning natural resources utilization, ecological conservation and rehabilitation, and the balancing of inter-regional development are waiting for solutions and call for the active participation of various stakeholders. Therefore, to foster widespread participation and cooperation in ecological restoration is the second biggest challenge in China.

The third biggest challenge originates in the limited understanding of ecological regimes and the interactions between humankind and nature. This is the most typical challenge concerning worldwide science and technological development in the field of ecology and environmental sciences. In our current state of knowledge about ecology and environmental sciences, it is hardly possible, if not impossible, to accurately understand, forecast and control the dynamics of the physical environment and human–nature interactions. This will be true even for quite a long time in the future because of the great complexity of nature and human–nature relations. Therefore, ecological restoration can at best be treated as a "real-world experiment" (Gross and Hoffmann-Riem, 2005). It requires the integration of ecological field work, large-scale implementations and wider socioeconomic contexts. And only through these long-term, real-world experiments can ecological restoration gradually grow into maturity both as a scientific pursuit and as a comprehensive practical engineering approach that will heal degraded lands and the abnormal human–nature relationships.

8.3.6 Opportunities and outlook

The opportunities for degraded land restoration lie in the following four broad respects. Firstly, the rapid economic growth of China continuously enhances the availability of investment funds for ecological restoration projects and the related scientific research. Secondly, average environmental awareness and concern of the general public is now high due to years of environmental education, environmental protection practices and the enforcement of environmental laws. Thirdly, international cooperation in the field of ecological restoration is improving continuously. Finally, ecology, resources and environmental sciences are given much importance in the mid and long-term (2006–2020) plan for science and technology development in China issued by the State Council. The scientific and technological advancement of ecological and environmental sciences will surely provide more opportunities for bringing ecological restoration projects towards successful ends. Accordingly, the future development of restoration-related ecological and environmental sciences may include the following aspects.

The establishment of theoretical and operational frameworks

Because of the large territory and the significant socioeconomic and physical differences between areas, a comprehensive regionalization scheme for ecological restoration needs to be formulated from the beginning. However, this regionalization should not start from scratch. It is rather an integration process that incorporates the already existing comprehensive physical geographical, economic and ecological regionalization schemes with special attention to the characteristics and requirements of ecological restoration. Priorities can be set for ecological restoration in a space based on the comprehensive regionalization scheme (Hyman and Leibowitz, 2000). Thus a rational "top-down" approach for selecting ecological restoration sites and projects can be adopted. At the same time, the theoretical and application issues, such as ecological restoration decision-making (Pastorok *et al.*, 1997; Cairns, 2000), the effective planning and implementation of ecological rehabilitation projects (Quon *et al.*, 2001), and methods for monitoring and evaluation of restoration success (Choi, 2004) in the Chinese context need to be addressed.

Development of experiment-based restoration research

All kinds of ecological restoration research, such as theoretical development, modelling, planning and design, need to be based on or verified by certain scientific data obtained by long-term monitoring, survey and experiments. Experiment-based research can help in solving problems such as the ecological implications of landscape metrics, the relationship between landscape patterns and ecological processes, and the right scaling for ecological restoration projects. The effects of restoration on landscapes and approaches to the regulation of restoration activities can only be detected and devised after experimental research on various spatio-temporal scales. Therefore, in order to make theoretical development scientifically sound and the research results widely applicable, experiment-based ecological restoration research should be enhanced.

Data are the basic "feeding stuff" in the process of scientific research. Models are only tools that represent the essential features of a system for analyzing the relationships between its inner components and with the external environment within the established boundary conditions (Dale, 2003). Data and modelling are both, therefore, indispensable for furthering the development of restoration ecology in China. Field experiment stations have been set up since the 1950s (B. W. Huang *et al.*, 1990). Many scientific data have been collected in these stations. Data are not equivalent to useful scientific information, but are merely the raw materials. Subsequently, simulation and modelling are required to understand the effects of ecological restoration on various scales. Restoration ecology can be advanced on the basis of multi-source, consistent data acquisition and through robust simulation using adapted models.

Monitoring, restoration and research

Besides the Chinese Ecological Research Network, there are some field stations supported by other ministries in China. The national-level field station system for long-term monitoring and research on regional ecological and environmental dynamics has been launched recently. This system, planed to cover one hundred field stations, will promote the research and applications of ecological restoration in China. The research planned can be summarized into five main categories, including research into an integrated inland river system, research into water resource dynamics under climate change and human activities, quantification of the natural and anthropogenic factors impacting environmental factors, and environmental impact assessments of large scale projects.

The focuses of the above research fields depend on the regional context. Integrated inland river system research may focus on water circulation, the hydro-ecological processes and interactions on multiple scales and management of inland river systems including water resources and ecological integrity. Research into water resource dynamics may focus on the influences of land use or land cover change on the hydrological cycle and water resources on the Loess Plateau, the mechanism, assessment and forecasting of mountain disasters, and the environmental and socioeconomic impacts of glacier change. The quantification of the natural and anthropogenic factors impacting environmental regimes focus mainly on the driving forces leading to soil and water loss, and their influence on regional vegetation dynamics, the processes and driving forces of desertification, and the interactions between desertification, climate fluctuation and human activities.

Various large-scale projects such as the Three Gorges Dam Project (TGDP), conversion of farmland to woodland or grassland, comprehensive development of inland rivers and lakes, the Qinghai-Tibet Railway, the South-North Water Transfer Project, and the West-East Gas Pipeline Project have been subjected to environmental impact assessments. TGDP is a typical case of a large-scale engineering project believed to have significant environmental impacts. It took eighty years for the project to develop from a theoretical idea to the initiation of construction in 1993. Then the preliminary Three Gorge Reservoir (TGR), covering an area of 58,000 km^2, was formed and started regulating water in 2003 (Wu *et al.*, 2003). The whole TGDP will be finished in 2009, according to the construction plan (Su, 2004). The project has several objectives including flood control, electricity generation, waterage transportation, fishery and tourism development (Y. B. Chen, 2004). The most direct environmental impact is the inundation of large areas of croplands and plantations (260,000 km^2) and settlements, with the inhabitants (already 418,000 before the end of 2001 and about 1.3 million by the end of its construction) having to relocate (G. J. Chen, 2003). Other important environmental impacts include various kinds of pollution degrading the water quality, landscape fragmentation, elevated risks of biodiversity loss, danger for human health, and geological disasters (Fu *et al.*, 2003; Wu *et al.*, 2003; W. L. Chen *et al.*, 2004; Fang *et al.*, 2006). These environmental problems can be alleviated, at least partly, through ecological restoration and effective management. Rational landscape planning, vegetation restoration and settlement eco-rehabilitation have been carried out in this respect (Z. Guo *et al.*, 2003; F. Q. Chen *et al.*, 2004; Z. H. Shi *et al.*, 2004; Xiao and Lei, 2004). However, the environmental change brought about by this kind of large-scale project is usually nonlinear (X. Q. Xu *et al.*, 2002) and long lasting. Therefore, ecological monitoring and environmental impact assessments are crucial for ecological security and the sustainable development of these project-affected regions. Other scientific issues, including landscape dynamics and ecosystem stability in key regions, and ecotones of different ecosystem or landscape types should also be researched as hotspots in the future (Plate 8.5).

The establishment and advancement of unified restoration ecology

The development of restoration ecology adapted to Chinese characteristics should be initiated and advanced. Firstly, future restoration ecology development in China must be based on the physical, environmental and socioeconomical situation of the country. Secondly, unified ecological restoration research includes research field unification, multidisciplinary unification, scale unification and stakeholder unification. Research field unification, the basis for comprehensiveness, is the unification of static (pattern) research, dynamic (processes) research and application research; multidisciplinary unification, the foundation of innovation, is the unification of the pertinent disciplines under a holistic landscape framework to build unified theories and methodologies of restoration ecology; scale unification, the methodological necessity, is the matching of spatio-temporal and management scales and scaling; stakeholder unification, the bridge between science and society, is the unification of researchers, decision-makers and the users of ecologically restored lands.

8.4 Concluding remarks

China is a developing country with a large territory, a long history of human development, a big population, and significant environmental and socioeconomic differentiation. Driven by physical and socioeconomic factors, and also as the result of land use history, land degradation problems and the related environmental impacts have became more and more prominent and have aroused much concern. The Chinese government, with the contribution of many international partners, has taken action in addressing the problems of land degradation and its negative environmental impacts since the 1950s, of various kinds. Among the actions taken, the establishment and management of a protected area system (J. G. Liu *et al.*, 2003), and the most recent programs including natural forest protection, desertification control, and sloped farmland revegetation (W. H. Li, 2004; J. Xu *et al.*, 2006) are national-scale efforts to conquer the problems of land degradation. However, the overwhelming complexity of the Chinese context has produced many challenges, including the heavy reliance on state finance, the lack of inter-agency cooperation, the insufficient consideration of local interests, the neglect of appropriate practices, and the rigidity and inconsistency of certain policies (J. Xu *et al.*, 2006). Therefore, even though great efforts have been made towards the resolution of these problems, their real-world effectiveness is still not as good as expected.

Because of the uniqueness of China in the world both physically and socio-economically, progress in the theory and practice of rehabilitation of degraded land will improve the situation of the whole world, especially the developing world, will advance the pertinent science and technological fields and will improve regional sustainable development capacity and human welfare. Many ambitious plans have been issued by different levels of governmental agencies in China to conquer the land degradation problems. Nevertheless, it is not sufficient to rely solely on the government. Support from the general public, scientists, entrepreneurs, and other local stakeholders is very important as well. Further collaboration with the international community will certainly provide great help for future ecological restoration campaigns. With all the above concerted efforts, there is hope that these campaigns will be driven to successful conclusions.

References

Allen, C. D., Savage, M., Falk, D. A., Suckling, K. F., Swetnam, T. W., Schulke, T., Stacey, P. B., *et al.* (2002). Ecological restoration of Southwestern ponderosa pine ecosystems: A broad perspective. *Ecological Applications*, **12** (5): 1418–1433.

Bai, Y. F., Han, X. G., Wu, J. G., Chen, Z. Z. and Li, L. H. (2004). Ecosystem stability and compensatory effects in the Inner Mongolia grassland. *Nature*, **431**: 181–184.

Banister, J. and Zhang, X. (2004). China, economic development and mortality decline. *World Development*, **33** (1): 21–41.

Banks, T., Richard, C., Ping, L. and Zhaoli, Y. (2003). Community-based grassland management in western China: Rationale, pilot project experience, and policy implications. *Mountain Research and Development*, **23** (2): 132–140.

Bian, Y. S. (2003). Establishment regulations of construction planning of ecological demonstration area, county, municipality and province. *Environmental Protection*, **31** (10): 22–26 (in Chinese).

Brown, R. T., Agee, J. K. and Franklin, J. F. (2004). Forest restoration and fire: Principles in the context of place. *Conservation Biology*, **18**: 903–912.

Cairns, J. Jr. (2000). Setting ecological restoration goals for technical feasibility and scientific validity. *Ecological Engineering*, **15**: 171–180.

Chang, Q., Li, H. Y. and He, Y. (2005). The model of resources and environment management of urban dry-up river in northern China – A case study of ecological restoration and reconstruction of Hutuo River. *Journal of Natural Resources*, **20**: 7–13 (in Chinese with English abstract).

Chen, F. Q., Zhang, L. P. and Xie, Z. Q. (2004). Vegetation restoration of waste land in the Three Gorges area. *Resources and Environment in the Yangtze Basin*, **13**: 286–291 (in Chinese with English abstract).

Chen, G. J. (2003). Situation and problems in the Three Gorges Reservoir area. *Resources and Environment in the Yangtze Basin*, **12** (2): 107–112 (in Chinese with English abstract).

Chen, L. Z. (1993). *Biodiversity in China: Status and Conservation Measures*. Beijing: Science Press (in Chinese).

Chen, L. Z. (1994). The status and conservation of biodiversity in China. In *Theories and methods of biodiversity research*, ed. Y. Q. Qian and K. P. Ma. Beijing: China Science and Technology Press (in Chinese), pp. 13–35.

Chen, M., Zhang, L. J., Wang, R. S. and Hai, B. G. (2005). Dynamics of ecological footprint of China from 1978 to 2003. *Resources Science*, **27** (6): 132–139 (in Chinese with English abstract).

Chen, W. L., Xie, Z. Q. and Xiong, G. M. (2004). Preliminary study on the ex-situ conservation of land plants in the Three Gorges Reservoir area. *Resources and Environment in the Yangtze Basin*, **13** (2): 174–177 (in Chinese with English abstract).

Chen, Y. B. (2004). Effect of Three Gorge Project (TGP) on the sustainable development in the Yangtze River Basin. *Resources and Environment in the Yangtze Basin*, **13** (2): 109–113 (in Chinese with English abstract).

China National Committee for the Implementation of the UN Convention to Combat Desertification (1992). *China National Action Program to Combat Desertification.* Beijing: Ministry of Forestry (in Chinese).

Choi, Y. D. (2004). Theories for ecological restoration in changing environment: Toward "futuristic" restoration. *Ecological Research*, **19**: 75–81.

Covington, W. W., Fulé, P. Z., Alcoze, T. M. and Vance, R. K. (2000). Learning by doing–Education in ecological restoration at Northern Arizona University. *Journal of Forestry*, **98** (10): 30–34.

Dale, V. H. (2003). Opportunities for using ecological models for resource management. In *Ecological Modeling for Resource Management*, ed. V. H. Dale. New York: Springer-Verlag, pp. 3–19.

Deng, F. T., Sun, P. S., Deng, F. S. *et al.* (2005). Research on the demonstration project of treatment wetland for polluted water purification near the Dianchi Lake. *Environmental Engineering*, **23** (3): 29–31.

Ding, Y. and Dai, X. (1994). Temperature variation in China during the last 100 years. *Meteorology*, **20** (12): 19–26 (in Chinese).

Du, Q. (2005). Limitation of major territory-based resource laws in ecological conservation. *Environmental Protection*, **33** (6): 29–32 (in Chinese).

Ellis, E. C., and Wang, S. M. (1997). Sustainable traditional agriculture in the Tai Lake Region of China. *Agriculture, Ecosystems and Environment*, **61**: 177–193.

Ellison, A. M. (2000a). Restoration of mangrove ecosystems. *Restoration Ecology*, **8**: 217–218.

Ellison, A. M. (2000b). Mangrove restoration: Do we know enough? *Restoration Ecology*, **8**: 219–229.

Fang, T., Fu, C. Y., Ao, H. Y. and Deng, N. S. (2006). The comparison of phosphorous and nitrogen status of the Xiangxi Bay before and after the impoundment of the Three Gorges Reservoir. *Acta Hydrobiologica Sinica*, **30** (1): 26–30 (in Chinese with English abstract).

Feng, S., Tang, M. and Zhou, L. (2000). Level fluctuation in Qinghai Lake during the last 600 years. *Journal of Lake Sciences*, **12**: 205–210 (in Chinese with English abstract).

Feng, Z., Yang, Y., Zhang, Y., Zhang, P. and Li, Y. (2005). Grain-for-green policy and its impacts on grain supply in West China. *Land Use Policy*, **22**: 301–312.

Field, C. D. (1998). Rehabilitation of mangrove ecosystems: An overview. *Marine Pollution Bulletin*, **37**: 383–392.

Foster, D., Swanson F., Aber J. *et al.* (2003). The importance of land-use legacies to ecology and conservation. *Bioscience*, **53**: 77–88.

Fu, B. J., Liu, G. H., Wang, X. K. and Ouyang, Z. Y. (2004a). Ecological issues and risk assessment in China. *International Journal of Sustainable Development and World Ecology*, **11**: 143–149.

Fu, B. J., Liu, G. H., Lu, Y. H., Chen, L. D. and Ma, K. M. (2004b). Ecoregions and ecosystem management in China. *International Journal of Sustainable Development and World Ecology*, **11**: 397–409.

Fu, B. J., and Lu, Y. H. (2006). The progress and perspectives of landscape ecology in China. *Progress in Physical Geography*, **30** (2): 232–244.

Fu, C., Wu, J., Chen, J., Wu, Q. and Lei, G. (2003). Freshwater fish biodiversity in the Yangtze River Basin of China: Patterns, threats and conservation. *Biodiversity and Conservation*, **12** (8): 1649–1685.

Fulé, P. Z., Waltz, A. E. M., Covington, W. W. and Heinlein T. A. (2001). Measuring forest restoration effectiveness in reducing hazardous fuels. *Journal of Forestry*, **99** (11): 24–29.

Fulé, P. Z., Covington, W. W., Smith, H. B. *et al.* (2002). Comparing ecological restoration alternatives: Grand Canyon, Arizona. *Forest Ecology and Management*, **170**: 19–41.

Gao, Q. H. (2003). The distributions of natural disasters and the divisional disaster-countermeasures for China. *Earth Science Frontiers*, **10** (Suppl): 258–264 (in Chinese with English abstract).

Gao, Z., Liu, J. and Zhuang, D. (1999). An analysis of eco-environmental quality conditions of China's land resources. *Journal of Natural Resources*, **14** (1): 93–96 (in Chinese with English abstract).

Ge, Q. S., Fang, X. Q., Zhang, X. Q. and Wu, S. H. (2005). Remarkable environmental changes in China during the past 50 years: A case study on regional research of global environmental change. *Geographical Research*, **24** (3): 345–358 (in Chinese with English abstract).

Geng, S., Zhou, Y., Zhang, M. and Smallwood, K. S. (2001). The model of resource and environment management of urban dry-up river in northern China: A case study of ecological restoration and reconstruction of Hutuo River. *Journal of Natural Resources*, **20**: 7–13 (in Chinese with English abstract).

Gross, M. and Hoffmann-Riem, H. (2005). Ecological restoration as a real-world experiment: Designing robust implementation strategies in an urban environment. *Public Understanding of Science*, **14**: 269–284.

Gundale, M. J., De Luca, T. H., Fiedler, C. E., Ramsey, P. W., Harrington, M. G. and Gannon, J. E. (2005). Restoration treatments in a Montana ponderosa pine forest: Effects on soil physical, chemical and biological properties. *Forest Ecology and Management*, **213**: 25–38.

Guo, Q. (1999). Mechanisms for sustainable agriculture: Effects of the garden-plantation system in Shaanbei, China. *Journal of Sustainable Agriculture*, **13** (4): 23–37.

Guo, Z., Xiao, X., Gan, Y. and Zheng, Y. (2003). Landscape planning for a rural ecosystem: Case study of a resettlement area for residents from land submerged by the Three Gorges Reservoir, China. *Landscape Ecology*, **18** (5): 503–512.

Han, Y. W., and Gao, J. X. (2005). Analysis of main ecological problems of grasslands and relevant countermeasures in China. *Research of Environmental Sciences*, **18** (3): 60–62 (in Chinese with English abstract).

Harrington, C. A. (1999). Forests planted for ecosystem restoration or conservation. *New Forests*, **17**: 175–190.

Hasenauer, H. (2002). Forest ecosystem restoration (Preface). *Forest Ecology and Management*, **159**: 1–2.

He, F. L. (2004). The countermeasures of mangrove ecosystem restoration in Futian, Shenzhen City. *Environmental Science and Technology*, **27** (4): 81–83.

He, Z. S. (2003). Basic theories and applications of restoration and re-establishment for degraded forest ecosystem. *Journal of Chongqing College of Education*, **16** (3): 59–62 (in Chinese).

Hu, A. (2006). The interactions of industrialization and urbanization development in China. *Statistics and Decision Making*, **22** (14): 99–101 (in Chinese).

Huang, B. W., Zuo, D. K. and Chen, F. Z. (1990). Field experimentation of physical geography in China. *Acta Geographica Sinica*, **45**: 225–234 (in Chinese with English abstract).

Huang, C. (2004). Discussion on wetland conservation and management in China. *Forest Resources Management*, **26** (5): 36–39 (in Chinese with English abstract).

Hyman, J. B. and Leibowitz, S. G. (2000). A general framework for prioritizing land units for ecological protection and restoration. *Environmental Management*, **25** (1): 23–35.

Jiang, J., Zhang, S., Huang, Q. and Deng, X. (2005). Analysis on the restoration and ecological recovery of Dongting lake by stopping cultivation. *Journal of Lake Sciences*, **16**: 325–330 (in Chinese with English abstract).

Jing, X. (2006). Analysis on China's urbanization and its political economic background. *Journal of Social Sciences*, **28** (4): 119–123 (in Chinese with English abstract).

Kettle, W. D., Rich, P. M., Kindscher, K., Pittman, G. L. and Fu, P. (2000). Land-use history in ecosystem restoration: A 40-year study in the prairie-forest ecotone. *Restoration Ecology*, **8**: 307–317.

Korn, M. (1996). The dike-pond concept: Sustainable agriculture and nutrient recycling in China. *Ambio*, **25**: 6–13.

Lei, K. (2005). Some thoughts on wetland park construction and development in China. *Forest Resources Management*, **27** (2): 23–26 (in Chinese with English abstract).

Li, C. H., and Zhou, Q. (2005). Environment worsen mechanism and restore solutions of wetland eco-system on the rivers sources area. *Nationalities Research in Qinghai*, **16** (1): 25–29 (in Chinese with English abstract).

Li, H. L., Li, X. N., Hou, X. M. *et al.* (2003). Technology of coastal ecological restoration: research progress and existing problems. *Urban Environment and Urban Ecology*, **16** (6): 36–37 (in Chinese with English abstract).

Li, J. W., Yuan, X. and Li, J. Q. (2006). Study on resources utilization and relative legislative issues of nature reserves of China. *Forest Inventry and Planning*, **31** (4): 55–60 (in Chinese with English abstract).

Li, M. S. (2006). Ecological restoration of mineland with particular reference to the metalliferous mine wasteland in China: A review of research and practice. *Science of the Total Environment*, **357** (1–3): 38–53.

Li, S. (2003). Study on the development stages of converting cropland for forest and grassland. *World Forestry Research*, **16** (1): 36–41 (in Chinese with English abstract).

Li, S. and Zhai, H. (2004). Comprehensive regionalization of the conversion of farmland to forests in China. *Journal of Mountain Science*, **22**: 513–520 (in Chinese with English abstract).

Li, W. H. (ed.) (2003). *Ecological Agriculture: Theories and Practices of Sustainable Agriculture in China*. Beijing: Publishing Center for Environmental Science and Technology, Chemical Industry Press (in Chinese).

Li, W. H. (2004). Degradation and restoration of forest ecosystems in China. *Forest Ecology and Management*, **201**: 33–41.

Li, X., Zhang, J., Zheng, W., Tang, C., Miao, Z. and Liu, K. (2004). Calculation and analysis of land consolidation potential in rural habitat during rapid urbanization process in China. *Transactions of CSAE*, **20** (4): 276–279 (in Chinese with English abstract).

Li, X. W., Luo, C. D., Hu, T. X. and Zhang, J. (2001). Suggestions on restoration and reconstruction of degraded forest ecosystem in the upper reaches of the Yangtze River. *Acta Ecologica Sinica*, **21**: 2117–2124 (in Chinese with English abstract).

Li, Y. (2005). The ecological conditions of grasslands in China have been improved significantly. *Pratacultural Science*, **22** (4): 52 (in Chinese with English abstract).

Li, Y. F. (2005). Cause of formation of China's land degeneration and improvement of legal control system. *Environmental Protection*, **33** (2): 24–27 (in Chinese).

Liao, B. W., Zheng, S. F., Chen, Y. J. and Li, M. (2005). Advance in researches on rehabilitation technique of mangrove wetland. *Ecologic Science*, **24** (1): 61–65 (in Chinese with English abstract).

Lin, N. F. and Tang, J. (2002). Geological environment and causes for desertification in arid-semiarid regions in China. *Environmental Geology*, **41**: 806–815.

Lin, P., Zhang, Y. H. and Yang, Z. W. (2005). Protection and restoration of mangroves along the coast of Xiamen. *Journal of Xiamen University (Natural Science)*, **44** (Sup.): 1–5 (in Chinese with English abstract).

Liu, G. (1999). Soil conservation and sustainable agriculture on the Loess Plateau: Challenges and prospects. *Ambio*, **28**: 663–668.

Liu, G. H., Fu, B. J., Chen, L. D. and Guo, X. D. (2000). Characteristics and distributions of degraded ecological types in China. *Acta Ecologica Sinica*, **20** (1): 13–19 (in Chinese with English abstract).

Liu, H. (1995). Types and characteristics of land degradation and countermeasures in China. *Natural Resources*, **17** (4): 26–32 (in Chinese with English abstract).

Liu, H. Y. (2005). Characteristics of wetland resources and ecological safety in China. *Resources Science*, **27** (3): 54–60 (in Chinese with English abstract).

Liu, J. G. and Diamond, J. (2005). China's environment in a globalizing world. *Nature*, **435**: 1179–1186.

Liu, J. G., Ouyang, Z., Pimm, S. L. *et al.* (2003). Protecting China's Biodiversity. *Science*, **300**: 1240–1241.

Liu, J. Y., Yue, T., Ju, H., Wang, Q. and Li, X. (2005a). *Integrated Ecosystem Assessment of Western China*. Beijing: China Meteorological Press (in Chinese).

Liu, J. Y., Zhan, J. and Deng, X. (2005b). Spatio-temporal patterns and driving forces of urban land expansion in China during the economic reform era. *Ambio*, **34** (6): 450–455.

Liu, Y. and Guo, H. C. (2003). Ecological restoration and landscape design of urban lakes. *Urban Environment and Urban Ecology*, **16** (6): 51–53 (in Chinese with English abstract).

Liu, Y. and Pen, H. (2003). Analysis on social economic driving forces for wetland ecosystem degradation in Boyang Lake. *Jiangxi Social Sciences*, **10**: 231–233 (in Chinese).

Lu, Y. C. and Zhang, S. G. (2003). Imperative technical problems to be solved and strategies recommended in regions of the protection program of Chinese natural forests. *Forest Research*, **16**: 731–738.

Luo, X. Z., Zhu, T. and Sun, G. Y. (2003). Wetland restoration and reconstruction in Da'an Paleochannel of Songnen Plain. *Acta Ecologica Sinica*, **23**: 244–250 (in Chinese with English abstract).

Ma, G. and Ning, D. (2006). The empirical correlation analysis between consumption and economic growth in China (1978–2004). *Shandong Economy*, **23** (3): 25–27 (in Chinese).

Ma, L. J. C. (2005). Urban administrative restructuring, changing scale relations and local economic development in China. *Political Geography*, **24**: 477–497.

Martin, L. M., Moloney, K. A. and Wilsey, B. J. (2005). An assessment of grassland restoration success using species diversity components. *Journal of Applied Ecology*, **42**: 327–336.

Miao, Z. and Marrs, R. (2000). Ecological restoration and land reclamation in open-cast mines in Shanxi Province, China. *Journal of Environmental Management*, **59**: 205–215.

Min, Q. W. and Cheng, S. K. (2002). Water resources security and countermeasures in China oriented to globalization. *Resources Science*, **24** (4): 49–55 (in Chinese with English abstract).

Ministry of Agriculture (2003a). Ecological conservation and rehabilitation planning of grasslands in China: Part one. *Pratacultural Science*, **20** (8): 68–69 (in Chinese with English abstract).

Ministry of Agriculture (2003b). Ecological conservation and rehabilitation planning of grasslands in China: Part two. *Pratacultural Science*, **20** (9): 80–84 (in Chinese with English abstract).

Mitchell, D. J., Fearnehough, W., Fullen, M. A. and Trueman, I. C. (1996). Ningxia desertification, reclamation and development. *China Review*, **5**: 27–31.

Mitchell, D. J., Fullen, M. A., Trueman, I. C. and Fearnehough, W. (1998). Sustainability of reclaimed desertified land in Ningxia, China. *Journal of Arid Environments*, **39**: 239–251.

Moore, M. M., Covington, W. W. and Fulé, P. Z. (1999). Reference conditions and ecological restoration: A southwestern ponderosa pine perspective. *Ecological Applications*, **9**: 1266–1277.

Pastorok, R. T., MacDonald, A., Sampson, J. R., Wilber, P., Yozzo, D. J. and Titre, J. P. (1997). An ecological decision framework for environmental restoration projects. *Ecological Engineering*, **9**: 989–1007.

Pavlikakis, G. E. and Tsihrintzis, V. A. (2000). Ecosystem management: A review of a new concept and methodology. *Water Resources Management*, **14**: 257–283.

Qin, D. H., Ding, Y. H, Su, J. L. and Wang, S. M (eds.) (2005). *Climate and Environmental Change in China*. Beijing: Science Press (in Chinese).

Qiu, D. and Wu, Z. (1996). On the degradation and restoration of shallow eutrophic Chinese lakes. *Resources and Environment in the Yangtze Valley*, **5**: 355–361 (in Chinese with English abstract).

Quon, S. P., Martin, L. R. G. and Murphy, S. D. (2001). Effective planning and implementation of ecological rehabilitation projects: A case study of the regional municipality of Waterloo (Ontario, Canada). *Environmental Management*, **27** (3): 421–433.

Ren, H. and S. Peng. (2003). The practice of ecological restoration in China: A brief history and conference report. *Ecological Restoration*, **21**: 122–125.

Ren, X. Y., Cai, S. M., Wang, X. L., Du, Y. and Wang, Q. (2004). Study of wetland restoration in the middle reaches of Yangtze River. *Journal of Central China Normal University (Nat. Sci.)*, **38** (1): 114–116 (in Chinese with English abstract).

Shang, S., Du, J., Li, X., Shen, Q. and Wang, L. (2003). Ecological restoration engineering technology of eutrophic lake–a case study of lake Wuliangsuhai in Inner Mongolia. *Chinese Journal of Ecology*, **22** (6): 57–62 (in Chinese with English abstract).

Shi, J. C. (2004). The Hani terraced-field at Honghe: Typical Chinese wetland. *Journal of Yunnan Nationalities University*, **21** (5): 77–81 (in Chinese with English abstract).

Shi, P. and Li, W. (1999). Rehabilitation of degraded mountain ecosystems in southwestern China: An integrated approach. *Ambio*, **28**: 390–397.

Shi, T. (2002). Ecological agriculture in China: Bridging the gap between rhetoric and practice of sustainability. *Ecological Economics*, **42**: 359–368.

Shi, Z. H., Cai, C. F., Ding, S. W., Wang, T. W. and Chow, T. L. (2004). Soil conservation planning at the small watershed level using RUSLE with GIS: A case study in the Three Gorges area of China. *Catena*, **55** (1): 33–48.

Song, S., Fan, T. and Wang, Y. (1997). Comprehensive sustainable development of dryland agriculture in Northwest China. *Journal of Sustainable Agriculture*, **9** (4): 67–84.

State Environmental Protection Administration of China (2003). Investigation report on eco-environmental situation in China's mid-east regions. *Environmental Protection*, **31** (8): 3–8 (in Chinese).

State Environmental Protection Administration of China (2004). Investigation report on eco-environment in China. *Environmental Protection*, **32** (5): 13–18 (in Chinese).

Stephenson, N. L. (1999). Reference conditions for giant sequoia forest restoration: Structure, process, and precision. *Ecological Applications*, **9**: 1253–1265.

Su, W. C. (2004). Main ecological and environmental problems of water-level fluctuation zone (WLFZ) in Three Gorges Reservoir and controlling measures. *Journal of Yangtze River Scientific Research Institute*, **21** (2): 32–34 (in Chinese with English abstract).

Swetnam, T. W., Allen, C. D. and Betancourt, J. L. (1999). Applied historical ecology: Using the past to manage for the future. *Ecological Applications*, **9**: 1189–1206.

Tan, Y. Z., Wu, C. F., Wang, Q. R., Zhou, L. Q. and Yan, D. (2005). The change of cultivated land and ecological environment effects driven by the policy of dynamic equilibrium of the total cultivated land. *Journal of Natural Resources*, **20** (5): 727–734 (in Chinese with English abstract).

Tong, C. H., Yang, X. E. and Pu, P. M. (2003). Degradation of soil and water ecological system of Moganhu catchment and control countermeasures. *Journal of Soil and Water Conservation*, **17**: 72–88 (in Chinese with English abstract).

Wang, B., Cui, X. and Yang, F. (2004). Chinese forest ecosystem research network (CFERN) and its development. *Chinese Journal of Applied Ecology*, **23** (2): 84–91 (in Chinese with English abstract).

Wang, G. M., Jiang, G. M., Peng, Y., Yu, S. L. and Li, Y. G. (2005). Community-based conservation strategy: A case study of degraded ecosystem restoration in Hunshandak sandy land. *Acta Ecologica Sinica*, **25**: 1459–1465 (in Chinese with English abstract).

Wang, R. S. (2005). Thought on plan and development of administrative eco-district. *Environmental Protection*, **33** (5): 28–33(in Chinese).

Wang, S. P. and Li, Y. H. (1999). Degradation mechanism of typical grassland in Inner Mongolia. *Chinese Journal of Applied Ecology*, **10** (4): 437–441 (in Chinese with English abstract).

Wang, T., Wu, W., Xue, X., Han, Z., Zhang, W. and Sun, Q. (2004). Spatial temporal changes of sandy desertified land during last five decades in northern China. *Acta Geographica Sinica*, **59**: 203–212 (in Chinese with English abstract).

Wang, W. G. (2005). Research on the legislative issues pertaining to ecological environment construction. *Environmental Protection*, **33** (8): 20–23 (in Chinese).

Wang, Z. and Li, A. (2005). Amelioration of soil media during ecological restoration in the mining waste land. *China Mining Magazine*, **14** (3): 22–23 (in Chinese with English abstract).

Wang, Z. L. (2005). Strategic thinking of protecting grassland ecology in China. *Grassland of China*, **27** (4): 1–9 (in Chinese with English abstract).

Wen, D. and Pimentel, D. (1992). Ecological resource management to achieve a productive, sustainable agricultural system in northeast China. *Agriculture, Ecosystems & Environment*, **41**: 215–230.

Wong, M. H. and Luo, Y. M. (2003). Land remediation and ecological restoration of mined land. *Acta Pedologica Sinica*, **40** (2): 161–169 (in Chinese with English abstract).

Wu, J., Huang, J., Han, X., Xie, Z. and Gao, X. (2003). Three Gorges Dam experiment in habitat fragmentation? *Science*, **300**: 1239–1240.

Xiao, H. L., Li, X. R., Duan, Z. H., Li, T. and Li, S. Z. (2003). Succession of plant-soil system in the process of mobile dunes stabilization. *Journal of Desert Research*, **23**: 605–611 (in Chinese with English abstract).

Xiao, W. F. and Lei, J. P. (2004). Spatial distribution, disturbance and restoration of forests in the Three Gorges Reservoir region. *Resources and Environment in the Yangtze Basin*, **13** (2): 138–144 (in Chinese with English abstract).

Xie, Z. H. (2005). Thought and key task of SEPA of the eleventh-five-year plan on the nature ecological conservation. *Environmental Protection*, **33** (8): 5–10 (in Chinese).

Xiong, Y., Wang, K. L., Lan, W. L. and Qi, H. (2004). Evaluation of the lake recovery area eco-compensation in Dongting lake wetland. *Acta Geographica Sinica*, **59**: 772–780 (in Chinese with English abstract).

Xu, F., Guo, S. Y. and Zhang, Z. X. (2003). The distribution of soil erosion in China at the end of the twentieth century. *Acta Geographica Sinica*, **58** (1): 139–146 (in Chinese with English abstract).

Xu, J., Yin, R., Li, Z. and Liu, C. (2006). China's ecological rehabilitation: Unprecedented efforts, dramatic impacts, and requisite policy. *Ecological Economics*, **57**: 595–607.

Xu, P. Z. and Qing, B. Q. (2002). Degraded reasons, restoration and re-establishment of ecosystem in Taihu shore. *Water Resources Conservation*, **3**: 31–36 (in Chinese).

Xu, S. X., Zhao, X. Q., Sun, P., Zhao, W. and Zhao, T. B. (2002). A serious menace to biological resources: Losses of biodiversity. *Resources Science*, **24** (2): 6–11.

Xu, X. Q., Xiao, B. D., Fang, T. and Zheng, L. (2002). Nonlinear and time-delayed environmental effect of the Three Gorges Reservoir Area. *Resources and Environment in the Yangtze Basin*, **11** (1): 73–78.

Xu, Y., Ding, Y. H. and Zhao, Z. C. (2003). Scenario of temperature and precipitation changes in northwest China due to human activity in the twenty-first century. *Journal of Glaciology and Geocryology*, **25**: 327–330 (in Chinese with English abstract).

Xu, Z., Xu, J., Deng, X., Huang, J., Uchida, E. and Rozelle, S. (2005). Grain for green versus grain: Conflict between food security and conservation set aside in China. *World Development*, **34**: 130–148.

Xu, Z. Q., Li, W. H., Min, Q. W. and Xu, Q. (2005). Researches on changes of value of ecosystem in Xilin River Basin. *Journal of Natural Resources*, **20**: 99–104 (in Chinese with English abstract).

Yang, X., Zhang, K., Jia, B. and Ci, L. (2005). Desertification assessment in China: An overview. *Journal of Arid Environments*, **63**: 517–531.

Ye, X. J., Wang, Z. Q. and Li, Q. S. (2002). The ecological agriculture movement in modern China. *Agriculture, Ecosystems and Environment*, **92**: 261–281.

Ye, Z. H., Wong, J. W. C., Wong, M. H., Baker, A. J. M., Shu, W. S. and Lan, C. Y. (2000). Revegetation of Pb/Zn mine tailings, Guangdong province, China. *Restoration Ecology*, **8** (1): 87–92.

Yuan, Q., Xu, Z., Shi, W. G. and Wu, X. H. (2004). Establishment of the sharing information system of grassland resources in China. *Grassland of China*, **26** (4): 16–20 (in Chinese with English abstract).

Yue, T. X., Liu, J. Y., Jørgensen, S. E., Gao, Z. Q., Zhang, S. H. and Deng, X. Z. (2001). Changes of the Holdridge life zone diversity in all of China over half a century. *Ecological Modelling*, **144**: 153–162.

Yun, J. F. (2002). The status of grassland resources in China. *Flowers and Horticulture in China*, **2** (17): 4–5 (in Chinese).

Zha, Y. and Gao, J. (1997). Characteristics of desertification and its rehabilitation in China. *Journal of Arid Environments*, **37**: 419–432.

Zhang, J., Zhao, Z. K. and Li, X. W. (2005). Wetland restoration and habitat rehabilitation in human dominated area: A case study in Zhangdu Lake, Wuhan city. *Resources Science*, **27** (4): 133–139 (in Chinese with English abstract).

Zhang, M. X., Yan, C. G., Wang, J. C. and Chen, F. (2001). Analysis on wetland degradation and its reasons in China. *Forest Resources Management*, **23** (3): 23–26 (in Chinese with English abstract).

Zhang, S. G., Xiao, W. F. and Jiang, Z. P. (2002). Sustainable forest management in China; the basic principles and practices. *Plant Biosystems*, **136** (2): 159–166.

Zhang, W. G. (2005). Evolution analysis and development approach of eco-province construction in China. *Environmental Protection*, **33** (2): 38–41 (in Chinese).

Zhang, X. S. (1994). Principles and optimal models for development of Maowusu sandy grassland. *Acta Phytoecologica Sinica*, **18**: 1–16 (in Chinese with English abstract).

Zhao, S. (2005). The degradation, conservation, and rehabilitation of wetlands in China. *Advances in Earth Science*, **20**: 701–704 (in Chinese with English abstract).

Zhao, X. (2001). The first symposium on restoration ecology held in Guangzhou, China. *Ecological Restoration*, **19**: 72–73.

Zhao, Z. (2006). The institutional background of urbanization in China and its system restrictions. *Urban Problems*, **25** (2): 9–11 (in Chinese with English abstract).

Zheng, Y. R. and Zhang, X. S. (1998). The diagnosis and optimal design on high efficient ecological economy ecosystem in Mu Us sandy land. *Acta Phytoecologica Sinica*, **22**: 262–268 (in Chinese with English abstract).

Zhou, G. and Niu, W (eds.) (2000). *Sustainable Development Strategies in China (Reading Material for Decision Makers)*. Beijing: Xiyuan Press (in Chinese).

Zhu, Z. and Liu, S. (1988). Desertification processes and their control in Northern China. *Chinese Journal of Arid Land Research*, **1**: 27–36.

9

Conservation, restoration and creation of wetlands: a global perspective

WILLIAM J. MITSCH

9.1 Introduction

The importance of wetland environments to the development and sustenance of cultures throughout human history is undeniable. Since early civilization, many cultures have learned to live in harmony with wetlands and have benefited economically from wetlands surrounding their settlements, whereas other cultures quickly drained the landscape. The propensity in Eastern civilizations was not to drain the landscape, as is frequently done in the West, but to work within the aquatic landscape, albeit often in a heavily managed way. Dugan (1993) makes an interesting comparison between *hydraulic civilizations* that controlled water flow through the use of dikes, dams, pumps, and drainage tiles, partly because water was only seasonally plentiful, and *aquatic civilizations* that better adapted to their surroundings of water-abundant floodplains and deltas and took advantage of nature's pulses, such as flooding. It is because the former approach of controlling nature rather than working with it is so dominant around the world today that we find such high losses of wetlands worldwide.

I refer to the people who live in proximity to wetlands and whose culture is linked to them as *wetlanders*. The ancient Babylonians, Egyptians, and the Aztecs in what is now Mexico developed specialized systems of water delivery involving wetlands. The Camarguais of southern France, the Cajuns of Louisiana, the Marsh Arabs of southern Iraq, and some Far Eastern cultures have lived in harmony with wetlands for hundreds of years. Native Americans in North America have, for centuries, harvested and reseeded wild rice (*Zizania aquatica*) along the littoral zones of lakes and streams. Domestic wetlands such as rice paddies feed an estimated half of the world's population. Cranberries are harvested from bogs, and the industry continues to thrive today in North America. Many aquatic plants besides rice, such as Manchurian wild rice (*Zizania latifolia*), are harvested as vegetables in China. Wetlands are known for their abundant wildlife and are

Ecological Restoration: A Global Challenge, ed. Francisco A. Comin. Published by Cambridge University Press.
© Cambridge University Press 2010.

therefore important for emerging ecotourism uses by natives in many parts of the world. Some of the most biologically rich wetlands in the world are found in the Rift Valley in Tanzania and in the seasonally flooded Okavango Delta in Botswana. Even where wetlands are artificially created, wildlife can become abundant in a relatively short time as with the Oostvaardensplassen in The Netherlands. The Russians, Finns, Estonians, Irish, and even New Zealanders, among other cultures, have mined their peatlands for centuries, using peat as a source of energy and for horticultural purposes. The production of fish in shallow ponds or rice paddies developed several thousands of years ago in China and Southeast Asia, and crayfish harvesting is still practiced in the wetlands of Louisiana and the Philippines. Coastal marshes in northern Europe, the British Isles, and New England were used for centuries and are still used today for the grazing of animals and hay production, and coastal mangroves are harvested for timber, food, and tannin in many countries throughout Indo-Malaysia, East Africa, and Central and South America. Reeds and even the mud from coastal and inland marshes have been used for wall construction, fence material and thatching for roofs in Europe, Iraq, Japan, and China (Mitsch and Gosselink, 2007).

This paper first provides some general definitions and describes the extent of wetlands in the world and how they are being lost to human development. The paper then turns its attention to the positive side: activities to restore and create wetlands on a large scale throughout the world, based on the principles of ecological engineering. Four case studies of large-scale wetland restoration and creation, three in North America and one in the Middle East, are then described with general conclusions in each case study of what needs to be done. These case studies are followed by general conclusions related to humans and wetlands and our ability and need to recreate these important ecosystems.

9.2 Wetland definitions and global extent

Wetland definitions often include three main components. Wetlands are distinguished by the presence of water, either at the surface or within the root zone, they often have unique soil conditions that differ from adjacent uplands, and they support vegetation adapted to the wet conditions (*hydrophytes*) and, conversely, are characterized by an absence of flooding-intolerant vegetation. Climate and geomorphology define the degree to which wetlands can exist, but the starting point is the *hydrology*, which, in turn, affects the *physicochemical environment*, including the soils, which, in turn, determines with the hydrology what and how much *biota*, including vegetation, is found in the wetland. And just as importantly, the biota can cause feedbacks that modify the hydrology and physicochemical environment.

A number of common terms such as swamp, marsh, and mire have been used over the years to describe different types of wetlands. The history of the use and misuse of these words has often revealed a decidedly regional or at least continental origin. Although the lack of standardization of terms is confusing, many of the old terms are rich in meaning to those familiar with them (González-Bernáldez, 1992). A *marsh* is known by most as a herbaceous plant wetland. A *swamp*, on the other hand, has woody vegetation, either shrubs or trees. There are subtle differences among marshes. A marsh with significant (>30 cm) standing water throughout much of the year is often called a *deepwater marsh*. A shallow marsh with waterlogged soil or shallow standing water is sometimes referred to as a *sedge meadow* or a *wet meadow*. Intermediate between a marsh and a meadow is a *wet prairie*. Several terms are used to denote peat-accumulating systems. The most general term is *peatland*, which is generally synonymous with *moor* and *muskeg*. There are many types of peatlands, the most general being *fens* and *bogs*.

Based on several estimates, the extent of the world's wetlands is generally thought to be from 7 to 9 million km^2, or about 4 to 6 percent of the land surface of the Earth (Mitsch and Gosselink, 2007). Almost 86 percent of the estimated total natural wetland area is found in tropical, subtropical, and boreal regions of the world. Temperate zone wetlands constitute only about 14 percent of the world's natural wetlands. Scientists have estimated there are between 1.3 and 1.5 million km^2 of rice paddies in the world. By way of comparison, a total of 1.45 million km^2 of wetlands has been registered with the Ramsar Convention on Wetlands of International Importance as of the end of 2006 (Ramsar Convention, 2006). This represents about 16 to 21 percent of the world's total wetlands although the Ramsar definition is quite liberal in including deepwater ponds and estuaries as wetlands.

9.3 Wetland losses

With all of the values of wetlands that have been identified (see Mitsch and Gosselink, 2007 for a complete description), not to mention the aesthetics of a landscape in which water and land often provide a striking panorama, one would expect wetlands to be revered by humanity; this has certainly not always been the case. Wetlands have been depicted as sinister and forbidding, and as having little economic value, throughout most of history. For example, in the *Divine Comedy*, Dante describes a marsh of the Styx in Upper Hell as the final resting place for the wrathful:

> Thus we pursued our path round a wide arc of that ghast pool,
> Between the soggy marsh and arid shore,
> Still eyeing those who gulp the marish [marsh] foul.
>
> *(Dante Alighieri)*

Centuries later, Carl Linnaeus, crossing the Lapland peatlands, compared that region to that same Styx of Hell:

Shortly afterwards began the muskegs, which mostly stood under water; these we had to cross for miles; think with what misery, every step up to our knees. The whole of this land of the Lapps was mostly muskeg, *hinc vocavi* Styx. Never can the priest so describe hell, because it is no worse. Never have poets been able to picture Styx so foul, since that is no fouler.

(Carl Linnaeus, 1732)

As a result of these and similar views of wetlands, particularly in Western society, the drainage and destruction of wetlands became accepted practice around the world and were even encouraged by specific government policies. Wetlands were replaced with agricultural fields and urban sprawl. There are a number of causes for loss of inland wetlands, the most notable being drainage for agriculture, forestry, and mosquito control; filling for residential, commercial, and industrial development; filling for solid-waste disposal; and mining of peat. With over 70 percent of the world's population living on or near coastlines, coastal wetlands have long been destroyed through a combination of excessive harvesting, hydrologic modification and seawall construction, coastal development, pollution and other human activities.

It is probably safe to assume that we are still losing wetlands at a fairly rapid rate globally and that we have perhaps lost as much as 50 percent of the original wetlands. There are a number of areas where the loss rate has been documented (Table 9.1). The estimate of about 50 percent loss of wetlands since European settlement in the lower forty-eight US states is fairly accurate as is the 90 percent loss of wetlands in New Zealand. Several regions of the world, for example, Europe and parts of Australia, Canada, and China, have lost an even higher percentage of regional wetlands.

9.4 A more optimistic approach: creating and restoring wetlands and watersheds

Because of their important values, wetland creation and restoration need to become fundamental approaches to our management of our water resources. But do we have sufficient scientific knowledge of wetland function to be able to create and restore wetlands? We are now in a position to make a substantial contribution to the "greening" of the planet through ecological engineering and ecosystem restoration. Ecologists are now refining the techniques of restoring function in degraded ecosystems and countless ecologists now call themselves restoration ecologists. Agricultural engineers, known for the efficiency with which they drained the landscape, are changing their names and their actions in many locations by restoring ditches to stream channels and farmlands to wetlands. Civil engineers, the nation's

Table 9.1 *Percent loss of wetlands in various geographic locations in the world.*

Location	Percentage loss	Reference
NORTH AMERICA		
United States	53	Dahl (1990)
Canada		National Wetlands Working Group (1988)
Atlantic tidal and salt marshes	65	
Lower Great Lakes – St. Lawrence Rivers	71	
Prairie potholes and sloughs	71	
Pacific coastal estuarine wetlands	80	
AUSTRALASIA		
Australia	>50	Australian Nature Conservation Agency (1996)
Swan Coastal Plain	75	
Coastal New South Wales	75	
Victoria	33	
River Murray Basin	35	
New Zealand	>90	Dugan (1993)
Philippine mangrove swamps	67	Dugan (1993)
CHINA	60	Lu (1995)
EUROPE	60	Revenga *et al.*(2000)

Source: Mitsch and Gosselink (2007).

top river straighteners, are busy removing dams and restoring river meanders. Restoration and creation of ecosystems is now an industry in many countries and those countries are showing the way for others to follow.

To do this restoration correctly, a new application of science is needed. I prefer to use the term *ecological engineering* because restoration is not simply a case of studying a system, making hypotheses, and testing those hypotheses as scientists are trained to do. Ecological engineering involves designing or redesigning ecosystems and the landscape (Mitsch and Jørgensen, 2004). Ecological engineering is defined as "the design of sustainable ecosystems that integrate human society with its natural environment for the benefit of both" (Mitsch, 1993; Mitsch, 1998; Mitsch and Jørgensen, 2004). This design or redesign must have basic scientific principles to guide it; a recently published book specifies nineteen principles (Mitsch and Jørgensen, 2004). Among the basic concepts that distinguish ecological engineering from other fields of engineering or science are its basis in the self-designing capacity of ecosystems and its potential for conserving nonrenewable energy resources while supporting ecosystem conservation.

There are countless examples around the world where ecological engineering is in progress in order to restore the watery landscape, with an emphasis on wetland

Table 9.2 *Comparison of restoration projects described in the paper.*

Restoration project	Area, ha
Delaware Bay	5,800
Mesopotamian marshlands	1,500,000
Everglades	4,600,000
Louisiana delta	2,500,000
Mississippi River Basin	2,200,000

and river restoration. Four examples of wetland restoration on a large scale that are being undertaken or planned are summarized here (Table 9.2).

9.4.1 Delaware Bay

A large coastal wetland restoration project in Eastern USA involves the restoration, enhancement and preservation of 5000 hectares of coastal salt marshes on Delaware Bay in the states of New Jersey and Delaware in northeastern USA (Figure 9.1). This estuary enhancement, being carried out by an electric utility, with advice from a team of scientists and consultants, was undertaken as mitigation for the potential impacts of once-through cooling from a nuclear power plant. The reasoning was that the impact of once-through cooling on fin fish, through entrainment and impingement, could be offset by increased fisheries production brought about by restoring salt marshes. Because of the uncertainties involved in this kind of ecological trading, the area of restoration was estimated as the salt marshes that would be necessary to compensate for the impacts of the power plant on fin fish, times a safety factor of four. The project was the subject of a special issue in the journal *Ecological Engineering* (Peterson *et al.*, 2005).

The most important type of restoration in this project involves the reintroduction of tidal inundation to about 1800 hectares of former diked salt-hay farms. Many marshes along Delaware Bay have been isolated by dikes from the bay, sometimes for centuries, and put into the commercial production of salt hay (*Spartina patens*). Hydrologic restoration was accomplished by excavating breaches in the dikes and, in most cases, connecting these new inlets to a system of recreated tidal creeks and existing canal systems. In some cases, restoration involved enhancing drainage by re-excavating some initial tidal creeks in these newly flooded salt marshes. After initial tidal creeks were established, the system "self-designed" more tidal channels and increased the channel density. Self-design works when the proper conditions for propagule disbursement are provided (Teal and Weinstein, 2002; Hinkle and

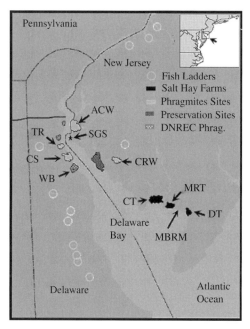

Fig. 9.1. Salt marsh restoration at Delaware Bay, eastern USA. Letters indicate sites where salt hay farms are converted to salt marshes or *Phragmites* cover is controlled (map courtesy of PSEG, Newark, New Jersey, USA; also published in Peterson *et al.*, 2005).

Mitsch, 2005). Allowing tidal and storm forces to self-design the marsh hydrology caused these marshes to develop natural tidal cycles and *Spartina* to establish itself.

9.4.2 Florida Everglades

Engineers, ecologists, resource managers and even politicians are now completely redesigning the "ecological plumbing" in south Florida to provide a more ecologically integrated system for the Florida Everglades (Plate 9.1). The restoration of the Florida Everglades, the largest wetland area in the United States, is actually several separate initiatives being carried out in the 4.6 million hectares Kissimmee-Okeechobee-Everglades (KOE) region in the southern third of Florida. As part of this project, the Kissimmee River in Florida is being "restored" to something resembling its former self before it was canalized thirty years ago at enormous cost. Overall, the Everglades restoration, as now planned by the US Army Corps of Engineers, will cost almost US$ 8 billion and will be carried out over twenty years or more. Problems in the Everglades have developed because of (1) excessive nutrient loading to Lake Okeechobee and to the Everglades itself, primarily from agricultural runoff, (2) loss and fragmentation of habitat caused by urban and agricultural development, (3) spread of *Typha* and other invasives and exotics to the Everglades, replacing native vegetation, and (4) hydrologic alteration due to an extensive system of canals and straightened rivers built by the US Army Corps of Engineers and others for flood protection, and maintained by several water management districts.

9.4.3 Louisiana delta and the Mississippi River Basin

Louisiana delta

The Mississippi River delta and coastline in Louisiana in south-central USA are disappearing into the sea and major efforts are underway to reduce land loss along that coastline. Louisiana is one of the most wetland-rich regions of the world with 36,000 km^2 of marshes, swamps and shallow lakes. Yet Louisiana is suffering a rate of coastal wetland loss of between 6,600 and 10,000 hectares per year as it converts to open water areas on the coastline, due to natural causes (land subsidence) and human causes such as river levee construction, oil and gas exploration, urban development, sediment diversion, and possibly climate change. There has been, since the early 1990s, a major interest in reversing this rate of loss and even gaining coastal areas, particularly freshwater marshes and salt marshes.

An ambitious ecological engineering project, now called the Louisiana Coastal Area (LCA) project, is being developed now in Louisiana to re-engineer the coastline to curtail the land loss. This project, estimated to cost US$ 14 billion, would be the largest ecological engineering project in the world if undertaken. Already, there has been considerable public interest in the project in Louisiana, which would like to convince the rest of the United States that the Mississippi delta is "America's Wetland." The plan is to carry out a suite of dozens of projects that are meant to build land-shoreline protection in critical areas, river diversions, to optimize the functioning of Mississippi River distributaries such as the Atchafalaya River, and to restore barrier islands. Because the Mississippi River is maintained for shipping purposes and the navigational channel is now maintained as a channel well downstream of New Orleans and out to the Gulf of Mexico deep waters, simply allowing the river to cut through its artificial levees would cause great economic harm to southern Louisiana. Rather, plans have to be made in a hydrologically complex situation.

River diversions are a large part of the delta restoration plan. For example, the Caernarvon freshwater diversion is the largest of several diversions currently in operation on the Mississippi River in Louisiana. The diversion structure is on the east bank of the river below New Orleans and 131 km upstream of the Gulf of Mexico. The Caernarvon diversion delivers water to the Breton Sound estuary, an 1100 km^2 area of fresh, brackish, and saline wetlands. Studies have already illustrated that sedimentation is significant and land building is occurring adjacent to these river diversions.

Some of the most recent information on the potential for restoration of the Louisiana Delta, particularly after the impact of Hurricane Katrina on New Orleans in late August 2005, are contained in a report by Boesch *et al.* (2006). In addition a special issue of *Ecological Engineering* on the restoration of the Gulf of Mexico and the Louisiana Delta has been published (Chapman and Reed, 2006).

The LCA plan is now in jeopardy as priorities shift to engineering solutions for levees and rebuilding of homes in flood-prone areas instead of ecologically engineered solutions of wetland restoration and sensible human settlements. The US$ 14 billion that was estimated to be needed for this ecological engineering may well be swallowed up by the reconstruction of the city and its levees. The US Army Corps of Engineers (2006), in their recent report to the US Congress in the aftermath of Hurricane Katrina, admits that hurricane defense of New Orleans includes the coastal wetlands and barrier islands that are, themselves, disappearing yet vital:

Coastal ecological features form the outer line of defense against storm waves. Barrier island systems absorb waves from approaching storms and help limit the amount of water that enters estuaries in advance of tropical systems. Back-barrier marshes and coastal fringe wetlands act as tidal and wave buffers protecting inland features.

... A better system approach would involve fighting storm surges on the outer fringe of populated areas with large surge barriers and armored levees fronted by natural coastal features.

Also, coastal populations should recognize the extreme storm dangers and plan accordingly by using better construction techniques to withstand storms and efficient evacuation plans to move out of the paths of harmful hurricanes.

To solve the problems in New Orleans and restore the valuable delta to a balanced landscape the following seven principles have been suggested (Costanza *et al.*, 2006a, 2006b).

1. Let the water decide. Building structures below sea level is asking for trouble.
2. Avoid abrupt boundaries between deepwater systems and uplands.
3. Restore natural capital of the coastal wetlands.
4. Use the Mississippi River to rebuild the coast, rather than keep the current system that constrains the river between levees, allowing the resources of freshwater, sediments, and nutrients to flow into the deeper waters of the Gulf and cause offshore hypoxia. Diversions of water and sediments from the Mississippi are a major component of the LCA plan.
5. Restore the built capital of New Orleans to the highest standards of high-performance green buildings and a car-limited urban environment with high mobility for everyone.
6. Rebuild the social capital of New Orleans to twenty-first century standards of diversity, tolerance, fairness, and justice.
7. Restore the Mississippi River Basin to minimize coastal pollution and the threats of river flooding in New Orleans. Upstream changes in the 3 million km^2 Mississippi drainage basin have significantly changed nutrient and sediment delivery patterns to the delta. Changes in farming practices in the drainage basin can improve not only the coastal restoration process, but also promote sustainable farming practices in the entire basin (see below).

The Mississippi River Basin

The Gulf of Mexico, south of the Louisiana Delta, continues to have annual "dead zones" that now spread well over 20,000 km^2 (Rabalais *et al.*, 2002) (Figure 9.2). Nitrogen, particularly nitrate-nitrogen, is the most probable cause; 80 percent of the nitrogen input is from the 3 million km^2 Mississippi River Basin (41 percent of lower forty-eight states of the USA). The excessive nitrate-nitrogen flowing from Midwestern USA is exacerbated by the fact that there has been excessive drainage of the upper Mississippi River Basin that causes waters to leave the fertilized farmland quickly and immediately be transported to the Gulf of Mexico. The control of this hypoxia is important in the Gulf of Mexico because the continental shelf fishery in the Gulf is approximately 25 percent of the USA total. Flooding and water pollution caused by excessive nitrogen within the Mississippi River Basin itself are additional reasons why a restoration approach would synchronize with a solution to the Gulf of Mexico hypoxia problem.

A number of approaches are being considered for controlling nitrogen flow into the Gulf; many of them involve large-scale modifications of land-use practices in Midwestern USA. Among the options are modifying agricultural practices, e.g., reduced fertilizer use, alternate cropping techniques; tertiary treatment (biological, chemical, physical) of point sources; landscape restoration, e.g., riparian buffers and wetland creation, to control non-point source pollution from farmland; stream and delta restoration; and atmospheric controls of NO$_x$. The approach that appears to have the highest probability of success with a minimum impact on farming in the Midwestern USA is wetland and riparian ecosystem restoration. It has been suggested that something of the order of 22,000 km^2 of restored and created wetlands and even

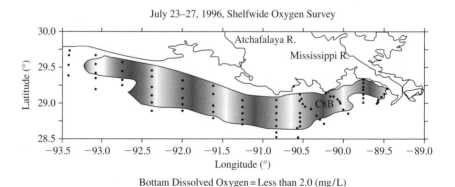

Fig. 9.2. Extent of the Gulf of Mexico hypoxia in July 1996. From Rabalais *et al.*, 1999.

more restored riparian buffers would be necessary to provide enough denitrification to substantially reduce the nitrogen entering the Gulf of Mexico (Mitsch *et al.*, 2001; Mitsch *et al.*, 2005; Mitsch and Day, 2006). This would involve approximately 3 percent of the Mississippi River Basin. Interestingly, Hey and Phillipi (1995) found that a similar wetland restoration effort would be required in the Upper Mississippi River Basin to mitigate the effects of very large and costly floods such as the one that occurred in the summer of 1993. Recent Federal programs that provide resources to farmers such as the Conservation Reserve Enhancement Program (CREP) of the US Department of Agriculture have been implemented with the hope that they will provide wetlands and riparian ecosystems that can help limit the amount of nitrogen fertilizer that is discharged from Midwestern USA (Plate 9.2).

9.4.4 The Mesopotamian marshlands

The crown jewel wetlands of the Middle East are the Mesopotamian marshlands of southern Iraq and Iran. These wetlands are in an arid region of the world and exist at the confluence of the Tigris and Euphrates Rivers. The watersheds of both the Euphrates and the Tigris lie predominantly in the countries of Turkey, Syria, and Iraq. The Tigris-Euphrates Basin has been subject to water control projects for over six millennia. The Mesopotamian wetlands are the largest wetland ecosystem in the Middle East, have been the home to the Marsh Arabs for 5,000 years and support a rich biodiversity (UNEP, 2001). Since 1970, the wetlands have been damaged dramatically (Plate 9.3). The Mesopotamian wetlands once were between 15,000 and 20,000 km^2 in area but were drained in the 1980s and 1990s to less than 10 percent of that extent. There was a 30 percent decline just in the period between 2000 and 2002. The draining of the wetlands was the result of human-induced changes. Upstream dams and drainage systems constructed in the 1980 and 90s drastically altered the river flows and have eliminated the flood pulses that sustained the wetlands (UNEP, 2001). Turkey alone constructed more than a dozen dams on the upper rivers. But the main cause of the disappearance of the wetlands was water control structures built by Iraq from 1991 to 2002.

These marshes, dominated by *Phragmites australis*, are located on the inter-continental flyway of migratory birds and provide wintering and staging areas for waterfowl. Two-thirds of West Asia's wintering waterfowl have been reported to live in the marshes. Globally threatened wildlife species which have been recorded in the marshes include eleven species of birds, five species of mammals, two species of amphibians and reptiles, one species of fish and one species of insect. The drying of the marshes has had a devastating effect on wildlife and of course on the Marsh Arabs living in Mesopotamia (UNEP, 2001).

Since the overthrow of Saddam Hussein's dictatorship in 2003 in Iraq, there has been a concerted effort by the Iraqis and the international community to restore the marshlands (Richardson *et al.*, 2005). The restoration has often taken the form of local residents breaking dikes or removing impediments to flooding. BBC News, on 24 August 2005 reported that images show that at least 37 percent of the wetlands were restored at that time while *Frontiers in Ecology and the Environment* (Vol. 3, No. 8, Oct 2005) reported that "at least 74 species of migratory waterfowl and many endemic birds have been sighted in a survey of Iraq's marshland…" It was also reported that as many as 90,000 Marsh Arabs have returned to the wetlands already (Alwash, Director of Eden Again, personal communication, 2006).

Dr Alwash has also suggested that perhaps as much as 75 percent of the marshlands will be restored. There are still several questions that remain unanswered about whether full restoration can occur, including whether adequate water supplies exist given the competition from Turkey, Syria, and Iran, and within Iraq itself, and whether the landscape connectivity of the marshes can be re-established (Richardson and Hussain, 2006).

9.5 Wetland restoration on a global scale

There is a need for restoration of degraded wetlands throughout the world. These restoration efforts are needed for the peatlands and tundra of northern climes and for tropical mangroves that continue to be degraded and destroyed around the world. There are now more reasons than ever to protect wetlands on a global scale because of the significant carbon stored in these systems, particularly in northern peatlands and tropical swamps. Coastal wetlands provide protection from coastal storm tides and even tsunamis. Many of these restoration efforts, such as that proposed for the Mississippi River Basin, require a major effort of cooperation between different regions, states, and even countries. There is also the necessity to train people on the correct principles to carry out the restoration correctly. Perhaps the large-scale restoration projects described above will provide these principles.

Large-scale water resource problems are complex because of both the hydrologic and ecological complications but also because of their connections to society. Solving these problems is even more complex. Complex ecological challenges such as the restoration of coastal Louisiana, the Mississippee River Basin, and the Mesopotamian Marshlands demand ecological, not only technological, solutions. Overall, the major wetland restoration projects described in this chapter are much more significant than current piece-meal regulations that provide support for small wetland restoration and creation.

9.6 Conclusions

Human culture can live in an environment where water is not rapidly drained from the landscape, as witnessed by major cultures that have existed in harmony with wetlands for centuries. Our recognition of the importance of wetlands, coupled with our increased sophistication in developing techniques for recreating and restoring natural aquatic ecosystems in the landscape should give us a sense of optimism that we can improve our watery world. The following conclusions can be obtained from the experiences presented above.

- Human utilization of wetlands has a strong historical context, especially in aquatic societies.
- Likewise human history has been filled with a negative perspective on wetlands, especially in hydraulic societies.
- Wetland utilization often borders between wetland function disturbance and destruction. Defining the line between is an international challenge.
- Wetland conservation and protection can usually be aided by various approaches to wetland valuation, but economics is a two-edged sword.
- The scale (time and space) of our watershed restorations will continue to increase in the coming years.
- Estuarine, coastal, and deltaic restoration will be more connected to watershed restoration in the future.
- The twenty-first century may finally be the century when the planet's rivers, watersheds, and wetlands are repaired on a large scale employing sound practices of ecological engineering and ecosystem restoration.
- Wetlands are the essence of life, human and otherwise.

References

Australian Nature Conservation Agency (1996). *Wetlands are Important* (two-page flyer). Canberra: National Wetlands Program, ANCA, ACT, Australia.

Boesch, D. F., Shabman, L. and Antle, L. G. (2006). *A New Framework for Planning the Future of Coastal Louisiana after the Hurricanes of 2005. Working Group for Post-Hurricane Planning for the Louisiana Coast*. Cambridge, MD: University of Maryland Center for Environmental Science.

Chapman, P. and Reed, D. (2006). Advances in coastal restoration in the northern Gulf of Mexico. Special Issue of *Ecological Engineering*, **26**: 1–84.

Costanza, R., Mitsch, W. J. and Day, J. W. (2006a). Creating a sustainable and desirable New Orleans. *Ecological Engineering*, **26**: 317–320.

Costanza, R., Mitsch, W. J. and Day, J. W. (2006b). A new vision for New Orleans and the Mississippi delta: Applying ecological economics and ecological engineering. *Frontiers in Ecology and the Environment*, **4**: 465–472.

Dahl, T. E. (1990). *Wetlands Losses in the United States, 1780s to 1980s*. Washington DC: US Department of the Interior, Fish and Wildlife Service.

Dugan, P. (1993). *Wetlands in Danger*. London: Michael Beasley, Reed International Books Limited.

González-Bernáldez, F. (1992). *Los paisajes del agua: Terminología popular de los humedales*. Madrid: J. M. Reyero Publ.

Hey, D. L. and Phillipi, N. S. (1995). Flood reduction through wetland restoration: The Upper Mississippi River Basin as a case study. *Restoration Ecology*, **3**: 4–17.

Hinkle, R. L. and Mitsch, W. J. (2005). Salt marsh vegetation recovery at salt hay farm restoration sites on Delaware Bay. *Ecological Engineering*, **25**: 240–251.

Lu, J. (1995). Ecological significance and classification of Chinese wetlands. *Vegetatio*, **118**: 49–56.

Mitsch, W. J. (1993). Ecological engineering – a cooperative role with the planetary life-support systems. *Environmental Science & Technology*, **27**: 438–445.

Mitsch, W. J. (1998). Ecological engineering – the seven-year itch. *Ecological Engineering*, **10**: 119–138.

Mitsch, W. J., Day, J. W., Gilliam, J. W. *et al.* (2001). Reducing nitrogen loading to the Gulf of Mexico from the Mississippi River Basin: Strategies to counter a persistent ecological problem. *BioScience*, **51**: 373–388.

Mitsch, W. J. and Jørgensen, S. E. (2004). *Ecological Engineering and Ecosystem Restoration*. New York: John Wiley & Sons, Inc.

Mitsch, W. J. Day, J. W.Jr., Zhang, L. and Lane, R. (2005). Nitrate-nitrogen retention by wetlands in the Mississippi River Basin. *Ecological Engineering*, **24**: 267–278.

Mitsch, W. J. and Day, J. W.Jr. (2006). Restoration of wetlands in the Mississippi-Ohio-Missouri (MOM) River Basin: Experience and needed research. *Ecological Engineering*, **26**: 55–69.

Mitsch, W. J. and Gosselink, J G. (2007). *Wetlands*. New York: John Wiley & Sons.

National Wetlands Working Group (1988). *Wetlands of Canada*. Ecological Land Classification Series, no. 24. Montreal: Environment Canada, Ottawa, Ontario, and Polyscience Publications, Inc.

Peterson, S. B., Teal, J. M. and Mitsch, W. J. (eds.) (2005). Delaware Bay salt marsh restoration. *Special Issue of Ecological Engineering*, **25**: 199–314.

Rabalais, N. N., Turner, R. E., Justic, D., Dortch, Q. and Wiseman, W. J. (1999). *Characterization of Hypoxia. Topic 1 Report for the Integrated Assessment on Hypoxia in the Gulf of Mexico*. NOAA Coastal Ocean Program, Decision Analysis Series No. 15. Silver Spring: NOAA Coastal Ocean Office.

Rabalais, N. N., Turner, R. E. and Scavia, D. (2002). Beyond science into policy: Gulf of Mexico hypoxia and the Mississippi River. *BioScience*, **52**: 129–142.

Ramsar Convention (2006). *The List of Wetlands of International Importance*. (www.ramsar.org).

Revenga, C., Brunner, J., Henninger, N., Kassem, K. and Payne, R. (2000). *Pilot Analysis of Global Ecosystems: Freshwater Systems*. Washington: World Resources Institute.

Richardson C. J., Reiss, P., Hussain, N. A., Alwash, A. J. and Pool, D. J. (2005). The restoration potential of the Mesopotamian marshes of Iraq. *Science*, **307**: 1307–1311.

Richardson, C. J. and Hussain, N. A. (2006). Restoring the Garden of Eden: An ecological assessment of the marshes of Iraq. *BioScience*, **56**: 447–489.

Teal, J. M. and Weinstein, M. P. (2002). Ecological engineering, design, and construction considerations for marsh restorations in Delaware Bay, USA. *Ecological Engineering*, **18**: 607–618.

UNEP-United Nations Environmental Programme (2001). *The Mesopotamian Marshlands: Demise of an Ecosystem (UNEP/DEWA/TR.01–3 Rev.1)*. Nairobi, Kenya: UNEP.

US Army Corps of Engineers (2006). *2006 Louisiana Coastal Protection and Restoration. Preliminary Technical Report to United States Congress*. New Orleans: US Army Corps of Engineers, New Orleans District.

10

Uses, abuses and restoration of the coastal zone

FRANCISCO A. COMÍN, JORDI SERRA
AND JORGE A. HERRERA

10.1 Introduction

Global changes in the coastal zones have created key challenges for humanity, both in the short and long term. During the last decades sea level rise has been considered one of the major problems related to climate change (IPCC, 2007). However, scientists are now calling for attention to be paid to other changes which have been taking place in the coastal zone, in the near-coastal sea and in the continents. These changes have caused extremely serious disturbances on a global scale both in the structure of the physical systems constituting the coast (Carter, 1995) and in the processes (geomorphological, biogeochemical) which regulate its dynamics (Schlesinger, 1997). In fact, there is a strong relationship between structure and function in the coastal zone, as in all the ecosystems, but here it is more intensive because of the huge dynamics of the relatively narrow piece of the Earth's surface where continents and seas interact intensive and continuously. The coastal zone can be considered, together with the atmosphere, as the most intensive interphase of the Earth on all spatial and temporal scales (Comín et al., 2004).

The coastal fringe is also a major ecotone of the Earth, the zone with more intensive changes between ecosystems of the Earth on a global scale. It is also the ecotone with more variety of changes in terms of space and time, since the most intensive ranges of physical, biogeochemical and biological changes of the Earth take place in the coastal zones. Coastal zones range from rocky vertical coastal fringes to flat lowland intertidal zones in the deltaic area of the mouths of rivers. Beyond doubt, changes in the coastal zones, if we include lowland sedimentary coasts, are the most intense compared to any other ecosystem.

It is in this ecotone where most matter and energy are exchanged between the oceans and the continents. Matter and energy are exchanged in both directions, although matter is hugely transported towards the seas and energy towards the continents, both directly through waves and tides (Nixon, 1988; Carter, 1995) and

Ecological Restoration: A Global Challenge, ed. Francisco A. Comin. Published by Cambridge University Press.
© Cambridge University Press 2010.

through the atmosphere (Rodó, 2003). So, the coastal fringe is a zone of gradients of all types, including the topographic gradient, which is very important because it regulates the transport of materials (Pernetta and Milliman, 1995), as is the salinity gradient, because it highly determines the biological structure of coastal ecosystems (Remane and Schlieper, 1971; Deegan and Garrit 1997, Attrill, 2002; Sangiorgio *et al.*, 2007).

Much human activity takes place in the coastal zones. They are the permanent living place for 50 percent of the human population (Small and Nicholls, 2003), and are highly influenced by human activity both on land and sea. (Smith *et al.*, 1999; Kleppel *et al.*, 2006). It is a major platform for the transport of people and products, including food distribution all around the world. But, at the same time, it is also a dumping zone for many residues of human activities. Because of these intense activities, the coastal fringes have suffered huge disturbances to their structure in three spatial dimensions and, consequently, these disturbances have affected the temporal changes and the functions performed by these zones. Disturbances range from changes in the spatial distribution of physical components (e.g., land cover changes into massive urbanization, mining, aquaculture) to changes in the natural biological populations (e.g., sites given over to agriculture, alien species introduced in coastal lagoons and marshes). These disturbances cause a loss of ecosytem services, including buffering against sea and air storms, water quality improvement and food provision (Martínez *et al.*, 2007). These direct disturbances have reached a global scale, and there is almost no coastal area free from the direct impact caused by human uses (Plate 10.1). In addition to this, indirect impacts are caused by the recent sea level rise originated by the burning of fossil fuels, changes of biogeochemical cycles and changes in biodiversity, all provoked by the human activity that developed during the industrial era. The global scale of these direct and indirect disturbances is to be widely observed, especially after recognition of the magnitude of the accelerated ice-melting process taking place in the Arctic Ocean and Antarctica, which is changing large and new coastal areas from ice-water into land-water interphases (Serreze *et al.*, 2003; http://www.nasa.gov/vision/earth).

So, ecological restoration of the coastal zones is required on a global scale, corresponding to the magnitude of coastal-zone degradation caused by direct and indirect human activities and natural phenomena impacting on coastal zones with lowered resilience after human impacts. It is an unavoidable challenge for humans during the twenty-first century. It requires a global approach, such is the scale of the degradation origin and its impact. Human migrations including tourism, coastal-zone urbanization including industrial developments, water and soil pollution, a rise in the sea level, and many other changes are taking place on a global scale. Both the origins of and the impacts caused by these changes are heterogeneously distributed in the coastal zone, but the whole human population is implicated and so the restoration of the degraded zones requires a global approach, and probably a heterogeneous distribution of efforts both in space and time.

All around the globe, intensive urbanization of large coastal areas is taking place without considering geographic and cultural differences among coastal zones. Some recent geophysical phenomena show the intense dynamics of processes taking place in the coastal zones, or affecting them, and the risk of not implementing adaptive strategies for the use and management of the coasts. Specific phenomena, such as the Indian Ocean tsunami (24th December 2004), the Katrina hurricane in the Mississippi delta (August 2005), and Tabasco floods on the coasts of the Gulf of Mexico (October 2007), caused the loss of many human lives and goods. It is clear from these cases, and many others taking place every year, that the coastal zones are at great risk because of the degradation of natural ecosystems. The frequency of these phenomena is inversely related to their intensity. Thousands of low intensity phenomena take place every year with similar impacts but on a lower scale. These impacts are amplified by the degradation of natural ecosystems and the loss of the functions performed by the coastal zones.

With these perspectives in mind, a review of the peculiarities of the coastal zone and its ecosystems, and a general review of the uses and abuses in the coastal zone are presented in this chapter, as well as an approach to the ecological restoration of coastal zones at different spatial scales. All of this is from the perspective of the challenge of extending integrated management of coasts (Christie, 2005; McVey *et al.*, 2006), including the restoration of degraded ecosystems, to the global scale.

10.2 Types of coastal ecosystems and their restoration

10.2.1 General types of coasts

The variety of coastal types and associated environments is very large and the factors shaping them are mainly derived from geology and climate. Coastal morphology as created by coastal processes is the main criterion on which to classify or distinguish coastal types. At the same time, the state of an individual coastal system can be measured by its meterological and climate-dependent morphology. From rocky cliffs and irregular coastlines to low-lying and smooth shorelines and deltas, any coastal type can change over different timescales, from decades to geological periods (Figure 10.1).

Major coastal systems or environments have been described and studied by geomorphologists and engineers over more than two centuries, beginning with Lomonosov (1759), the first to describe marine coastal erosion, Lyell (1830) and his "uniformitarism" concept, Darwin (1842), whose Beagle cruise inspired him to study fringing reefs and coastal changes, followed by modern authors such as Johnson

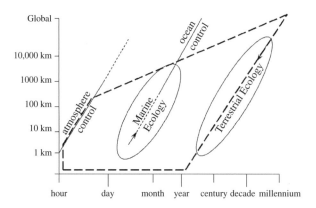

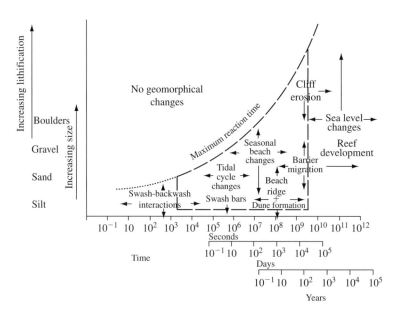

Fig. 10.1. Top: spatial and temporal scales at which most terrestrial and aquatic processes take place. The discontinuous line limits the spatial and temporal scales of interest for most coastal aquatic processes. (Modified from Steele, 1995). Bottom: spatial and temporal scales at which major physical coastal processes take place. (Modified from Carter, 1995).

(1919), who wrote about the "beach equilibrium profile" and cyclicity of coasts, Shepard (1948), who made the widest coastal classification, Davis and Duncan (2004) and many other contemporary scientists working on coastal morphodynamics.

From a general perspective, both the spatial and temporal scales must be considered in accordance with the objective of the classification and the geomorphological and dynamic characteristics of the area of interest. As an example, Sherman (1995) differentiates for coastal dunes a macroscale (greater than decades), mesoscale (month to decades) and microscale (second to months). For coastal management purposes, more specific classifications can be applied but they may require the study of past coastal changes to provide insights into rates of changes which cannot be observed directly or frequently. So, both the spatial and temporal scales of change must be considered in order to plan and implement management and restoration works in a coastal zone. The sedimentary budget and coastal morphodynamics approach is very useful (Carter, 1995). It considers such parameters as the coastal cell and drift current, its length and flow, as the basics for classifying a particular coast. The result is a classification into five types of coast, from cliffs to low coasts, and a number of intermediate cases, which are the result of the processes regulating coastal morphodynamics. Another type of coast can be considered: those areas intensively modified by anthropogenic actions. Likewise, different types of such coasts can exist close to each other in a small coastal zone. Similar approaches, using historical factors, geomorphological processes, and the type of materials constituting the coastal zone can also be useful for planning coastal management and restoration (Fairbridge, 2004).

10.2.2 Ecosystem types in the coastal zone

Geological history, weather and climatic patterns, terrestrial controls such as freshwater and sediment inputs, ocean controls such as currents and tides, and natural events such as storms and hurricanes, among many others, are variables that contribute to the mosaic of coastal ecosystems observed in the coastal zones of the Earth (Plates 10.2 and 10.3). A summary description of the major characteristics of the typical types of ecosystems in the coastal zone is presented below.

Coral reefs

Coral reefs are unique ecosystems and key components of large coastal zones. As ecosystems, they absorb much of the incoming wave energy, contribute to beach formation and provide a habitat for a rich diversity of marine life. Many coral organisms build their skeletons and shells out of calcium carbonate. When these organisms die, their skeletal remains are transported to the beach or are cemented into the framework of the reef. Most of the light-colored sand on beaches derives from coral reefs. Coral reefs are sensitive environments that require very good water quality conditions and are very efficient marine organisms that thrive in nutrient-poor environments. Corals can survive occasional short-term siltation events, however, repeated or chronic silt plumes or a single large event will kill them. Nutrient loading is also harmful to coral reefs, due to favored eutrophication that has as a symptom a reduction in transparency; however, climate change could have a major impact through temperature raising which damages the coral producing their bleaching. These and other causes of coral reef degradation, including physical disturbance by intensive recreational activities, are global phenomena. Therefore a global strategy is needed for the restoration of coral reefs on adequate spatial and temporal scales. Efficient restoration of coral reefs requires deep knowledge of the species forming the reef and the processes regulating the functioning of the ecosystem, but techniques such as transplantation have been proven to work efficiently (Edwards and Gomez, 2007).

Sandy beaches

A beach is formed by the accumulation of sediment, usually sand or gravel, which occupies a portion of the coast; a beach includes any dunes present. This ecosystem is on a dynamic equilibrium, where the net influx of sediment (or source), equals the net loss of sediment (or sink). Sources of beach sediment include skeletal material from coral reefs, transport of sand onto the shore, longshore transport, headland erosion, volcanic glass, river input, and erosion (scarping) of the coastal upland. Where there is an imbalance between sources and sinks, the beach either erodes or accretes. Coastal processes such as erosion and accretion are site-specific, season-specific, and interannual. Beach processes can vary dramatically from one end of a particular beach to the other, due to connectivity with other ecosystems such as sand dunes. Dunes are accumulations of wind-blown sand. Although some dunes are bare, most are vegetated with coastal plants, which help stabilize the dune. Vegetation traps wind-blown sand and then grows up through the accumulation of sand. This process is repeated to build larger dunes. Many dunes are host to burial sites and are legitimate environmental systems that support specific ecosystems.

Because of their recreational and economic importance (Honey and Krantz, 2007) many sand beaches have been destroyed by urbanization and strong efforts are being made now to restore them. In addition to a general approach which must consider the dynamics of this ecosystem type, specific actions for every site must be taken, based on a deep knowledge of the local characteristics of the sea and adjacent land sites, particularly for coastal sand dune ecosystems (Ley *et al.*, 2007), and also of general sea and atmospheric currents regulating the dynamics of a larger coastal zone.

Seagrasses

Seagrasses are an unusual group of marine angiosperms, all having a somewhat grass-like appearance. They grow just in soft substrates, often forming extensive underwater meadows. Despite their low species richness (fifty-seven species in the entire world), they remain of critical importance and, in many areas, account for a large proportion of inshore marine productivity. Moreover, they act as an important habitat, adding structural complexity, as well as a source of nutrition for many species in the coastal zone.

Unlike mangroves, seagrass communities are widely distributed in both tropical and temperate seas. The complex and often deep root structures, combined with the surface layer of leaves, serve to stabilize sediments, contributing to coastal protection and shoreline stability. They provide more directly tangible economic benefits through their importance to many artisanal and commercial fisheries. The seagrass habitat is vital as the feeding ground for a number of threatened species; however, due to lack of information on habitat distribution and changes, it is not possible to map with great accuracy the actual distribution of seagrass species. This ecosystem is mainly threatened by coastal eutrophication and those activities associated with coastal tourism and navigation. Its physical degradation is mainly caused by sand extraction to feed eroded beaches and the use of trawling gear over the seabed. Transplantation has become a common technique for seagrass recolonization. This may be a good start for ecosystem restoration if other requirements related to sediment, water quality, light availability and animal integration are met accordingly (Fonseca *et al.*, 1998).

Mangroves

The term mangrove is used both to describe a group of plants and the communities in which these plants occur. Mangrove plants are shrubs or trees that live in or adjacent to the intertidal zone in tropical and subtropical coastal zones. They are adapted to a regime of widely varying salinities, and to periodic and sometimes prolonged inundation. Typically, they are located along sheltered shores and in estuarine environments. Their role in fisheries has been widely recognized: many fish species use mangroves as breeding and nursery grounds. They are a source of timber and wood for fuel and play a critical role in coastal protection, and are also highly productive, typically exporting large quantities of carbon to neighboring systems, but also acting as important carbon sinks, both from their own biomass and also from the nutrients delivered from upstream ecosystems.

Among the main threats to mangrove ecosystems are the land-cover change into aquaculture systems (mostly shrimp farms) and into touristic developments, which took place at high rates all around the world in recent decades. Maintaining the integrity of well-preserved mangrove systems and restoring a basic mixture of water flows without too much stress (e.g., desiccation, salinization) for mangrove plants is essential for keeping the high number of services mangroves can provide, if well-preserved or restored with an ecological perspective. Restoring mangroves with the short-term perspective of planting trees is not an efficient strategy

(Plate 10.4). In general, favoring natural hydrological processes and spontaneous recolonization by mangrove propagules is more efficient than planting by hand, although the involvement of local people is always a step forward towards the efficient restoration of large degraded zones (Turner and Lewis, 1997; Lewis and Streever, 2000).

Marshes

Marshes, like mangroves, are transitional areas between land and water, occurring along the intertidal shore of estuarine zones and sounds, where salinity ranges from fresh in upriver marshes to ocean strength, which itself may vary in coastal zones subject to much evaporation. Salinity, frequency and extent of flooding of the marsh determine the types of plants and animals found there. Salt marshes are ecosystems typically located in temperate coastal zones, especially around estuaries, deltas, and coastal plains, and are characterized and dominated by herbaceous vegetation. These valuable wetlands have frequently been written off as wasteland, and drained in order to make them "useful." Salt marshes are some of the most productive ecosystems in the world as they convert large amounts of carbon dioxide into reduced carbon compounds, which may accumulate in the marsh soil acting as a carbon sink. The carbon compounds that the marsh plants create form the basis of the food web supporting uncountable species of aquatic life, and also contribute sand and silt to coastal accretion. Undoubtedly, their primary productivity is greater than that of agricultural crops, a sad irony when one considers that some marshes have been drained in past years for the purpose of farming. They are not only very important habitats for wildlife, but also the fisheries they support are of extremely high economic importance for coastal areas. They even play an important role in reducing coastal erosion as coastal marshes help to keep beaches from eroding away. Those restoring coastal marshes must consider the scale of the coastal zone as well as most of the impacts affecting these areas (reduction of flows, pollution of waters, plant and animal invasions, etc). Existing and newly created coastal marshes play a critical role in counteracting sea-level rises and climate change as they may present high accretion rates, particularly in sites exposed to tidal currents, and accumulate high amounts of organic matter (Day and Templet, 1990). These can be included as objectives for the restoration of degraded marshes if restoration is approached with the perspective of recreating the mosaic of habitats a coastal marsh should offer, rehabilitating water flows and linked ecological processes (Zedler, 2000a,b; O'Brien and Zedler, 2006).

Mud flats

Tidal flats are non-vegetated areas protected from wave action and composed primarily of mud transported by tidal channels. An important characteristic of the tidal flat environment is its alternating tidal cycle of submergence and exposure to the atmosphere. They are harsh, unpredictable environments. Tidal flats appear to be barren wastelands but they are, in fact, highly productive areas that support large numbers of animals, particularly macroinvertebrates and shorebirds. The extent to which tidal flats contribute to the flow of energy within an estuary is mostly determined by how frequently they are flooded with seawater and the length of time they are exposed. The components of the sediments are decided by the physical characteristics of each individual area, and subsequently, greatly influence both the biological diversity and productivity of the habitat. Tidal flats are key areas for food production for humans in many parts of the world. Disturbances in the land nearby (erosion by deforestation and forest fires and urbanization) and the sea (oil spills) have been the major factors degrading tidal flats for years. The conservation and restoration of degraded tidal flats requires a typical strategy based on a combination of scientific and technical, economic and social approaches. Providing an adequate topography, habitat conditions for living organisms, and recruitment and settlement conditions for marine organisms are basic elements of restoration programs for degraded tidal flats (Steyer *et al.*, 2003); mud flats can account for more wetland acreage than other types of coastal wetlands (Field *et al.*, 1991).

Rocky shores

Rocky shores are areas of bedrock exposed between the extreme high and extreme low tide levels on the seashore. This coastal type is located in all latitudes. Plants and animals are distributed by height according to the tolerance of the species to either exposure to air or submergence in water during the tidal cycle. This zonation can be very clearly marked and abrupt. In this environment, tidal pools often have rich communities of organisms normally associated with the lower shore or subtidal habitats. Rocky shores form as a result of marine erosion on the bedrock, due to a combination of a rising sea level and wave action, in areas where there is little sediment. Rocky shores are not usually the subject of ecosystem restoration projects, compared to other types of coastal ecosystems. However, impacts from excessive plant and animal harvesting (Castilla, 1999) and from oil spills (Hawkins *et al.*, 2002) are among the most frequent degradation causes addressed by restoration.

Coastal lagoons

Coastal lagoons are formed by the enclosing of a piece of the coastal sea between a sand barrier and the continent. They form part of the mosaic of ecosystems providing diversity as the result of the dynamism of coastal zones. A mixture of fresh and marine waters via surface and groundwater circulation, as well as precipitation and evaporation, may result in a gradient of salinity from fresh to hypersaline environments. Hypersaline environments are more common in tropical and temperate latitudes because of high evaporation. Accordingly, large changes in the biological community take place along the spatial and temporal scales. Coastal lagoons follow a geomorphological history from very open systems to closed accreted zones (Ayala-Castañares and Phleger, 1969). It is important to know the stage reached and which functions (mostly hydrogeomorphological factors) are regulating these changes to decide about how to restore degraded coastal lagoons. Most degradation of coastal lagoons is caused by disturbance of water flows, land cover changes onshore (agriculture, urbanization) and water eutrophication. Restoration of degraded coastal lagoons requires the rehabilitation of processes (water flows and geomorphological processes), and this may also require large physical changes at the coastal-zone scale. Coastal lagoons are the habitat for huge numbers of animals and provide food for many people in the world. Their restoration should be part of integrated coastal zone management plans (Coastal Zone Management Authority, 1999).

10.2.3 Forecast of changes after climate change; relative sea level rise

The climate change happening now is the most intense ever identified and will have serious effects on the coastal zones. The fourth report of the IPCC (2007) concludes that there is strong evidence that the global sea level gradually rose in the twentieth century, it is currently rising and it is also forecast to rise at a greater rate in this century. The two major processes causing the global sea level rise are thermal expansion of the oceans and land-based ice melting (IPCC, 2007). It should be noted that the origins of the present sea level rise are the impacts caused by human activities such as fossil fuel burning, land use and land cover changes and biodiversity loss.

The degradation of the coastal zone is mostly caused by direct impacts derived from human uses of it. Human activities have altered 28 percent of the coastal area of the world, considered as the area ranging from the continental shelf to a depth of 200 m, to intertidal areas and adjacent land within 100 km of the coastline (Martínez et al., 2007). In spite of this degradation, coastal ecosystems contribute 77 percent of the total economic value of global ecosystem services (Costanza et al., 1997). Three major factors are responsible for 55 percent of the variability among countries in terms of ecological, economic and social attributes of their coasts: the degree of conservation, the ecosystem service product (the total value of ecosystem services and products of the coasts), and the demographic trend (Martínez et al., 2007).

The recent effects of climate change on sea levels are clear. A sea level rise of 1.8 ± 0.5 mm/yr was observed between 1961 and 2003, as estimated from tide gauge records. There is evidence that the rate of sea level rise accelerated between the mid-nineteenth and the mid-twentieth centuries based upon tide gauge and geological data (Church and White, 2006). The global average rate of sea level rise measured by means of TOPEX (Topography Experiment for Ocean Circulation) and Poseidon satellite altimetry from 1993 to 2003 was 3.1 ± 0.7 mm/yr. Precise satellite measurements since 1993 provide unambiguous evidence of regional variability of sea level change (Cazenave and Nerem, 2004; Leuliette *et al.*, 2004). In some regions and coasts in the western Pacific Ocean, the rates of rise since 1993 are up to several times the global mean, while in other regions, on the coasts of the eastern Pacific Ocean, the sea level is falling. However, a general sea-level rise trend has been observed for most of the coasts in the world between 1955 and 2003. Observations of sea-level rise are consistent with observed changes in the ocean heat content and the cryosphere. However, other processes may have also contributed to the global sea-level rise (Bindoff *et al.*, 2007).

The projection for the future is a continuous increase in the sea level of between 20 and 50 cm during the twenty-first century, depending on the world's socioeconomic scenario and the measures adopted with respect to climate change. The report of the IPCC (2007) refers to the shortage of studies that relate the loss of coastal zones by erosion to sea-level rise. However, it is well documented that coastal erosion continues and accelerates all around the world as a result of the combination of sea-level changes and direct human intervention (Bird and Schwartz, 1985; Wolters *et al.*, 2005) and because of global warming (Zhang *et al.*, 2004).

The impacts of global changes and global warming on the coastal zones are part of a network of positive and negative feedbacks, which include other forcing functions and direct human activities. The number and proportion of high-strength cyclones has increased in all the oceans over the last thirty-five years, although the number of cyclones and days with cyclones has decreased in all the oceans, except the North Atlantic, during the last decade (Webster *et al.*, 2005). In any case, the consequences of extreme climatic events are more related to the vulnerability of the human population than to the danger inherent in the phenomena. Hurricanes Katrina (August 23–31 2005) and Rita (September 24–30 2005) impacted the north coast of the Gulf of Mexico, caused the death of 2000 people, and devastated the city of New Orleans and other areas of southern USA, while hurricanes of similar magnitude, such as Wilma (October 19–22 2005) and Gilbert (September 14–16 1988) killed 63 and 341 people, respectively, on the coasts of the Yucatan Peninsula. The damage caused by such hurricanes is related to several factors, including the number of inhabitants and land use on the coast, but is also related to exposure to risk, a lack of forecasts and poor coordination (SBC, 2006).

Other important global impacts have been foreseen. Very negative effects of global warming on coral reefs have been forecast, even in relatively conservative scenarios (CO_2 atmospheric concentration: 500 ppm by the end of the twenty-first century). There are predictions for a 50 percent decrease in the calcification of many marine organisms, particularly corals, due to acidification of the sea surface caused by a decrease in pH related to the increase of CO_2 (Hoegh-Guldberg *et al.*, 2007). Acidification progresses as atmospheric CO_2 increases. A 0.1-unit decrease of the ocean average pH has been already observed, with predictions for an additional decrease of 0.5 units during the twenty-first century. Water acidification reduces the calcification rate required for coral growth (Turley *et al.*, 2006). A water temperature that is higher than normal takes endosymbiotic algae, which provide nutrients for corals, out of the coral structure (Donner *et al.*, 2005). Furthermore, the combination of these two impacts and other related impacts (biodiversity decline, relative sea-level rise) creates a network of feedback processes which result in a decrease of corals and biodiversity in reef systems, and a general decline of the area covered by this type of ecosystem (Kleypas *et al.*, 2001; Kleypas *et al.*, 2006; Guinotte *et al.*, 2003). In fact, it has been shown that these factors have caused a 14 percent growth reduction of the Australian Great Barrier since 1990 (De'ath *et al.*, 2009). These changes will bring serious consequences for humans, as coral reef systems support fisheries and coastal recreation, nourish beach sand, and protect the coast, among other functions.

Relative sea-level fluctuations on a geological time scale are the basis for morphological changes of coasts on a large spatial scale. Long-term erosion is intensified when the sea-level fluctuations accelerate and sediment supply changes (Zhang *et al.*, 2004). Also accretion rates in coastal wetland zones could follow at similar rates if water and sediment transport is not altered or if it is favored jointly with processes leading to plant colonization and growth (Valiela, 2006). Again, indirect factors (global warming, relative sea-level rise) and forcing functions (water and sediment transport, coastal urbanization, etc) interact in a network of positive and negative feedbacks with diverse results which depend on the adaptive and mitigating capacity of the society towards this new challenge. Anthropogenic uses are responsible for up to one third of the loss of wetland and mangrove forest areas in the world (Nicholls *et al.*, 1999; Valiela *et al.*, 2001), but other perspectives should be taken into account, since a high spatial heterogeneity is linked to coastal processes. Many land coastal areas are the result of land reclamation activities carried out in the past in shallow bays and coastal zones. Many of these types of coastal sites are now at risk of disaster. A small increase in sea level would rehabilitate many coastal ecosystems and would increase the area of shallow coastal zones, which are also very productive and attractive for human uses. So, it is clear that serious global negative impacts are caused by global warming and the related sea-level rise, but the network of positive and negative effects at different spatial and temporal scales that small changes of the sea level can provoke in the coastal zones is not so clear-cut. This is crucial analysis which must be done in order to decide the type of strategy and actions to be performed, if any, concerning sea-level rises.

Nicholls (2004) simulated the impacts of a sea-level rise on coastal population and wetlands for different scenarios, which were defined taking into account economic and population growth and adoption of mechanisms by humans to adapt to sea-level rises. He concluded that human-induced impacts on coastal wetlands will be greater than the impacts of a sea-level rise, based on existing trends, that small islands are and will be particularly vulnerable to increased flooding during the twenty-first century, that vulnerability can be decreased significantly if human societies develop a high environmental consciousness and adopt adaptive mechanisms to sea-level rises, and that the impacts of sea level rises on coastal flooding could be larger in the twenty-second century, in a scenario with low environmental concern and a high increase in coastal population.

However, it is still debated what will be the net result of a sea-level rise in terms of the area covered by land-based coastal ecosystems. Much work is still to be done to elucidate the positive effects of sea-level rises and the resulting recovery of coastal ecosystems. For example, when coastal marshes and terrestrial zones are flooded, levels of nutrients will rise and this will trigger a network of processes, including enlarging fisheries. No doubt a strategy based on limiting the impacts and adapting to them is the less costly alternative to global changes, including climate change (Hekstra, 1990).

10.3 Uses of the coastal zone

10.3.1 Tangible uses of the coastal zone

Direct human benefits

The coastal zone contributes more than one third of the global value provided by the ecosystems of the Earth (Costanza *et al.*, 1997). One of its most outstanding values is its role as a buffer for physical disturbances caused by the sea and the continents. This was made clear after the human disasters caused by the tsunami that devastated the coasts of the Indian Ocean in December 2004 (150,000 people died) and by hurricanes Katrina and Rita, which destroyed most of New Orleans in August and September 2005. The magnitude of these disasters made it evident that removing and destroying mangroves, marshes and dunes also removes the capacity of the coastal zone to act as a buffer against the energy discharged by waves, tides, winds and river floods. This was well known some time ago (Lugo *et al.*, 2000) but unfortunately humans like to live in the coastal zone and that is why they have destroyed many of these ecosystems. Fifty per cent of the world's population live in the 100-km coastal fringe and about 500 million live in a 5-km wide coastal strip (Nicholls and Small, 2002; Miyazaki *et al.*, 2006). Urbanization has been one of the major causes of land-cover change in the coastal zones, contributing to the destruction of natural ecosystems. The forecast is that people living on the coastal zones will double by 2025, increasing both urbanization and the density of the human population on the coasts (UNDP, 2001; Tibbetts, 2002).

In addition, coastal ecosystems, including coral reefs, are a key source of sand to maintain beach and other lowland dynamics (Carter, 1995; Finkl, 2004). On the other hand, the coastal zone has an intense water circulation which contributes to regulate global biogeochemical cycles. In fact, the coastal zone is the major marine source of some of the greenhouse gases (nitrous oxide, carbonyl sulfide, methane), and it is, at the same time, an important carbon sink because of the extremely high productivity of many of its ecosystems. It is also well known that the most important fisheries of the world are located in the coastal zone, as 90 percent of the world harvest comes from the marine environment (FAO, 1999) and nearly two thirds of all fish harvested depend upon coastal wetlands, seagrasses and coral reefs in various stages of their life cycles (Hinrichsen, 1998). This is because of the direct fertilizing effect of the freshwater discharges from the land and the indirect fertilizing effect caused by the mixing of different water masses in the coastal zone.

The concept, dimensions and importance of the coastal zone may change depending on the type of approach adopted (Crowell *et al.*, 2007). However, it is true that human uses are increasing in coverage and intensity in areas where the sea and the continents interact directly (UN, 2005).

Integrative uses of the coastal zone and interactions among ecosystems

Human activities affecting the coastal zone are so abundant that it is quite difficult to observe sites with purely natural characteristics and processes, particularly after the expansion of tourism on a global scale. The energy and sediment unbalances created by human impacts on coastal systems lead to the accelerated erosion of the most fragile zones and the fragmentation of habitats. It is necessary to have natural littoral zones to provide sand for coastal dynamics. Disturbed coastal systems are made resilient by proximity to areas that can provide energy and materials for self-organization (Figure 10.2).

Most coastal protection and exploitation actions have been based on a conventionally engineered approach consisting of establishing hard structures in the land-water interphase and extracting near-shore sand and gravel for construction and beach nourishment purposes. These practices must be reformulated if we wish to face efficiently the impacts of climate and global changes. Use of the coastal zone should not be planned or performed by separating different zones but by following an integrated plan of acceptable uses in different parts of a given coastal zone. In the

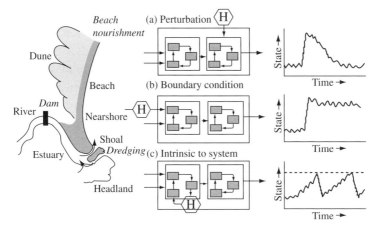

Fig. 10.2. Human action (H) and the coastal system: a) viewed as a perturbation, for instance beach nourishment; b) altering boundary conditions, e.g., river damming altering sediment input; and c) as an intrinsic component of the system, e.g., dredging a channel. Schematic response of morphology is sown on the right: returns to its origin (up), remains altered changing boundary conditions (central) and repeated negative feedback where human action is intrinsic to the system. Modified after Woodroffe (2002).

same way, management of the coastal zone will not be efficient if performed separatedly in zones which are geomorphologically, hydrologically or ecologically connected (Born and Miller, 1988; Sorensen, 1997).

10.3.2 Non tangible uses of the coastal zone

Protection against storms, microclimate, land-ocean interactions

Erosion is a net effect over time when there is less new sediment entering a coastal system than leaving it. Erosion by itself is not harmful, as long as the shape of the shoreline can change and allow for such losses, following an erosion-transport-sedimentation cycle. What happens to coastlines when the sea-level rises? Many coastal geologists believe that the sea-level is rising at a rate that is dependent on the coastline. A seemingly small amount of change in water level can cause extensive changes to the shoreline over time as gently sloping and flatter beaches are inundated. The most dramatic effects are seen during extreme storms, when the tidal range swings to great extremes. Increases in levels of bay waters as well as ocean waters will narrow barrier islands from both sides, causing flooding and erosion. Storms are often the most obvious precipitant of severe erosion, but much of the sand that does not travel too far off shore in the storm will return to the beach with favorable weather.

Dunes are an excellent natural flood barrier and the best sand reservoir in coastal zones; at the same time, salty marshes absorb wave energy during storm surges, thereby counteracting erosion. Theory and a considerable body of field studies indicate that a wide beach provides significant benefits in the form of storm damage reduction. During storms with elevated water levels and high waves, a wide beach performs as an effective energy absorber with the wave energy dissipated across the surf zone and the beach rather than impacting on the upland structures. So it is clear what will happen if part of the beach area is occupied or urbanized with rigid structures constructed on the beach, wetland and dune habitats.

Typical seasonal changes to the beach profile are the response to annual climatic wave energy changes. The same adaptation process occurs in shorter or longer period changes, from microclimate and high frequency conditions up to longer periods as climatic, glacio-eustatic and geological periods. Transgressive and regressive periods (processes giving a geological record from low to high sea level and vice versa) are well documented for the last thousand million years (Figure 10.1). Coastal processes during changes in sea level on a lower scale have also been recognized and recorded as transgressive facies (transgressive sand beach, barrier island, etc). The main difference between late Quaternary and older geological processes and the present trend is a changed "land-ocean" interaction situation. During the past century and before, man used and influenced the coastal zone structure and dynamics by constructing dams, harbors, sea walls, groynes etc., extracting granulates from coastal, river and marine areas and decreasing the amounts of continental water and sediment reaching the coastal zone. Coastal works can alter the movements of local sediments and can also affect the regular supply of solid material from rivers. The result is a highly coastal system in which sediments are not balanced and a general tendency towards erosion. For this reason, the first recommendation from a comparative research exercise on the erosion of the European coasts was to "restore the sediment balance and to provide space for coastal processes" in order to protect biodiversity and mitigate the impact of climate change on coastal morphodynamics (Doody *et al.*, 2004).

Present and forecast sea level rises and the increase in the frequency and force of coastal storms resulting from climate change are likely to mean that new strategies will be required for protection of coasts and adaptation to coastal erosion and flooding over a long time.

The impacts of conventional coastal defences vary widely according to the techniques used, their specific design and the characteristics of the local environment. Some generalized impacts of physical coastal defences include disturbances of natural ecosystem processes and biotope structures of beaches, dunes, cliffs and the nearshore zone by partial or complete modification of landforms and sedimentary processes, both on a local and regional scale; continuous loss of characteristic marine-influenced ecosystems, such as episodically flooded coastal and riverine wetlands, coastal wet-forests and active cliffs; an increasing threat to the biodiversity of coastal areas; and visual deterioration. The impacts of hard engineering are usually more severe than soft engineering. Hard engineering generally results in long-term changes in coastal morphology, particularly erosion, alongside protected areas. It also often leads to a reduction in the width of the shoreline as low-lying backshore areas are reclaimed behind defences. This leads to a decrease in the size of shore habitats, a phenomenon called "coastal squeeze." Soft engineering is generally a more environmentally friendly approach which works towards providing a dynamic equilibrium at the coastline whereby erosion and flooding are kept to a minimum. It also generally requires more space to be used, thereby reducing coastal squeeze.

Defensive structures that are designed to reduce wave energy at the shoreline often result in the build-up of sediment in the wave shadow of the structure. In some situations this may lead to the covering of existing shoreline ecosystems or other changes. Hard defence techniques that reduce upper shore and cliff erosion also disrupt longshore sediment movements which often leads to the accelerated erosion of adjacent shorelines (Carter, 1995; MESSINA, 2006).

Biogeochemical processes in the coastal zone

The coastal zone is the area where a great amount of matter circulates towards the sea, but is mostly deposited on the coast, where it undergoes intense biogeochemical transformations that are very important in the global cycles of elements (Smith and Hollibaugh, 1993; Smith *et al.*, 2000; Siefert, 2004). Every year, around 11–27 petagrams (10^{15} g) of sediments are transported by rivers and reach the sea (Beusen *et al.*, 2005). These sediments contain particles associated with small amounts of carbon, nitrogen and phosphorus of the order of a few hundreds of teragrams (10^{12} g) per year (Schlesinger, 1997; Crossland *et al.*, 2005). These materials contribute to the creation of coastal habitats (i.e., deltas and marshes) and the fertility of nearby marine ecosystems (Valiela *et al.*, 2000), including the rich coastal fisheries where most fishing around the world takes place, even though one third of fisheries production comes from aquaculture (FAO, 2007).

However, the coastal zone also exports a part of the organic matter produced to the atmosphere, as several respiratory processes (methanogenesis, nitrate and sulfate reduction) operate in the anaerobic sediments that are common in the organically rich coastal aquatic ecosystems and these facilitate the production of reduced forms of carbon, nitrogen and sulfur (Capone and Kiene, 1988). But a considerable reduction in the quantity of sediment transported from the continents to the coastal zones took place during the twentieth century because of the damming of rivers, and a further reduction is expected in the twenty-first century. These unbalanced conditions for the roles of the coastal zone in the biogeochemical cycles are part of the global changes taking place all around the world. Restoration at watershed scale is a challenge which will ultimately contribute to restore and manage the coastal zone and to involve people at large scale for this purpose (Kier, 1995).

10.4 Abuses in the coastal zone

10.4.1 Geomorphological approach: impacts which modify the coastline

Human actions in the coastal zone are generally persistent and resulting coastal landforms are progressively less free to change geomorphologically as they would in a natural environment. Coastal defence strategies have been heavily based on a static engineered response with solid coastal structures protecting the built environment; often the problem is to relocate downdrift to another part of the coast.

As has been mentioned above, the impact of this type of intervention, which modifies the sediment balance, is a new scenario for the geomorphological evolution of the coastal zone. Before decisions about land use (e.g., urban planning) of parts of the coastal zone are taken, projections of geomorphological changes related to natural dynamics and climate change on large spatial and long temporal scales must be considered in order to adopt some guidelines for general use, which may require cooperation between different municipalities and administrations.

Coastal erosion management should move away from piecemeal solutions to a planned approach based upon accountability principles, by optimizing investment costs, increasing social acceptability and keeping options open for the future. Preserving and restoring coastal resilience and a favourable sediment status must be the main objectives of coastal management plans (Doody *et al.,* 2004). It is therefore important to recognize the value of coastal sedimentary habitats (notably tidal flats, wetlands and dunes) in providing natural defence. Management plans must identify coastal segments or cells which will be protected and preserved from erosion and other changes. Four generic policy options identified in the UK are now classic points to be considered for the management of coastal zones: hold the line, move seaward, manage realignment, no active intervention (or "do nothing").

The first option can be applied where present infrastructures of ecological value require the maintenance of current conditions. Nevertheless, the first and second options can be changed if they become non-sustainable. A few examples of these types of changes are provided by some European cases, for example the lowland coasts in the Netherlands, where forecasts for sea level rise and climate conditions make it impossible to continue with plans to increase the dyking of the coastal zone, which was the strategy adopted after the large floods of 1954. As the European Environment Agency recognizes, "a further increase in flood risk is expected due to the rise in sea level, climate change, and further economic and social development. Technical solutions no longer form the sole answer to this increase" (Voight *et al.,* 2004). The cases of Venice, New Orleans, the Ebro Delta and many others make it evident that lowland coastal zones are under extreme risk of morphological changes in the short term (Dunbar *et al.*, 1992; Valiela, 2006).

It is impossible to solve these problems just with conventional engineering methods. It is necessary to adopt approaches that consider the re-establishment of a sedimentary balance according to dynamic processes and that incorporate decisions to adapt the socioeconomic system to the dynamic coastline.

10.4.2 *Ecosystem approach: impacts causing changes at ecosystem scale*

The coastal zone accounts for only 20 percent of all the land area in the world. Of this, 19 percent of all lands within 100 km of the coast (excluding the Antarctica and water bodies) are classified as altered, and 10 percent are semi-altered, involving a mosaic of natural and altered vegetation (Plate 10.1). A large percentage of this least modified category includes many uninhabited areas in northern latitudes. Many important coastal habitats, such as mangroves, wetlands, seagrasses, and coral reefs, are disappearing at a fast rate. Anywhere from 5 to 80 percent of the original mangrove area in various countries, where such data are available, is believed to have been lost. Extensive losses have occurred particularly in the last fifty years.

The coastal ecosystems do not just cover and protect the continents around the world, they change continuously, provide the habitat for a wide biodiversity, are highly productive ecosystems, and all are scenically beautiful. However, they are, unfortunately, under much pressure and severely threatened.

The main threats to the ecosystems of the coastal zone include coastline and channel modifications, chemicals and toxin dumping, channelization and groundwater withdrawal, and altered trophic webs. Many of the above-mentioned issues constitute direct and indirect threats to the natural assets of the coastal ecosystems. Prioritization of threats for the coastal zone is not a simple procedure, as the cumulative impacts from multiple uses of coastal areas might not be adequately addressed if issues are simplified. Furthermore, there are future and potentially dramatic threats not yet seen because of a delay to the effects, interaction among the various threats and new threats appearing over time.

Some of the major concerns for the coastal ecosystems are:

– *Eutrophication:* Declining water quality, algal blooms, death of sea-grass meadows and decreased fisheries are some of the impacts of coastal eutrophication. Increasingly, catchment management is beginning to include the identification of catchment–coast–ocean implications for coastal management and water quality. The Mississipi watershed–Gulf of Mexico approach shows the strong relationship between the massive use of nutrients in the watershed, the loss of wetlands, and the extensive dead zone in the Gulf of Mexico, just in front of the Mississippi Delta. It is

a clear example of watershed scale disturbance affecting negatively both the continental and the marine watersheds (Mitsch and Day, 2006; Rabalais *et al.*, 2007).
– *Urbanization:* A variety of recreational activities is undertaken in the coastal zone,
   many of which perform important functions for community health, nutrition,
   income and general welfare. The impacts and issues associated with the recreational use of the coastline, in particular habitat destruction and modification,
   through huge tourism developments, is the main concern, especially in developing tropical areas. Nature-based tourism with low impact facilities could be a
   strategy to minimize impacts on the coastal ecosystems.
– *Sea and land-based aquaculture*: Coastal aquaculture is a growing industry and
   has great potential for economic investment and returns. However, issues associated with aquaculture include coastal impacts from land-cover change and
   offshore impacts on water quality, introduction of pest organisms, disturbance
   of trophic relationships, attraction of predators and changes of sediments and
   benthic communities near and under the grow-out facilities.
– *Marine pollutants and oil spills*: Toxic contaminants find their way into the coastal
   and marine environment through numerous routes, entering from catchment
   washout of herbicides and pesticides, polluted or contaminated stormwater out-
   falls, inappropriate coastal sewage systems, anti-fouling paints, oil spills and
   bilge pumping, rubbish blown out from land (including plastic bags from popular
   coastal fishing spots and campsites) and illegal dumping at sea (Moore *et al.*,
   2001; Moore, 2003).

10.4.3 *Specific processes: dumping, eutrophication, urbanization, pollution, exotic species, fragmentation*

Illegal and accidental dumping of pollutants in the coastal zone is an impact that
causes global concern. Most dumping takes place during or following the direct
use of the products. However, the biggest impact is caused by spills from vessels near
the coast (NRC, 2003). Part of the oil evaporates and is decomposed by photoche-
mical oxidation and microorganisms (up to 50 percent), and the rest remains dispersed in the sea waters and deposited in the sea sediments and coasts (Valiela, 2006).

Volunteer participation in oil removal from impacted coastal areas is increasing over
time and contributes significantly to mitigate damage to both coastal habitats and
species (SPREEX, 2007). However, the impacts still affect negatively the whole trophic
network and persist for a long time (Samiullah, 1985). It is interesting to note that the
number of oil spills from oil vessels decreased from between fourteen and twenty-nine
per year during the 1970s to between three and five per year during the present decade, in
spite of the increased world oil demand (ITOPF, 2007). Levels of concern about oil
spills, particularly from vessels, are rising as marine oil transport increases both in

quantity of vessels and number of routes. A decrease in huge oil spills from vessels could be achieved rapidly if rules for transport (navigation routes, company management, the professional skills of employees) and materials used (double-hulled tanks, security systems) are immediately adopted and followed (Rodríguez, 2007).

Obviously, a global reduction in oil produced, which can be achieved by energy saving and diversification programs, is a commendable objective to be promoted for countries with huge oil demand rates (Oliver, 2006). The accomplishment of these objectives will lead to a decrease in the risk of oil vessel accidents, as well as in the impacts of oil dumping in the coastal zone. Oil-contaminated sites can be treated using a variety of techniques and, most probably, a combinination of different techniques is required to optimize resultss (NRC, 2003). Nevertheless, the restoration of ecosystems impacted by oil spills is very challenging, as the coastal zone is very complex, and the effects of oil spillages may last for decades (Irvine *et al.*, 2006).

Eutrophication has been a major subject of research for much of the second half of the last century (Nixon, 1995; Cloern, 2001). Major factors causing coastal eutrophication are increased discharges of nutrients, mostly nitrogen, from anthropogenic uses, either in the coastal zone or inland but transported into the coastal zone by surface and groundwater flows (ESA, 2000; Valiela and Bowen, 2002; Aranda *et al.*, 2006). Any efficient corrective action to combat coastal eutrophication must be related to the reduction of the nitrogen discharge from continental watersheds, which mostly comes down to the reduction of discharges derived from the use of fertilizers and urban wastewater disposal. This could become a global strategy, which can be implemented regionally or country by country, aimed at adopting new practices of agriculture and building more efficient wastewater treatment plants, but also at recovering the nitrogen recycling capacity of ecosystems, which is related to their good ecological functioning (Cloern, 2001).

Land use and cover changes are taking place at a very high rate on the coasts. Urbanization of the coastal zone is a major global change which provides housing and business locations to human populations but also impacts negatively on the coastal zone. In some regions of the world, urban areas occupy long coastal zones that bridge frontiers, when observed on a large scale. Not surprisingly, these types of coastal changes contribute to facilitate the occurrence of epidemic sicknesses, such as cholera (Colwell and Huk, 1994).

On a smaller spatial scale, habitat destruction and reduction of the natural ecosystem area, together with disturbance of surface and groundwater flows is a common occurrence (Kempe, 1988; Doody, 2004; Ekercin, 2007). In addition, discharges of continental wastewater and solid wastes from urban and intensively fertilized agricultural zones are the reasons for the loss of water and sediment quality and changes in the structure of their trophic webs, including a decrease in fisheries and recreational values (Smith *et al.*, 2003; Rabalais *et al.*, 2007).

In addition to urbanization, mangrove and salt marsh destruction in favor of aquaculture and other commercial uses (e.g., oil extraction, salt production) are also major land-use changes responsible for global habitat loss and degradation in coastal zones, although positive initiatives indicate that much more could be done to buffer and integrate these uses on a large scale (SEAFDEC-AQD, 2006). Urbanization is one of the major impacts causing degradation of the coastal zone on a global scale. Large areas of southern Europe and other developed countries, including some in tropical latitudes, have been transformed into urban zones (CO-DBP(99)11–Committee, 1999). Up to 60 percent of the world's historical mangrove resources has been lost due to population pressures and clearing for agriculture, urban development, logging and fuel, while in contrast, shrimp farming is said to be responsible for just 5 percent of total mangrove loss (Global Aquaculture Alliance, 2001). However, in some parts of the world this pattern can be different from the global one, e.g., in Ecuador 27 percent of the mangrove loss was due to shrimp farming between 1965 and 1995; a timid mangrove recovery, 5 percent of the area lost, is taking place. Sound estimates by FAO (Wilkie and Fortuna, 2003) conclude that the world has lost 5 million hectares of mangroves over the last twenty years, or 25 percent of the mangroves existing in 1980. The current mangrove area worldwide has now fallen below 15 million hectares, from 19.8 million hectares in 1980. It must be also remembered that the net benefit provided by a hectare of mangrove in a healthy state, which is a benefit to all humanity, is five times higher than the economic benefit of a hectare of mangrove transformed into a shrimp farm, which is a benefit only to a reduced group of people (Millennium Ecosystem Assessment, 2005). Whatever the numbers are, there is a general agreement on the necessity and utility of establishing regulations for their conservation and use on a regional scale (Coastal Aquaculture Authority Government of India, Ministry of Agriculture), and on a global scale (McVey *et al.*, 2006).

In addition to coastal water eutrophication, and habitat destruction and fragmentation, invasions by alien species and pollution cause a decrease in water and landscape quality for recreational purposes, as well as health risks for the people using the coastal zone (Jiang *et al.*, 2001; Williamson *et al.*, 2005; ISSG, 2007). In any case, the demand for implementation of integrated sustainable plans for the use of the coastal zone which cannot be ignored (Yunis, 2006).

10.5 Restoration of the coastal zone: what is being done and what is necessary

10.5.1 *Approach on large spatial scales and long time scales*

In order to avoid degradation, coastal systems must be always in a dynamic equilibrium with climatic and hydrologic processes. Morphological stability, the shape and position of individual elements or whole systems depend on interplaying factors associated with the coastal environment along the spatial and time dimensions. As an example, an undisturbed system of coastal foredunes is in a complex state of dynamic equilibrium resulting from the effects of wind, waves, tides and vegetation. The dune system retreats under wave attacks, when sand is lost to form off-shore bars, and advances during calmer weather, acting at the same time as a sand reservoir and a flexible coastal barrier against sea erosion and flooding. This cyclical process can be continuous in time, over seasonal, annual or longer periods, until a change in the sediment feedback occurs, caused, for example, by coastal works, river diversion or damming. Changes in sea level, as predicted for the present century, will act progressively on foredune systems, landwards migration and adaptation. This includes spatial changes of plant communities (Huiskes, 1990). In the cases of a negative net accretion rate or a lack of space for new colonizations or a displacement of habitat, the system will disappear.

Mediterranean deltas are a good example of unbalanced systems, but at the same time, they show a high capacity of adaptation by retrogression, if there is space available. A typical example is the Ebro Delta, which faces the impact of the present null river sediment transport by dam effect (Guillén and Palanques, 1992; Serra 1997). It shows a rapid change in the perimetrical sandy coast and dune system, but at the same time, the overall surface is maintained because the sediments at the front of the delta are moved to shaoling zones at the terminal spit and bays (Rodríguez *et al.*, 2008). In the long term, this change will be followed by a perimetrical growth in height coupled by a sea-level rise and wave overwash effects, but the delta plain will remain at the same or at a lower (subsidence) level. To prevent this expected change, the Spanish Coastal Administration is planning a managed realignment project following recommendations from teams of scientists and technicians. This project's objectives are to take a 500 m coastal zone (mainly rice fields) and implement a new 100 m wide beach and a 400 m dune-wetland system (Galofré and Montoya, 2007).

The resilience or ability of the coast to accommodate changes induced by any anthropic activity and specifically by present meteorological extreme events and sea-level rise, is the main factor contributing to the maintenance of the functions of the coastal system over a long time scale. Key features for resilience are a net positive sediment budget and sufficient space for coastal processes to operate (Doody *et al.*, 2004).

Other coastal systems, for example extensive salt marshes in temperate estuaries, and mangroves in tropical areas, have decreased in size through erosion over the last few decades. In this context, the link between elevation, sea level and sedimentation represents just one aspect of overall transitional system morphodynamics. The case of Blyth estuary, Suffolk (eastern England), indicates a slightly different pattern of salt marsh adjustment than that which has been reported for other estuaries in the region. Vertical sedimentation everywhere within the salt marsh outpaces the post-1964 sea-level rise, whilst the total area of marsh has actually increased by around 14 percent since 1887. However, whilst sediment supply is clearly adequate to maintain the elevations of established marshes and to drive sedimentary infilling of the tidal frame, overall marsh morphodynamic behavior − which includes temporal and spatial variation in the foci of accretion and erosion − appears to be governed more by gross changes in estuary morphology and process regime, rather than by the regional sea-level rise (French and Burningham, 2003).

Sustainable development of the coastal zone and the conservation of its habitats require respect for the natural functioning of the coastal system in all senses, and hence its natural resilience to dynamic and frequent changes such as erosion and flooding, and to infrequent changes such as sea-level rises. Coastal resilience or the inherent ability of the coast to adapt its shape to changes is highly dependent on coastal geodiversity and processes; a deep understanding of the system is the best foundation for formulating good restoration and protection policies.

10.5.2 *Approach on the ecosystem scale*

Coastal ecosystems are some of the world's most biologically important areas. Rich in species and genetic diversity, coastal ecosystems are essential in storing and cycling nutrients, protecting shorelines, and filtering pollutants (Bryant *et al.*, 1995). Many different species rely on the habitats provided by these ecosystems for foraging, nesting, spawning, and nursery grounds. These areas are also very much appreciated by humans. Recreation and industry use coastal areas heavily; half of the world's coastal ecosystems are threatened by human activity (Bryant *et al.*, 1995).

Restoring coastal ecosystems has proven to be complicated: these ecosystems provide extra challenges due to their dynamic nature. Human activity has affected coastal ecosystems more heavily than any other marine habitat. Altered around the world by development, pollution, and habitat destruction, ecosystems have been restored to rebuild populations of plants and animals, and to support communities. The goal of coastal restoration projects should be primarily to recover the processes working as forcing functions and, associated with this, improving the structure of communities.

Simulation models of hydrology, nutrient biogeochemistry, and vegetation dynamics have been developed to forecast patterns in mangroves in the Florida Coastal Everglades. These models provide an insight into the way mangroves respond to specific restoration alternatives, and test the causal mechanisms of system degradation. Arguably, these models can also assist in selecting performance measures for monitoring programs that evaluate project effectiveness. This selection process, in turn, improves model development and calibration for forecasting mangrove response to restoration alternatives (Twilley and Rivera, 2005).

Over time, restoration goals have trended from habitat for a specific species to restoration of historical ecosystem states. This trend is especially seen in projects on the coastal wetlands for waterfowl; however, restoration objectives must focus on the functioning of entire ecosystems. The ability of these projects to succeed in the long-term has been increased by the amount of monitoring that is now being done. Monitoring pre- and post- restoration has had an impact on how successful and adaptive management is in restorations. Data collection has made the need for mid-course corrections more easily recognized and implemented, increasing the long-term success of a restoration action.

The importance of the dynamic nature of water in coastal systems is exhibited by efforts to restore coastal wetlands, such as mangrove areas. However, the failure of a connecting channel may cause inadequate tidal flushing and result in lower functioning of the site (Plate 10.4). Stagnant water and anaerobic conditions caused by this failure do not provide the environmental conditions needed for mangrove forest establishment.

Another approach to restoring coastal ecosystems is through adequate hydrology management, which involves rehabilitating water-level fluctuations in harmony with meteorological and hydrologic forcing functions to favor the establishment of the characteristic flora and fauna, reaching a minimum functioning of major ecological processes. In the case of wetlands such as mangroves and salt marshes the rehabilitation of natural water flows across the landscape should be enough for recolonization by key species which could drive on the recovery of the ecosystem.

One of the most important challenges in the restoration of the natural dynamics of coastal areas and the movement of water demand for coastal wetlands is that restorations cannot be isolated – the connectivity of coastal ecosystems to areas surrounding them is so critical for their functioning. Cooperation with surrounding landowners, residents, and others responsible for uses of the water (aquaculture, fisheries, tourism) is essential to meet the objectives of ecological restoration. Because of this, restoring coastal ecosystems is not possible if natural hydrology cannot be returned.

10.5.3 Other approaches: urban zones

Restoration required in urbanized coastal zones ranges from site-level actions such as removal of buildings constructed in wetland and dune zones to integrated management and restoration of large coastal zones (e.g., eastern and southern coastal zones of Spain, Indonesian coasts damaged by the tsunami in 2004). Whatever its scale, the ecological restoration of degraded, damaged or destroyed coastal zones requires the re-establishment of functional water flows because these constitute the major forcing function to recreate geomorphological and biogeochemical processes (Day *et al.*, 2007). However, wind and deep sea currents are also major factors which must be taken into account for coastal zone restoration. It is clear that artificial structures on the seashore may determine the present and future evolution of the coastline. So, restoration of coastal urbanized zones must take into account the potential effects of and limitations from the characteristics of the sea watershed. The same is valid for the restoration of coastal zones degraded by impacts from the continental watershed. To be efficient, the capacity of restoration at watershed scale should be considered, as proposed for some eroded coastal zones, and decreased sediment transport and pollution from the continental watershed (Costanza *et al.*, 2006a,b).

Obviously an approach that integrates scientific, technical, social, and economic perspectives is the most adequate for the planning and implementation of the restoration of urbanized coastal zones. Large urban areas in the coastal zone have received great economic investment and are valuable social settlements. Their restoration should be based on the demand of security for humans and their interests, but it should be linked to the roles that natural coastal ecosystems play in the preservation of human life and their contribution to global environmental cycles. This principle may be applied to the restoration of geographically defined coastal sites or zones, but also to generic plans related to political and social aspects (Pethick, 2002).

The improvement of degraded and damaged urban coastal zones is usually linked to the construction of artificial structures. A new approach, based on the principles of ecological restoration, should be adopted, particularly in those areas where the artificial solution has proven to be inefficient both for human protection and fund investment, as was proposed for New Orleans after the Katrina hurricane (Costanza *et al.*, 2006b). Restoration of specific coastal urban sites may be easily linked to their economic revitalisation (Cunningham, 2002). For this purpose, the major principles to be incorporated as part of urban planning in urban coastal zones include: leaving freely evolving natural areas (reefs, beaches, dunes, vegetated habitats) at the sea front which can adapt to major forcing functions (sea storms, currents and waves, wind), leaving enough free natural ecosystems (estuarine zones) to drive energy between the sea and the continent, both surface and groundwater, adapting urban building to buffer surface run-off and groundwater level changes and establishing ecological corridors to improve renewal of biodiversity in the urban coastal zone (Purcell *et al.*, 2002). Compensating mechanisms should be established in order to obtain space for naturally functioning coastal zones. These could include giving free land for urban developments in safe areas in the same urban zone or close to it (Aronson and Vallejo, 2006).

In contrast, the restoration of large coastal zones or generic habitats in a coast will require criteria and methods to recover geomorphological and biogeochemical processes that regulate seashore dynamics and those of coastal ecosystems, which themselves regulate further development of biological structures (Konisky *et al.*, 2006). Also, re-establishing the connectivity between functional ecosystems is a major objective in fragmented coastal landscapes (Marzluff and Swing, 2001). Otherwise, restoration action may have very limited success or even negative consequences for the safe future of the urban coastal zone.

10.6 Conclusions

As has been stated above, a large part of the coastal zone is still in good condition; most of it is located in coasts that are least urbanized. But many coastal areas have been transformed and degraded by direct and indirect human activities. Huge numbers of people are moving to coastal cities and tourist resorts, which occupy large areas of the coastal zone and affect negatively a wide belt of shallow coastal seas, reefs, beaches and dunes, estuaries, marshes and mangroves, as well as continental watersheds that feed the coastal zones with freshwater. Also, forecasts of the effects of sea storms and sea-level changes on the human systems established in the coastal zones are negative under any socioeconomic scenario considered on a global scale for this century and beyond (Nicholls, 2004). However, there seems to be a gap in the literature neglecting the positive effects of marine reflooding of large coastal areas, which would promote the recovery of lost habitats, the export of nutrients to the sea and an increase in coastal fisheries, together with the re-equilibration of damaged coastlines. Further insights are also required if economic valuation is used as a tool for decision support systems, since the relationships between ecosystem services and habitat size are not linear, as is the case for biological and ecological mediated processes (Barbier *et al.*, 2008).

Balancing coastal developments that are at risk of natural disaster, e.g., New Orleans, tourist resorts on the Mediterranean and Mexican coasts (Costanza *et al.*, 2006a; Herrera *et al.*, 2006), with enhanced nearshore estuarine and marine ecosystems (i.e., net ecosystem improvement) should be a top priority for coastal researchers and planners for this century (Thom *et al.*, 2005). Furthermore, integrated coastal zone management (Cican-Sain, 1998) should be incorporated as a basic step for coastal development. While this approach is a good start compared to previous practices based on occupation and explotaition of the coastal zone, it is not yet clear that the philosophy of governance by partnership with civil society incorporated by the integrated coastal zone management (ICZM) approach (COI, 2001) is enough, as abuses of the coastal zone usually proceed faster than rational recovery plans are put into practice. To practice integrated management efficiently, it is best to combine the mitigation of causes and impacts and adaptation to present and future coastal changes (including sea level rise) (Hekstra, 1990). Also, the mitigation of current impacts will be required to preserve human lives and goods. However, the preservation of natural coastal ecosystems and restoration of degraded ones is essential to develop sustainable human societies in the coastal zone and to contribute to world sustainability (Baird, 2005).

The ecological restoration of the coastal zone is a global issue as the impacts of sea level changes, human migration to the coastal zone and use of coastal areas for human transport and recreation are increasing on a global scale. So, restoration of degraded and damaged coastal zones should follow common rules worldwide. These must include the recovery of geomorphological and biogeochemical processes on large spatial scales and long time scales and the exchange of water, sediment and organisms between the sea and coastal ecosystems, as regulated by natural forcing functions on small spatial scales and short time scales. Removal of artificial structures that hinder these processes and facilitation of the biological activity rather than continued building in coastal areas is a basic rule for the detailed specification of restoration plans and actions.

With respect to people's involvement, volunteer participation can be very positive for restoration of damaged coastal zones (Walters, 2006) as has been shown after recent oil spills. However, establishing guidelines based on sound research and practice is essential for the effective restoration of coastal zones on all scales (Peterson *et al.*, 2003; NOAA-Univ. New Hampshire, 2004; Novoa, 2006).

In any case, what is required is a worldwide move from the present general cultural perception of the coastal zone as a place to be occupied with artificial structures and to be used intensively towards a perception of it as the zone of free and dynamic interactions between the sea and the continents which provide benefits for human society. This change can be accomplished by establishing general principles for integrated coastal zone management and by performing ecological restoration projects as a global network of actions that will demonstrate the benefits of developing human societies in the coastal zone on the basis of conserving, adapting and using the dynamics of the coastal ecosystems as an essential part of our life support system.

It has been accepted that the best strategy for the management of the coastal zone is one that integrates physical and biological aspects, including human interests. This can be implemented by integrating the huge amount of knowledge already developed on the functioning of the coastal zone and the experience of its management, particularly now that after developing integrated coastal zone management strategies have been developed (Clark, 1995). The challenge now is to develop integrated approaches which incorporate ecological restoration as part of a global strategy. It is the homage we must pay to those who have worked and work now to improve the natural functioning of the coastal zone: it is the duty we owe to future generations.

References

Aranda, N., Herrera, J. A. and Comín, F. A. (2006). Nutrient water quality in a tropical coastal zone with groundwater discharge, NW Yucatan, Mexico. *Estuarine, Coastal and Shelf Science*, **68**: 445–454.

Aronson, J. and Vallejo, R. (2006). Challenges for the practice of ecological restoration. In *Restoration Ecology*, ed. J. Van Andel and J. Aronson. Malden: Blackwell Publishing, pp. 234–247.

Attrill, M. (2002). A testable linear model for diversity trends in estuaries. *Journal of Animal Ecology*, **71**: 262–269.

Ayala-Castañares, A. and Phleger, F. B. (1969). *Coastal lagoons. Memoirs of the International Symposium on coastal lagoons*. Mexico DF: Universidad Nacional Autónoma de México.

Baird, R. C. (2005). On sustainability, estuaries and ecosystem restoration. The art of the practical. *Restoration Ecology*, **13**: 54–158.

Barbier, E. B., Koch, E. W., Silliman, B. R. *et al.* (2008). Coastal ecosystem based management with nonlinear ecological functions and values. *Science*, **319**: 321–323.

Beusen, A. H. W., Dekkers, A. L. M., Bowman, A. F., Ludwig, W. and Harrison, J. (2005). Estimation of global river transport of sediments and associated particulate C, N and P. *Global Biogeochemical Cycles*, **19** (4): 1–19.

Bindoff, N. L., Willebrand, J., Artale, V. *et al.* (2007). Observations: Oceanic climate change and sea level. In *IPCC 2007. Climate Change.The Physical Science Basis. Contribution of Working Group I to the Fourth Assessment. Report of the Intergovernmental Panel on Climate Change*, ed. S. Solomon, D. Qin, M. Manning, *et al*. Cambridge, UK and New York: Cambridge University Press, pp. 385–432.

Bird, E. C. F and Schwartz, M. L. (1985). *The World's Coastline*. New York: Van Nostrand Reinhold.

Born, S. and Miller, A. (1988). Assessing networked coastal zone management programmes. *Coastal Management*, **16**: 229–243.

Bryant, D., Rodenburg, E., Cox, T. and Nielsen, D. (1995). *Coastlines at Risk: An Index of Potential Development-Related Threats to Coastal Ecosystems. WRI Indicator Brief*. Washington DC: World Resources Institute.

Capone, D. G. and Kiene R. P. (1988). Comparison of microbial dynamics in marine and freshwater sediments: Contrasts in anaerobic carbon metabolism. *Limnology and Oceanography*, **33** (4–2): 725–749.

Carter, R. W. (1995). *Coastal Environments: An Introduction to the Physical, Ecological and Cultural Systems of Coastlines*. San Diego: Academic Press.

Castilla, J. A. (1999). Coastal marine communities: Trends and perspectives from human-exclusion experiments. *Trends in Ecology and Evolution*, **14**: 280–283.

Cazenave, A. and Nerem, R. S. (2004). Present-day sea level change: Observations and causes. *Reviews of Geophysics*, **42** (3): 1–20.

Christie, P. (2005). Is integrated coastal management sustainable? *Ocean & Coastal Management*, **48**: 208–232.

Church, J. A. and White, N. J. (2006). A 20th century acceleration in global sea-level rise. *Geophysical Research Letters*, **33**: L01602–L01604.

Cican-Sain, B. (1998). *Integrated Coastal and Ocean Management: Concepts and Practices*. Washington DC: Island Press.

Clark, J. R. (1995). *Coastal Zone Management Handbook*. Boca Raton, FL: CRC Press.

Cloern, J. E. (2001). Our evolving conceptual model of the coastal eutrophication problem. *Marine Ecology Progress Series*, **211**: 223–253.

Coastal Zone Management Authority (1999). *The National Integrated Coastal Zone Management Strategy for Belize*. Belize City: Coastal Zone Management Authority and Institute. (www.coastalzonebelize.org).

COI-Comisión Oceanográfica Internacional (2001). *Instrumentos y personas para una gestión integrada de la zona costera*. UNESCO, Manuales y guías 42. Paris: COI-UNESCO.

CO-DBP(99)11–Committee, CE for the Activities of the Council of Europe in the Field of Biological and Landscape Diversity. (1999). *European code of conduct for coastal zones*. Strasbourg: Secretariat General Direction of Environment and Local Authorities, Council of Europe (http:\strategy\co-dbp\docs\cdp11E.99).

Colwell, R. R. and Huk, A. (1994): Environmental reservoir of *Vibrio cholerae. Annals of New York Academy of Sciences*, **15**: 44–54.

Comín, F. A., Menéndez, M. and Herrera, J. A. (2004). Spatial and temporal scales for monitoring coastal aquatic ecosystems. *Aquatic Conservation: Freshwater and Marine Ecosystems*, **14** (S1): 5–17.

Costanza, R., d'Arge, R., de Groot, R. *et al.* (1997). The value of the world's ecosystem services and natural capital. *Nature*, **387**: 253–259.

Costanza, R., Mitsch, W. J. and Day, J. W. (2006a). A new vision for New Orleans and the Mississippi delta: Applying ecological economics and ecological engineering. *Frontiers in Ecology and the Environment*, **4**: 465–472.

Costanza, R., Mitsch, W. J. and Day, J. W. (2006b). Creating a sustainable and desirable New Orleans. *Ecological Engineering*, **26**: 317–320.

Crossland, C. J., Kremer, H. H., Lindeboom, H. J., Marshall Crossland, J. I. and Le Tissier, M. D. A. (eds.) (2005). *Coastal Fluxes in the Anthropocene. The Land-Ocean Interactions in the Coastal Zone Project of the International Geosphere-Biosphere Programme*. Global Change–The IGBP Series. New York: Springer.

Crowell, M., Edelman, S., Coulton, K. and McAfee, S. (2007). How many people live in coastal areas? *Journal of Coastal Research*, **23** (5): iii–vi.

Cunningham, S. (2002). *The Restoration Economy*. San Francisco: Berrett-Koehler Publishers.

Darwin, C. (1842). *The Structure and Distribution of Coral Reefs*. London: Smith Elder & Co.

Davis, R. A., and Duncan, M. F. (2004). *Beaches and Coasts*. Malden: Blackwell Publishing.

Day, J. W., and Templet, P. H. (1990). Consequences of sea level rise: Implications from the Mississippi Delta. In *Expected effects of climatic change on marine coastal ecosystems*, ed. J. J. Beukema, W. J. Wolff, and J. J. W. M. Brouns. Developments in Hydrobiology, 57. Dordrecht: Kluwer Academic Publishers, pp. 155–166.

Day J. W. Jr., Boesch, D. F., Clairain, E. J. *et al.* (2007). Restoration of the Mississippi Delta: Lessons from Hurricanes Katrina and Rita. *Science*, **315**: 1679–1684.

De'ath, G., Lough, J. M. and Fabricius, K. E. (2009). Declining coral calcification on the Great Barrier Reef. *Science*, **323**: 116–119.

Deegan, L. A. and Garrit, R. H. (1997). Evidence for spatial variability in estuarine food webs. *Marine Ecology Progress Series*, **147**: 31–47.

Donner, S. D., Skirving, W. J., Little, C. M., Oppenheimer, M. and Hoegh-Guldberg, O. (2005). Global assessment of coral bleaching and required rates of adaptation under climate change. *Global Change Biology*, **11**: 2251–2265.

Doody, P. (2004). "Coastal squeeze," a historical perspective. *Journal of Coastal Conservation*, **10**: 129–138.

Doody, P., Ferreira, M., Lombardo, S., Lucius, I., Misdorp, R., Niesing, H., Salman, A. and Smallegange, M. (2004). *Living with Coastal Erosion in Europe. Sediment and Space for Sustainability. Results from the Eurosion Study.* Brussels: Directorate General Environment of the European Commission.

Dunbar, J. B., Britsch, L. D. and Kemp, E. B. III. (1992). *Land Loss Rates: Louisiana Coastal Plain.* Technical Report GL-90–2. New Orleans: US Army Corps of Engineers.

Edwards, J. and Gomez, E. D. (2007). *Reef Restoration: Concepts and Guidelines.* Launceston, Tasmania: AT&M-Sprinta.

Ekercin, S. (2007). Coastline change assessment at the Aegean Sea coasts in Turkey using multitemporal Landsat imagery. *Journal of Coastal Research*, **23** (3): 691–698.

ESA-Ecological Society of America (2000). Nutrient pollution of coastal rivers, bays and seas. *Issues in Ecology*, **7**: pp. 1–17.

Fairbridge, R. W. (2004). Classification of coasts. *Journal of Coastal Research*, **20** (1): 155–165.

FAO-Food and Agriculture Organization of the United Nations (1999). *The State of World Fisheries and Aquaculture 1998.* Rome: FAO.

FAO-Food and Agriculture Organization of the United Nations (2007). *The State of World Fisheries and Aquaculture 2006.* Rome: FAO.

Field, D. W., Reyer, A. J., Genovese, P. V. and Shearer, B. D. (1991). *Coastal Wetlands of the United States.* Washington DC: National Oceanic and Atmospheric Administration.

Finkl, C. W. (2004). Coastal classification: Systematic approaches to consider in the development of a comprehensive scheme. *Journal of Coastal Research*, **20**: 166–213.

Fonseca, M. S., Kenworthy, W. J. and Thayer, G. W. (1998). *Guidelines for the Conservation and Restoration of Seagrasses in the United States and Adjacent Waters.* NOAA Coastal Ocean Program Decision Analysis Series No. 12. Silver Spring, MD: NOAA Coastal Ocean Office.

French, J. R. and Burningham, H. (2003). Tidal marsh sedimentation versus sea-level rise: A Southeast England estuarine perspective. In *Proceedings of Coastal Sediments 03*, Sheraton Sand Key, Clearwater, Florida. May 18–23 2003. American Association of Civil Engineers, pp. 1–14.

Galofré, J. and Montoya, F. J. (2007). Master plan for the sustainability of the Spanish coast: Tarragona and Castellon case study. *Proceedings of Coastal Zone 07*. Portland, Oregon, July 22–27 2007.

Global Aquaculture Alliance. (2001). *La crevette d'aquaculture et la forêt de mangrove.* St. Louis: Global Aquaculture Alliance (www.gaalliance.org).

Guillén, J. and Palanques, A. (1992). Sediment dynamics and hydrodynamics in the lower course of a river regulated by dams: The Ebro river. *Sedimentology*, **39**: 567–579.

Guinotte, J. M., Buddemeier, R. W. and Kleypas, J. A. (2003). Future coral reef habitat marginality: Temporal and spatial effects of climate change in the Pacific basin. *Coral Reefs*, **22**: 551–558.

Hawkins S. J., Allen, J. R., Ross, P. M. and Genner, M. J. (2002). Marine coastal ecosystems. In *Handbook of Ecological Restoration*, ed. M. R. Perrow and A. J. Davy. Cambridge University Press, pp. 121–148.

Hekstra, G. P. (1990). Man's impact on atmosphere and climate: A global threat? Strategies to combat global warming. *Developments in Hydrobiology*, **57**: 5–22.

Herrera, J. A., Comín, F. A. and Capurro Filograsso, L. (2006). Landscape, land-use, and management in the coastal zone of Yucatan Peninsula. In *The Gulf of Mexico: Ecosystem-Based Management*, ed. J. W. Day and A. Yáñez-Arancibia. The Gulf of Mexico: Its Origins, Waters, Biota, Economic & Human Impacts Series. Ecosystem

Based Management, Bulletin 89 of Institute for Gulf of Mexico Studies TAMUCC. Dallas: Texas A & M University Press, pp. 238–287.

Hinrichsen, D. (1998). *Coastal Waters of the World Strategies*. Washington DC: Island Press.

Hoegh-Guldberg, O., Mumby, P. J., Hooten, A. J. *et al.* (2007). Coral reefs under rapid climate change and ocean acidification. *Science*, **318**: 1737–1742.

Honey, M. and Krantz, D. (2007). *Marine Program*. Washington DC: World Wildlife Fund.

Huiskes, A. H. L. (1990). Possible effects of sea level changes on salt marsh vegetation. *Developments in Hydrobiology*, **57**: 167–172.

IPCC-Intergovernmental Panel on Climate Change (2007). The Physical Science Basis. Contribution of Working Group I to the *Fourth Assessment Report of the Intergovernmental Panel on Climate Change*, ed. S. Solomon, D. Qin, M. Manning *et al.* Cambridge (UK) and New York: Cambridge University Press.

Irvine, G. V., Mann, D. H. and Short, J. W. (2006). Persistence of 10-year old Exxon Valdez oil on Gulf of Alaska beaches: The importance of boulder-armoring. *Marine Pollution Bulletin*, **2**: 1011–1022.

ISSG-Invasive Species Specialist Group (2007). *Global Invasive Species Database*. Gland: Species Survival Commission of the IUCN-International Union for Conservation of Nature (www.issg.org/database).

ITOPF-The International Tanker Owners Pollutions Federation Ltd. (2007). *Oil tanker spill statistics 2007*. London: ITOPF.

Jiang, S. C., Nobel, R. and Chu, W. (2001). Human adenoviruses and coliphage in urban runoff-impacted coastal waters of southern California. *Applied and Environmental Microbiology*, **67**: 179–184.

Johnson, D. W. (1919). *Shore process and shoreline development*. Upper Sadle River: Prentice Hall.

Kempe, S. (1988). Estuaries. Their natural and anthropogenic changes. In *Scales and Global Change*, ed. T. Rosswall, R. G. Woodmansee and P. G. Risser. Paris: ICSU-Scientific Committee On Problems of the Environment-SCOPE 35, pp 189–217.

Kier, W. M. (1995). *Watershed Restoration. A Guide for Citizen Involvement in California*. Sausalito, CA: US Department of Commerce, National Oceanic & Atmospheric Administration.

Kleppel G. S., Richard De Voe, M. and Rawson, M. V. (2006). *Changing Land Use Patterns in the Coastal Zone: Managing Environmental Quality in Rapidly Developing Regions*. New York: Springer.

Kleypas, J. A., Buddemeier, R. W. and Gattuso, J. P. (2001). The future of coral reefs in an age of global change. *International Journal of Earth Sciences*, **90**: 426–437.

Kleypas, J. A., Feely, F. A., Fabry, V. J., Langdon, C., Sabine, C. L. and Robbins, L. L. (2006). *Impacts of Ocean Acidification on Coral Reefs and Other Marine Calcifiers: A Guide for Future Research*. NSF-NOAA-USGS, Report of Workshops. St. Petersburg, FL: U.S. Geological Survey (USGS) Integrated Science Center.

Konisky, R. A., Burdick, D. M., Dionne, M. and Neckles, H. A. (2006). A regional assessment of salt marsh restoration and monitoring in the Gulf of Maine. *Restoration Ecology*, **14**: 516–525.

Ley, C., Gallego, J. B. and Vidal, C. (2007). *Manual de restauración de dunas costeras*. Madrid: Ministerio de Medio Ambiente–Dirección General de Costas.

Leuliette, E. W., Nerem, R. S. and Mitchum, G. T. (2004). Calibration of TOPEX/Poseidon and Jason altimeter data to construct a continuous record of mean sea level change. *Marine Geodesy*, **27** (1–2): 79–94.

Lewis, R. R., and Streever, W. (2000). Restoration of mangrove habitat. Tech Note ERDC TN-WRP-VN-RS-3.2. Vicksburg, MS: US Army, Corps of Engineers.

Lomonosov, M. V. (1759). Meditationes de via Navis in Mari Certius Determinanda Praelectaein Publico Conventu Academiae Scientiarum Imperalis Petropolitanae die VIII Mai, A.C. 1759 Auctore Michaele Lomonosow Consilario Academico. In *Memoirs in Physics, Astronomy and Instrument*. M. V. Lomonosov (1955). Leningrad: USSR Academy of Sciences.

Lugo, A. E., Rogers, C. S. and Nixon, S. W. (2000). Hurricanes, coral reefs and rainforests: Resistance, ruin and recovery in the Caribbean. *Ambio*, **29**: 106–114.

Lyell, C. (1830). *Principles of Geology*. London: John Murray.

Martínez, M. L., Intralawan, A., Vázquez, G., Pérez-Maqueo, O., Sutton, P. and Landgrave, R. (2007). The coasts of our world: Ecological, economic and social importance. *Ecological Economics*, **63** (2–3): 254–272.

Marzluff, J. M. and Swing, K. (2001). Restoration of fragmented landscapes for the conservation of birds: A general framework and specific recommendations for urbanizing landscapes. *Restoration Ecology*, **9**: 280–292.

McVey, J., Lee, C. S. and O'Bryen, P. J. (2006). *Aquaculture and Ecosystems: An Integrated Coastal and Ocean Management Approach*. Baton Rouge, FL: The World Aquaculture Society.

MESSINA-Managing European Shoreline and Sharing Information on Near-shore Areas (2006). *Engineering the Shoreline: Introducing Environmentally Friendly Engineering Techniques throughout the World*. EU-Interreg IIIC.

Millennium Ecosystem Assessment (2005). *Ecosystems and Human Well-being: Synthesis*. Washington DC: Island Press.

Mitsch, W. J., and Day, J. W. Jr. (2006). Restoration of wetlands in the Mississippi-Ohio-Missouri (MOM) River Basin: Experience and needed research. *Ecological Engineering*, **26**: 55–69.

Miyazaki, N., Adeel, Z. and Ohwada, K. (eds.) (2006). *Mankind and the Oceans*. Tokyo: United Nations University Press.

Moore, C. J., Moore, S. L., Leecaster, M. K. and Weisberg, S. B. (2001). A comparison of plastic and plankton in the North Pacific central gyre. *Marine Pollution Bulletin*, **42**: 1297–1300.

Moore, C. (2003). Trashed across the Pacific Ocean, plastics, plastics, everywhere. *Natural History Magazine*, **112** (9) (www.naturalhistorymag.com).

Nicholls, R. J. (2004). Coastal flooding and wetland loss in the 21st century: Changes under the SRES climate and socio-economic scenarios. *Global Environmental Change*, **14**: 69–86.

Nicholls, R. J., Hoozemans, F. M. J. and Marchand, M. (1999). Increasing flood risk and wetland losses due to global sea level rise: Regional and global analysis. *Global Environental Change*, **9**: 69–87.

Nicholls, R. J. and Small, C. (2002). Improved estimates of coastal population and exposure to hazards released. *EOS Transactions*, **83** (2): 301–305.

Nixon, S. W. (1988). Physical energy inputs and the comparative ecology of lake and marine ecosystems. *Limnology and Oceanography*, **33** (4, Part 2): 1005–1025.

Nixon, S. W. (1995). Coastal marine eutrophication. A definition, social causes and future concerns. *Ophelia*, **41**: 109–219.

NOAA-National Oceanic and Atmospheric Administration and University of New Hampshire (2004). *Research & Development Priorities: Oil Spill Workshop*. Durham, NH: University of New Hampshire.

Novoa, X. (2006). The Prestige and international cooperation. Response onshore and site restoration. Madrid: CEPRECO-Centro para la Prevención y Lucha contra la Contaminación Marítima y del Litoral. Ministerio de Medio Ambiente. (www.cedre.fr/uk/publication/jourinfo06/).

NRC-National Research Council (2003). *Oil in the Sea III: Inputs, Fates and Effects*. Washington DC: National Academies Press.

O'Brien. E. and Zedler, J. B. (2006). Accelerating the restoration of vegetation in a southern California salt marsh. *Wetlands Ecology and Management*, **14**: 269–286.

Oliver, H. (2006). Reducing China's thirst for foreign oil: Moving towards a less oil-dependent road transport system. *China Environment Series*, **8**: 41–58.

Pernetta, J. C. and Milliman, J. D. (1995). *Land-Ocean Interactions in the Coastal Zone Implementation Plan*. IGBP Report No. 33. Stockholm Sweden: International Geosphere-Biosphere Programme.

Peterson, C. H., Rice, S. D., Short, J. W. *et al.* (2003). Long-term ecosystem response to the ExxonValdez oil spill. *Science*, **302**: 2082–2086.

Pethick, J. (2002). Estuarine and tidal wetland restoration in the United Kingdom: Policy versus practice. *Restoration Ecology*, **10**: 431–437.

Picó, M. J., Dolz, J., Pallisé, J., Prat, N. and Serra, J. (2005). Un delta amb futur: les mesures urgents per salvar l'ecosistema ric en biodiversitat. /NAT/, abril 2005: 24–30.

Purcell A. H., Friedich, C. and Resh, V. H. (2002). An assessment of a small urban stream restoration project in Northern California. *Restoration Ecology*, **10**: 685–694.

Rabalais, N. N., Turner, R. E., Sen Gupta, B. K., Platonand, E. and Parsons, M. L. (2007). Sediments tell the history of eutrophication and hypoxia in the northern Gulf of Mexico. *Ecological Applications*, **17** (5) Supplement 2007: 129–143.

Remane, A. and Schlieper, C. (1971) *Biology of Brackish Water*. Stuttgart: E. Schweiserbart'sche Verlagsbuchhandlung.

Rodó, X. (2003). Interactions between the tropics and extratropics. In *Global Climate. Current research and uncertainties in the Climate System*, ed. X. Rodó and F. A, Comín. Berlin: Springer, pp. 237–274.

Rodríguez, C. (2007). Los buques petroleros. *Revista Naval*, Año VII, Octubre. (www.revistanaval.com/contidos.php?ID=petroleros_i)

Rodríguez, I., Sánchez, M. J., Montoya, I., Gómez, D., Martín, T. and Serra, J. (2008). Internal structure of the aeolian sand dunes of El Fangar spit, Ebro Delta (Tarragona, Spain). *Geomorphology*.

Samiullah, Y. (1985). Biological effects of marine oil pollution. *Oil and Petrochemical Pollution*, **2**: 235–264.

Sangiorgio, F., Basset A., Pinna, M. *et al.* (2007). Ecosystem processes: Litter breakdown patterns in Mediterranean and Black Sea transitional waters. *Transitional Waters Bulletin*, **1**(3): 51–55. (http://siba2.unile.it/ese/twb).

SBC-Select Bipartisan Committee (2006). *A failure of initiative. Final Report of the Select Bipartisan Committee (Tom Davis, Chairman) to Investigate the Preparation for and Response to Hurricane Katrina*. Washington DC: US Government Printing Office.

Schlesinger, W. H. (1997). *Biogeochemistry. An Analyisis of Global Change*. San Diego, CA: Academic Press.

SEAFDEC AQD (2006). Mangrove friendly shrimp cultura: An ASEAN-SEAFDEC project (www.mangroveweb.seafdec.org.ph).

Serra, J. (1997). El sistema sedimentario del delta del Ebro. *Revista de Obras Públicas*, **3368**: 15–22.

Serreze, M. C., Maslanik, J. A., Scambos, T. A. *et al.* (2003). A record minimum arctic sea ice extent and area in 2002. *Geophysical Research Letters*, **30**: 10 (1)–10(4).

Shepard, F. P. (1948). *Submarine Geology*. New York: Harper & Row.

Sherman, D. J. (1995). Problems of scale in the modelling and interpretation of coastal dunes. *Marine Geology*, **124**: 339–349.

Siefert, R. L. (2004). The role of coastal zones in global biogeochemical cycles. *EOS*, **85** (45): 470–471.

Small, C. and Nicholls R. J. (2003). A global analysis of human settlement in coastal zones. *Journal of Coastal Research*, **19** (3): 584–599.

Smith, S. V. and Hollibaugh, J. T. (1993). Coastal metabolism and the oceanic organic carbon balance. *Reviews of Geophysics*, **31**: 75–89.

Smith, S. V., Dupra, V., Marshall Crossland, J. I. and Crossland, C. J. (2000). *Estuarine Systems of the South China Sea Region: Carbon, Nitrogen and Phosphorus fluxes*. LOICZ Reports & Studies No. 14. Texel, The Netherlands: LOICZ-Land-Ocean Interactions in the Coastal Zone.

Smith, S. V., Swaney, D. P., Talaue-McManus, L. *et al.* (2003). Humans, hydrology, and the distribution of inorganic nutrient loading to the ocean, *BioScience*, **53** (3): 235–245.

Smith, V. H., Tilman, G. D. and Nekola, J. C. (1999). Eutrophication: Impacts of excess nutrient inputs on freshwater, marine, and terrestrial ecosystems. *Environmental Pollution*, **100** (1–3): 179–196.

Sorensen, J. (1997). National and international efforts at integrated coastal management: definitions, achievement, and lessons. *Coastal Management* **25**: 3–41.

SPREEX-Spill Response Experience Coordination Action (2007). *Spill Response Experience and Research for Preparedness. Final Conclusions*. Brussels: European Commision, FP6 Priority Sustainable Surface Transport.

Steele, J. H. (1995). Can ecological concepts span the land and ocean domains? In *Ecological Time Series*, ed. T. M. Powell, and J. H. Steele. New York: Chapman & Hall, pp. 5–19.

Steyer, G. D., Sasser, C. E., Visser, J. M., Swenson, E. M., Nyman, J. A. and Raynie, R. C. (2003). A proposed coast-wide reference monitoring system for evaluating wetland restoration trajectories in Louisiana. *Environmental Monitoring and Assessment*, **81**: 107–117.

Thom, R. M., Williams, G. W. and Diefenderfer, H. L. (2005). Balancing the need to develop coastal areas with the desire for an ecologically functioning coastal environment. Is net ecosystem improvement possible? *Restoration Ecology*, **13**: 193–203.

Tibbetts, J. (2002). Coastal cities: Living on the edge. *Environmental Health Perspectives*, **110** (11): 674–81.

Turley, C., Blackford, J., Widdicombe, S., Lowe, D. and Nightingale, P. (2006). Reviewing the impact of increased atmospheric CO_2 on oceanic pH and the marine ecosystem. In *Avoiding Dangerous Climate Change*, ed. H. J. Schellnhuber, W. Cramer, N. Nakićenović, T. M. L. Wigley and G. Yohe. Cambridge University Press, pp: 65–70.

Turner, R. E. and Lewis, R. R. (1997). Hydrologic restoration of coastal wetlands. *Wetlands Ecology and Management*, **4** (2): 65–72.

Twilley, R. R. and Rivera-Monroy, V. H. (2005). Developing performance measures of mangrove wetlands using simulation models of hydrology, nutrient biogeochemistry, and community dynamics. *Journal of Coastal Research*, **40**: 79–93.

UNDP-United Nations Development Programme (2001). *World Resources 2000–2001– People and Ecosystems: The Fraying Web of Life*. Washington DC: World Resources Institute.

UN-United Nations (2005). *Human Development Report. International Cooperation at a Crossroads: Aid, Trade and Security in an unequal world*. New York: United Nations.

Valiela, I. (2006). *Global Coastal Change*. Malden: Blackwell Publishing.

Valiela I. and Bowen, J. L. (2002). Nitrogen sources to watersheds and estuaries: Role of land cover mosaics and losses within watersheds. *Environmental Pollution*, **118**: 239–248.

Valiela, I., Bowen, J. L. and York, J. K. (2001). Mangrove forests: One of the world's most threatened major tropical environments. *BioScience*, **51**: 807–815.

Valiela, I., Cole, M. L., McClelland, J., Hauxwell, J., Cebrián, J. and Joyce S. B. (2000). Role of salt marshes as part of coastal landscapes. In *Concepts and Controversies in Tidalmarsh Ecology*, ed. M. P. Weinstein, and D. A. Kreeger. Dordrecht: Kluwer Academic Publishers, pp: 23–38.

Voight, T., van Minnen, J., Erhard, M., Zebisch, M., Viner, D. and Koelemeifer, R. (2004). *Indicators of Europe's Changing Climate*. Copenhagen: European Environment Agency.

Walters, B. B. (2006). Local mangrove planting in the Philippines: Are fisherfolk and fishpond owners effective restorationists? *Restoration Ecology*, **8**: 237–246.

Webster, P. J., Holland, G. J., Cury, J. A. and Chang, H. R. (2005). Changes in tropical cyclone number, duration and intensity in a warming environment. *Science*, **309**: 1844–1846.

Williamson, K. E., Radosevich, M. and Wommack, K. E. (2005). Abundance and diversity of viruses in six Delaware soils. *Applied and Environmental Microbiology*, **71** (6): 3119–3125.

Wilkie, M. L., and Fortuna, S. (2003). *Status and Trends in Mangrove Area Extent Worldwide*. Forest Resources Assessment Working Paper No. 63. Rome: Forest Resources Division-FAO (www.fao.org/docrep).

Wolters, M., Bakker, J. P., Bertness, M. D., Jefferies, R. L. and Moller, I. (2005). Salt marsh erosion and restoration in south-east England: Squeezing the evidence requires realignment. *Journal of Applied Ecology*, **42**: 844–851.

Woodroffe, C. D. (2002). *Coast: Form, Process and Evolution*. Cambridge University Press.

Yunis, E. (2006). *Tourism in SIDS: A Key Factor for Economic, Social and Environmental Sustainability*. London: World Tourism Organization.

Zhang, K., Douglas, B. and Leatherman, S. P. (2004). Global warming and coastal erosion. *Climatic Change*, **64**: 41–58.

Zedler, J. B. (2000a). *Handbook for Restoring Tidal Wetlands*. Boca Raton, FL: CRC Press.

Zedler, J. B. (2000b). Progress in wetland restoration ecology. *Trends in Ecology and Evolution*, **15**: 402–407.

11

Spatial ecological solutions to mesh nature and people: Boston suburb, Barcelona region and urban regions worldwide

RICHARD T. T. FORMAN

11.1 Introduction

The overriding vision and objective, "to mesh nature and people on the land so they both thrive long-term," every week becomes more compelling, as society increasingly recognizes the massive human-caused degradation of the land and its powerful consequences for our future. The old ideas that the earth is infinite or that environmental problems are minor or that technology will take care of the problem are rapidly disappearing, as the public sees satellite and GIS images around the globe and senses that sprawl, water shortages, loss of natural areas, and much more are engulfing our neighborhoods.

What central paradigm or model is most promising as a solution? Water resource planning? Economics? Sustainable agriculture? Nature conservation? Social institution building? Other? Each can include nature and people, but none is likely to effectively address both dimensions. The land mosaic model which emerged over the past two decades from landscape ecology (Forman, 1995; Hobbs, 1995; Bennett, 2001; Gutzwiller, 2002; Opdam *et al.*, 2002) is the only promising paradigm I have been able to find. In essence, its principles point to spatial patterns and changes in the land that support ecology as well as culture on landscape or region scales. It does not promise a great increase in, for example, wealthy communities, farmland or grizzly bear habitat, but rather an effective spatial mesh of land uses to sustain both nature and people.

Thus the purpose of this chapter is to outline a series of analyses and, especially, solutions that further the stated overriding objective. All are for urban regions, more specifically the rapidly changing area of conflicting land uses surrounding the nearly continuous metropolitan built area. Paradoxically and unfortunately, this is the area largely avoided by both ecologists and urban planners. Yet the changing ring around the city offers unusual promise for implementing a mesh of nature and people that could have benefits for both the city and natural or rural areas outside the urban region. Three spatial scales provide complementary insights:

Ecological Restoration: A Global Challenge, ed. Francisco A. Comin. Published by Cambridge University Press.
© Cambridge University Press 2010.

1. Two solutions are outlined for a local unit (town, municipality, etc.) in the Boston region (USA): (a) a relatively novel regional approach; and (b) application of landscape ecology principles for the local unit as a whole.
2. An array of solutions is outlined for an urban region, Barcelona (Spain), organized around five themes: (a) nature; (b) food production; (c) water; (d) development; and (e) repetitive small places.
3. Finally, certain patterns of both ecological and human importance are identified by comparing urban regions, large to small, worldwide. Insights from two analyses are presented: (a) distance from city center to nearest water body; and (b) number and length of greenspace wedges projecting into a metropolitan area.

11.2 Local unit in the Boston region

The town of Concord, Massachusetts has 16,000 residents in an approximately 8-km diameter area located in the outer suburbs 35 km from downtown Boston, a medium-large US city. The town is widely known for an eighteenth century skirmish that effectively started the American Revolution, and as the nineteenth century literary center of America, where Emerson, Thoreau, Alcott and Hawthorne lived and wrote (Gross, 1976; Maynard, 2004). Today the town is about 40 percent built area, 45 percent forest and 15 percent farmland (Forman *et al.*, 2004). Approximately 35 percent of the land is protected greenspace (federal, state, town, non-profit organization, and private land), and a quarter of the land remains uncommitted and potentially available for built development.

Land planning is important in the town, in the present case to set priorities for open space (greenspace) protection. A *1992 Open Space Plan* (Ferguson *et al.*, 1993) applied certain landscape ecology principles, effectively creating a land mosaic model for town planning, which was enhanced in the most recent *Open Space and Recreation Plan 2004* (Forman *et al.*, 2004). In reaching this solution, the planning groups mainly worked from aerial photographs rather than from property-ownership boundary maps, in order to understand land-use patterns, plus major water, wildlife, and human movement routes. Also, to see the big picture, i.e., the fundamental landscape-wide or town-wide pattern, small "special sites" (of geological, ecological, historic, recreation, scenic, etc. value) were initially put aside, and later superimposed on maps.

Three types of large intact patches or areas, and also three types of major corridors, were obvious from the initial landscape-wide analysis. Six large natural areas, five large agricultural areas, and eight large built areas were prominent in the aerial photographs. Ten major water-protection corridors along streams and rivers (which were also wildlife corridors), four additional wildlife corridors, and many human corridors of various types were present. The network of large natural and agricultural areas connected by major water and wildlife corridors emerged as the key land-use

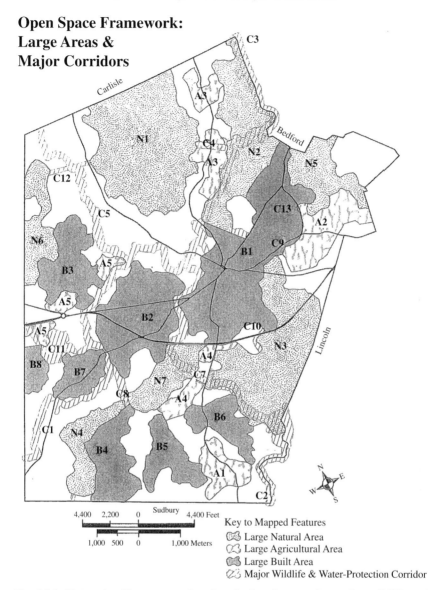

Open Space Framework: Large Areas & Major Corridors

Fig. 11.1. Network of large natural and agricultural areas plus major wildlife and water-protection corridors in a town. N = large natural area; A = large agricultural area; B = large built area; C = major wildlife and/or water-protection corridor. Concord, Massachusetts, USA. See Forman *et al.* (2004). Produced by M. Barrett; courtesy of Town of Concord.

planning model for the town. In the 2004 plan, one large natural area was added and adjustments were made to the water and wildlife corridors. Decision-makers and residents alike appreciated the resulting open space network (Figure 11.1), a big-picture solution that addressed both nature and human issues, from recreation and flood control to farmland, clean water, and biodiversity protection. As a test of this land-mosaic model approach, from 1992 to 2004, a remarkable eight percent of the entire land surface was protected for greenspace.

In recognition that many or most issues facing a local unit involve areas beyond its boundaries, the 2004 plan introduced a new regional approach to stimulate collaboration and action with surrounding communities. First, a list of about sixty ways in which Concord interacts with other towns was developed, including the particular towns involved in each issue. Both positive and negative interactions were included, such as supporting a high school or nursing home with another town, sharing a river corridor with several towns, addressing chronic problems of an airport with three towns, trying to collaborate on problems of a highway with several towns, and so forth. The list of interacting towns showed that Concord frequently interacts with fifteen other towns. Three towns with somewhat fewer interactions were added to these to make a relatively smooth oval around the central town. The result was a nineteen-town functional region based on interactions (Figure 11.2). This is effectively a "locality-centered region" (or town-centered region or simply Concord's region). The diameter is about five times that of the central town and encompasses approximately two towns beyond the central town in every direction.

Ten major land uses, including large natural areas, main roads and residential land, significantly affect Concord's greenspace, recreation, wildlife and water resources. Several key dimensions of these were mapped, including regional open space, rare species, and existing and proposed walking/bicycling trails. The regional open-space map shows a strong linkage of the central town with towns to the south and north, as well as the scarcity of greenspace adjacent to the town to the southwest (Figure 11.2). These striking patterns, and indeed those for the other major dimensions mapped, would not be evident without the regional approach. Planning for the town's greenspace thus is noticeably enhanced by meshing it with this regional pattern.

Regional collaboration acts as a catalyst, bringing an additional goal and benefit. Familiar problems with the usual regional approaches include: (1) a state or other extensive area divided into several planning regions, leaving most local units near a region boundary and with significant interactions outside their designated region; and (2) (good-intentioned) regional planning authorities with limited budgets, finite lives, and perceived as threats to local decision-making. In contrast, the local region approach here: (1) retains "home rule" (e.g., local government, taxes, budgets, voters, and land-use decisions); (2) highlights interactions with surrounding local units; and (3) recognizes the state (Massachusetts) requirement for local planning at regular intervals as a promising way to mesh the open space plans of surrounding towns.

To help accomplish the last item, the 2004 plan with regional (GIS) maps and data sources was provided to all towns in the region, with encouragement for usage and collaboration. Since each town is expected to do its own open space and recreation

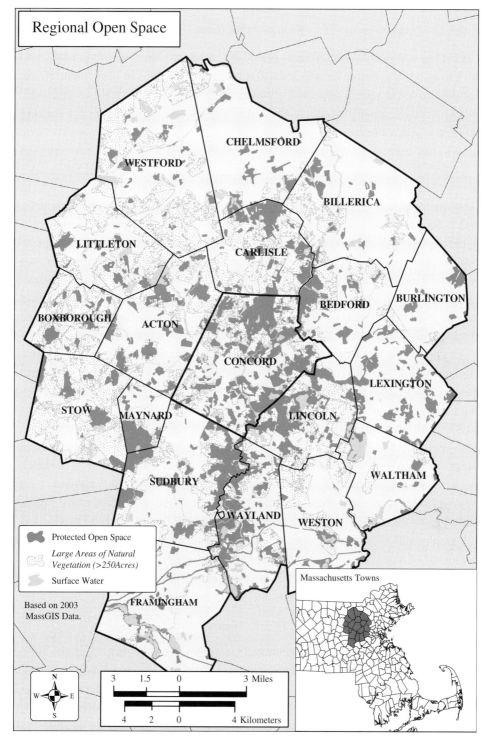

Fig. 11.2. Open space in a locality-centered region. Dark shading = protected areas; speckled areas = other natural vegetation >100 ha. Suburban Boston, USA. See Forman *et al.* (2004). Produced by M. Barrett; courtesy of Town of Concord, Massachusetts.

plan every five years, within a short period the towns in Concord's region are likely to be using the same regional data sources and maps. In this manner, the locality-centered region approach (for towns, counties, municipalities and other locales) should greatly stimulate the opportunities and expectations for regional collaboration and action.

11.3 Greater Barcelona region

Now the focus broadens to an entire urban region, in this case an area with a radius of about 65 km around Barcelona, a major European city. An array of ecological planning solutions encapsulated here for the region emerged from a 15-month conceptual planning project (Forman, 2004), with the broad objective to outline promising spatial arrangements and solutions that enhance natural systems and associated human land uses for the long-term future of the Greater Barcelona region. In effect, natural systems and human uses thereof in the urban region are designed to improve over ten and fifty years even though population growth and development continues.

The approach was to extensively visit and study the area with regional experts, consult with a wide range of leading specialists and thinkers, analyze numerous aerial photographs and maps for spatial patterns, identify functional flows and movements plus land-use changes, list important assumptions, identify basic principles from relevant fields, mesh principles and spatial models with the specific land-use patterns of the region, and outline and compare three alternative, feasible plans for the region as a whole. The objective of this urban region project closely parallels the vision (Forman, 2002) outlined at the beginning of this article.

Four themes or categories of issues were addressed with distinctive spatial solutions for the region as a whole (Forman, 2004): (1) nature; (2) food production; (3) water; and (4) development. In addition, one example ("edge of town") is presented of a generic solution for a type of small space repeated by the hundreds or thousands across the region.

11.3.1 Nature

The overriding threat to sustaining nature in an urban region is the loss, degradation, and fragmentation of natural habitat (Forman, 1995; Opdam *et al.*, 2002; Tess *et al.*, 2004; Forman, 2005). Without land planning and protection, large natural areas tend to be fragmented into small areas, resulting in a significant loss of biodiversity, clean groundwater or water from aquifers, and other values. Also major wildlife and stream or river corridors tend to be degraded or broken.

Eight large protected areas often called conservation parks exist in the Barcelona urban region (Plate 11.1) (Patronat del Parc de Collserola, 1990; Acebillo and Folch

2000). They provide numerous values to society, including rainwater infiltration and groundwater protection in a dry climate, recreational hiking, biking, hunting and nature education, wood production, and ecological protection of important habitats, species and biodiversity (Forman, 1995, Roda *et al.*, 1999; Gutzwiller, 2002; Opdam *et al.*, 2002; Peterken, 2002; Groom et al. 2005). These large protected "emeralds" are the most valuable natural resources in the region. Two additional large protected areas were recommended, with the number and locations based on landscape ecology principles, including the diversity of vegetation and rock types and the connectivity gained for individual emeralds and the overall network (Table 11.1) (Forman, 2004). The ten-emerald solution accomplished several goals: (1) integrated and protected a cluster of five small-to-medium parks (area no. 9); (2) protected a new large area with ecologically valuable remoteness and disturbance attributes, plus several European-Community-listed rare habitats (no. 10); and (3) enhanced the potential for connectivity with one existing isolated emerald (no. 1), both new emeralds (nos. 9 & 10), and the other large protected areas (especially nos. 4, 6, 7, & 8).

Landscape ecology principles highlight the importance of connectivity among the conservation parks, preventing the protected areas from becoming surrounded and isolated by future development (Saunders and Hobbs, 1991; Forman, 1995; Bennett, 2001; Jongman and Pungetti, 2004); some leaders and residents in the region also want this connectivity among the parks. The goals of connectivity here are for movement of local residents (including farmers), hikers, and wildlife. In an urban region, existing development forms a complex and difficult template for such connectivity. No single type of green corridor fits all cases. Thus five types of corridors with overlapping forms and functions were used in the plan (Table 11.1) (Forman, 2004). These types were tailored to the distance between protected areas, distribution of streams and rivers, and density of urbanization in areas traversed. Protection of the resulting network should noticeably enhance and sustain movements of walkers and wildlife across this urban region.

Natural vegetation is especially appropriate in the corridors, though in many cases farmland or intensive-use people-parks may also be present. Other types of green corridors presented in sections below include edge-of-town parks and parks that help to separate and maintain the distinctiveness of adjacent growing municipalities. Where natural corridors cross major transportation routes, wildlife underpasses or overpasses are required (Rosell Pages and Velasco Rivas, 1999; Forman *et al.*, 2003).

In short, the resulting emerald network is composed of large protected natural areas connected by green corridors. This solution, like that at the finer-scale locality near Boston (Figure 11.1), should sustain key characteristics of nature, as well as socioeconomic benefits, for decades ahead.

Table 11.1. Emerald network solution for nature. (a) Ten large protected areas are
listed with predominant vegetation types and rock types (plus variants) (letters
indicate types described in the cited reference); also listed is the number of likely
corridor linkages to surrounding large protected areas, both without and with
addition of the proposed new areas 9 and 10. (b) Five corridor connection types are
listed with their most typical form and function. *Source*: Forman (2004).

a) *Large protected natural areas or "emeralds"*

	Area no.	Veg. type	Rock type	Links without	Links with
Existing:	1	A	M	0	1
	2	A	var. of M	1*	1*
	3	B	N	3	3
	4	B	N	3	3
	5	C	O	2*	2*
	6	D	P	2	4
	7	D	var. of P	3	3
	8	E	Q	1	2
Proposed:	9	var. of D	var. of P	-	3 (to areas no. 6, 8, & 10)
	10	var. of D	var. of P	-	3 (to areas no. 1, 6, & 9)

* Additional linkage with a large protected area outside the urban region.

b) *Corridor connection types for emeralds*

Type	Form and function
1. Reconnection zone	Wide area; connects two smaller parks into one larger park
2. Green ribbon	Wide overland corridor; enhances movement along its length (commonly connecting emeralds)
3. Blue–green ribbon	Wide corridor along stream or river; protects watercourse and enhances movement along its length
4. Ribbon of pearls	Wide corridor with attached small-vegetation patches; provides "rest stops" to enhance movement along its length
5. String of pearls	Narrow vegetation-lined walkway with attached small patches; enhances movement through developed area.

11.3.2 Food production

Fragmentation and disappearance of farmland due to urbanization are especially
conspicuous in urban regions. This leaves the city increasingly dependent on food-
producing landscapes elsewhere, and on transportation systems and costs. Farmland
areas not only support agricultural communities and economies with food production close to metropolitan markets, but also support a range of distinctive important
wildlife and natural communities.

Six large productive areas (excluding forest and industry) are noteworthy in this urban region (Acebillo and Folch, 2000; Forman, 2004). Three are large farm and field areas (Plate 11.1), two are composed of small fields and plots on lower river floodplains or deltas, and one is a concentrated greenhouse area east of Barcelona. These six areas are well separated and produce quite different products, as listed below:

1. Grape and wine production
2. Grain production
3. Livestock and grain production from small fields and plots
4. Fruits and vegetables from small-market farming
5. Fruits and vegetables from family-food gardening and small-market farming
6. Concentrated greenhouses area for flower and vegetable production

Each large productive area also supports a somewhat different array of habitats, birds, plants and other ecological attributes. Although many small farmlands exist, the plan highlights sustaining these large productive areas as a priority.

The family-food gardening and small-market farming area (called an "agricultural park") adjacent to the Barcelona metropolitan area is especially significant, because it helps protect the region's most important aquifer against urban-related development. Also this productive landscape provides fresh vegetables and fruits with minimal transportation cost to markets and restaurants across the metropolitan area.

Greenhouses (glasshouses) for flower growing and vegetables tend to be concentrated in one area of valleys, where much woody vegetation has been removed and gully floods and erosion are severe. A partial solution is the establishment of numerous plantings and bits of natural woody vegetation in strategic spots to reduce erosion and flooding effects. This would be supplemented by limited recreational trail and picnic opportunities (with marketing benefits for growers) in this greenhouse area close to population and tourist centers.

The emerald network should be effective in protecting forest-related and even fire-dependent habitats and species in hills and mountains (Patronat del Parc de Collserola, 1990; Roda *et al.*, 1999). However, other key species such as many rare migratory birds concentrate in abandoned fields, woodland edges and other successional habitats, where seeds and insects typically abound year-round (Pino *et al.*, 2000). These habitats tend to be in valleys where urbanization commonly removes and fragments farmland. The solution proposed was to establish seven "agriculture-nature parks" on areas of viable small farms, and adjacent to the large protected areas (Forman, 2004). In these parks, rather than succumbing to urbanization, farmers would continue to own their land, with the constraint that some 10 to 20 percent of the productive area would remain in successional stages (old-field, shrubland, and young woods), the target habitats to be maintained for important biodiversity.

A consequence of the maintenance of the set of six large different productive areas, plus several agriculture-nature parks of small farms, is the provision of important economic flexibility and stability in the face of changes and surprises ahead. Of course the areas also provide farming communities, aesthetics, wildlife populations, biodiversity, and much more for the urban region's future.

11.3.3 Water

Solutions related to water in this Mediterranean-climate urban region are more difficult than in the two preceding cases. The major issues to be addressed, as documented by Patronat del Parc de Collserola,1990; Acebillo and Folch, 2000; Prat *et al.*, 2002, include: (1) too little water most of the time, i.e., scarcity; (2) too much water periodically, i.e., floods; (3) old industries with dams and pollutant discharges along streams and rivers; (4) stormwater and human sewage combined in pipes leading to overtaxed sewage-treatment facilities; (5) most stream and river lengths contaminated by sewage; (6) floodplain riparian vegetation commonly degraded or missing; (7) one of the four reservoirs polluted by pig-farm wastes, and another somewhat degraded by grape-growing farm runoff; (8) groundwater degraded by pig-farm wastes, and in another area by fertilizers from greenhouses; (9) watersheds for reservoirs inadequately protected and managed; (10) wetlands largely removed; (11) fish populations severely degraded and freshwater fishermen rare; (12) coastal beach, marsh, and woodland habitats degraded; and (13) saltwater intrusion along the coast.

No single solution provides for all or even most of these. Instead a package of solutions is provided, each addressing two or more of the issues (Forman, 2004). In this manner most issues are addressed from a number of perspectives. In addition, since many of the issues are interrelated, improvements in one or two are apt to improve others. In all cases, a solution is targeted at priority strategic locations (this is because a broad brush recommendation, e.g., for all riparian areas or stormwater and sewage systems or polluting industries, would politically have little chance of implementation). The net effect of this multi-issue solutions approach should be a major overall improvement in water resources and consequently their human uses.

The following partial list illustrates the overlap of both issues and solutions to water-related issues in the urban region (from Forman, 2004):

1. *Separate pipe systems for stormwater and sewage.* Enhance effectiveness of sewage treatment plants. Noticeably clean up water in streams and rivers. Enhance fish populations and fishermen. Support wetlands at ends of stormwater pipes.

2. *Create small wetlands where stormwater-pipes discharge.* Reduce flooding. Absorb stormwater pollutants and reduce amount entering surface water bodies. Increase seasonal habitat diversity and wetland species in widely dispersed locations.

3. *Disconnect hard surfaces.* Reduce flooding by preventing some stormwater from reaching streams and rivers. Recharge groundwater, which improves streamflow and generally enhances woody plants during dry periods.

4. *Create basin parks.* Absorb water from nearby areas to reduce downstream flooding. Provide at least seasonal wetlands and associated biodiversity. Provide recreational or visual amenities for neighborhoods.

5. *Remove livestock in floodplains.* Prevent degradation or destruction of riparian vegetation. Increase woody vegetation that reduces flooding on floodplain. Enhance floodplain habitat and species diversity.

6. *Enhance or replant riparian vegetation.* Provide stream and river protection against impacts from bank erosion and surrounding lands. Increase shade for cool-water fish. Provide branches and logs for fish habitat.

7. *Remove structures in floodplains.* Reduce impermeable surface and associated flooding. Eliminate fixed objects that during high water cause accelerated flow, erosion and scour.

8. *Limit construction on slopes exceeding a certain angle.* Decrease slope erosion. Reduce sediment impacts on streams. Reduce fire hazard.

9. *Create wetland complexes for birds.* Increase biodiversity by supporting many rare species and natural communities. Support migrant bird populations. Absorb water against floods.

10. *Ecotourism objectives in river valley.* Clean up a connected stream–river system. Support a diversity of recreational opportunities. Strengthen nature-based communities and economy.

11. *Address agricultural sources of nitrogen.* Plan, mitigate, and treat wastes in pig farming area. Reduce input and control runoff of agricultural chemicals in grape-growing area. Decrease fertilizer use in greenhouse area. Produce more clean water in a dry climate.

12. *Protect and manage land around reservoirs.* Permanently protect watershed areas. Enhance species and processes characteristic of large protected areas. Reduce human access, thus minimizing fires and erosion on steep slopes around reservoirs. Provide clean water in a dry climate.

13. *Create and support farming, parks, and natural areas on lower floodplains and deltas.* Absorb rainwater and minimize saltwater intrusion. Minimize pollution of aquifer due to urbanization and industrial pollutants. Protect distinctive coastal beach, wetland, and woodland habitats and species, thus enhancing regional biodiversity.

Although the improvements are presented here in abbreviated form (and only a few illustrated in Plate 11.2), essentially all are familiar to freshwater biologists and other water-resource experts. These solutions should improve all of the water issues listed at the outset, and produce a cumulative benefit for water in the region.

Water-related issues could be the Achilles heel that reduces Barcelona's influence, or they could be the savior against over-development, thus providing a bright future for the region. Certainly hydrologists, engineers and planners working closely with aquatic biologists and landscape ecologists are central to effective solutions for water in an urban region.

11.3.4 Development

Development, industry and transportation in an urban region, if inadequately planned or inappropriately located, represent a special threat to natural systems and their human uses (Browder and Godfrey, 1997; Calthorpe and Fulton, 2003; Tess *et al.*, 2004; Kowarik and Korner, 2005). The spread of urbanization removes, degrades and fragments natural habitat (Opdam *et al.*, 2002; Greenberg, 2002; Forman, 2005). Open space adjacent to a metropolitan area is particularly at risk (Ozawa, 2004). Also in some areas growing towns threaten to coalesce. Industries tend to degrade nearby locations, but more importantly, often send out plumes of chemical pollutants downstream and downwind. Transportation, in contrast, degrades habitat next to its extensive road network, and on a broad scale, fragments natural habitat and blocks wildlife movement corridors (Rosell Pages and Velasco Rivas, 1999; Forman *et al.*, 2003; Forman, 2005). A handful of solutions address at least in part this complex of topics (see also Plate 11.2):

1. *Development*
   (a) Growth and development encouraged in low-environmental-impact areas
     - Five small satellite cities
     - One commutable area near the big city
     - One strategic area for heavy industry center and truck transportation hub
   (b) Growth and development limited
     - Areas for limited, directional growth
     - Areas for no growth
     - Scattered areas for removal of buildings
   (c) Great park for large greenspace adjacent to the big city
2. *Transportation*
   (a) Vehicle commuter traffic discouraged and walking and biking encouraged with nodes of light and medium industry on edges of towns
   (b) Increased commuter rail and bus and transit-oriented development, with creative technologies to move people between town centers and transit stations
   (c) Enhanced rail, highway, and public transport capacity, between targeted growth areas
   (d) Truck transportation terminal
   (e) Underpasses and overpasses for people and wildlife crossing of major highways
3. *Industry*
   (a) Light and medium industry relocated from streamsides, and concentrated in nodes on edges of towns and small cities

    (b) New heavy-industry center with efficient fuel, electricity, water, waste treatment, diversity of employment, and transportation access

    (c) Heavy industry gradually relocated away from two riversides, facilitating water clean-up, connected riverside parks, and new investment opportunities

    (d) New "clean" heavy industry and employment encouraged

4. *Edge-of-town: repeated small places*

    (a) Benefits for biodiversity and wildlife movement.

    (b) Benefits for neighbors, future neighbors, and commuters

An especially effective strategy is to channel growth and development into areas where environmental damage would be slight. Thus, after evaluating conditions across the Barcelona Region, seven areas were targeted for future growth, if it occurs (Forman, 2004). Five are satellite cities between 40 and 50 km from downtown Barcelona. One is near the city, though separated by a large open space, to accommodate metropolitan area growth within close commuting distance. And one is a relatively flat strategic location for a proposed heavy industry center and truck transportation hub. In the six population-growth locations, the initial step is to establish a system of connected parks, a transportation system including public transport, and areas for building that minimize water and other environmental impacts.

A parallel inverse solution is to target areas for limited growth, no growth, and even building removal. Several no-growth-and-development villages and towns were pinpointed to protect prime agricultural landscapes, a major corridor, a reservoir, and so forth. Also several limited-growth communities were highlighted, where development could occur in certain directions but not in other directions due to an adjacent valuable resource. Scattered locations, such as floodplains, steep slopes and the center of an emerald, were also pinpointed for removing buildings or small communities (an ongoing process occurring at a low rate) that are especially damaging to natural systems and human uses thereof. Finally, a "green net" of greenspace strips between 100 and 200 m wide was proposed for connectivity and to help separate and maintain the identity or distinctiveness of adjoining growing municipalities that threaten to coalesce.

Large open space adjacent to, or being engulfed by, a metropolitan area is at special risk and requires a different solution (Patronat del Parc de Collserola, 1990; Greenberg, 2002; Ozawa, 2004). Such a space next to Barcelona (Acebillo and Folch, 2000; Boada and Capdevila, 2000; Prat *et al.*, 2002) only exists today because it has the most valuable aquifer in the region, has periodic mega-floods, and is enormously productive in providing fresh fruits and vegetables to markets and restaurants across the metropolitan area. A stunning stupendous park, to be one of the great parks of the world, connected by convenient walking and biking overpasses from the city and surrounded by productive small-market farms, was proposed (Plate 11.2) (Forman, 2004). Overlooking the cleaned-up river, the park is enhanced by walkways, picnic areas and ballfields, marked by rich wetlands and an abundance of waterbirds, laced with jogging paths and bikeways, enriched with highlights of history and heritage, sprinkled with reflection ponds and rushing water, surrounded by skyrocketing real estate value and economic opportunity, designed for big floodwaters to pass harmlessly by, and visible by day and night to highway travellers, commuter and high-speed rail lines, and air travelers from international airport to city. This is a Great Park highlighted by an inspiring symbol or flagship for the people and the city.

Transportation in an urban region raises another set of interlocking key issues relative to natural systems (Browder and Godfrey, 1997; Forman *et al.*, 2003). Solutions range from region-wide to single-location actions. Rapidly growing traffic is addressed in part with light and medium industry in nodes on the edge of towns to enhance walking and biking and reduce commuter traffic. Indeed, such industries along streams should be relocated near communities, both to restore the streams and provide jobs near residents. Also, more commuter rail and bus options with transit-oriented development (Cervero, 1998; Calthorpe and Fulton, 2003) should be coupled with creative technologies to move people between town centers and transit stations. It is recommended that transportation routes between expected growth areas include public transport and spatially mesh with the critical arrangement of natural systems in this plan. As mentioned above, a truck transportation terminal should be strategically located for the loading and unloading of goods from large long-distance trucks, as well as small trucks entering city streets. Products from the region's agricultural landscapes may also be efficiently handled in the hub.

Since the many highway corridors in an urban region frequently cross natural corridors, more underpasses and overpasses for the movement of local residents (including farmers), hikers, and wildlife are particularly important (Rosell Pages and Velasco Rivas, 1999; Forman *et al.*, 2003; Jongman and Pungetti, 2004). Several overpasses exist and twelve additional key locations were identified, where a pair of underpasses or overpasses should establish effective connectivity for the emerald network (Plate 11.2).

Heavy industry, traditionally located along rivers and seriously polluting them (Prat *et al.*, 2002), should also be appropriately and gradually relocated. A heavy industry center mentioned above is proposed for the region, which would provide the economic and environmental benefits of efficient fuel, electricity, water and waste treatment, a diversity of employment, and transportation access. Existing industries along stretches of two rivers should be targeted for clean-up, while the new center should accommodate new clean heavy industries providing employment. Relocating elsewhere an existing heavy-industry area immediately upwind of the metropolitan area should also have public health benefits, and permit the development of connected riverside parkland and new-investment land.

11.3.5 Repeated small places

The Greater Barcelona region plan takes the big picture, highlighting major patterns across the area (Forman, 2004). Yet certain small patterns are repeated by the hundreds or thousands, such as gullies, roads, and edges of towns, for which generic solutions (tailored to specific situations), if multiplied many-fold, would provide a large cumulative region-wide gain.

To illustrate, edges of towns are often little planned, at least for natural systems and their human uses, while urbanization tends to spread outward. The concept solution is a linear edge-of-town park with "outdoor rooms" that connects nodes of small and medium industry. In this model, a 50 m wide strip of parkland for intensive use by people separates the built area from farmland, which is likely to be developed in the future since it surrounds a growing town. The green edge-strip connects to stream corridors and walkways that cut into or across the town (Bennett, 2001; Greenberg, 2002; Jongman and Pungetti, 2004; Kowarik and Korner, 2005). The inner 15 m is an *allée* composd of a wide path with a row of sycamores (*Platanus*) either side and benches. The outer 35 m is a series of separate spaces, each at least partially lined with trees and shrubs, such as children's playground, basketball court, small football field, cultural or historical symbol, semi-natural trees, shrubs and rocks, and so on. Overall, the edge park provides three major benefits: an amenity and some nature for the existing neighborhoods; an amenity and some nature for the future neighborhoods that develop on its outer side; and a connected link for movement of some species from attached stream or walkway corridors. The nodes of light and medium industry provide jobs near to residents of the town to encourage walking and biking and discourage vehicular commuting and traffic.

In summary, this analysis of the Barcelona region addresses issues of nature, food production, water, and development for an urban region (Plate 11.2). Recognizing that implementation is likely at best to be in separate pieces and stages over time, a rich array of finite feasible solutions that fit together as a whole (rather than a large heavy imprint all at once) was provided to strengthen natural systems and associated human benefits for the region's future.

11.4 Urban regions compared

Finally, numerous patterns of both ecological and human importance are identified by comparing urban regions, large to small, worldwide. Such patterns are illustrated with two analyses: (1) distance from city center to nearest water body; and (2) number and length of greenspace wedges projecting into a metropolitan area.

Thirty-five urban regions on all continents, including small population (0.25–1.0 million), medium (1–4 million), and large (4–12 million) cities, are compared. For instance, the city of Santiago in Chile (4.7 million people) covers the central portion of a distinct, nearly continuous built metropolitan area, which is at the core of an urban region between the Andes to the east and the Pacific Ocean to the west (Plate 11.3). The urban regions examined generally have an average radius of between 70 and 100 km, based on several major linkages with the city, such as commuter rail lines, major reservoir water-sources, and one-day recreation areas.

The analytic approach is to measure spatial patterns of the built metropolitan area and of natural systems in the urban region (e.g., rivers, agricultural landscapes, highways, parks, towns and large forests). Measurements were made on broad satellite GIS images of 1:200,000 scale, which emphasizes regional rather than finer landscape or local patterns. The goal is to estimate how the region is structured and functions relative to natural systems and their human uses. Comparative spatial analyses of broad-scale patterns produce insights or probable patterns, which, like modelling results, are effectively hypotheses to be tested in part with detailed studies.

In the first analysis, cities of different size are related to distance to the nearest major freshwater and saltwater bodies (Figure 11.3). This provides the following insights:

1. Several cities are far (more than 50 km) from a major freshwater source (Figure 11.3, top rows left), suggesting that they have a major water transport system and maintenance concern, and are dependent on distant communities and conditions (in some cases outside the region) to minimize pollution and watershed degradation, and to maintain adequate and clean water sources. This may be a particular problem for large-population cities.
2. Some cities are adjacent to a major river (bottom left), suggesting that the downriver portion is polluted by the city, that controlling pollution immediately upriver and in the metropolitan area is a problem, and that to provide clean water further upriver requires some land protection around the river and its tributaries. This may be a particular problem for large cities.
3. Cities adjacent to or near a major reservoir (bottom left) have a problem adequately protecting its watershed and water against urbanization and pollution effects. A limited-size lake would pose a similar problem.
4. Cities adjacent to saltwater (lower right) typically have a port with protected harbor, an associated land-transportation system with environmental impacts crossing the region, considerable coastal strip development, and a large area of polluted coastal waters (with associated effects on biodiversity, near-shore fisheries, recreation and tourism).
5. Cities near saltwater (center right) have a separate port, with heavy environmental impacts likely between city and port.
6. Cities more than 30 km from saltwater (top right) may have relatively clean nearby coastal areas particularly suitable for terrestrial and marine habitats and species, near-shore fisheries, recreation and tourism.

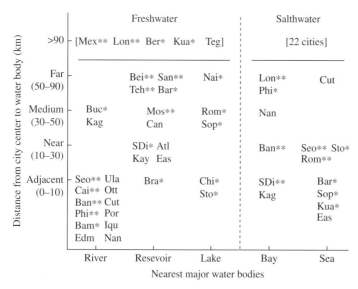

Fig. 11.3. Distance between cities of different size and major surface-water bodies. ** large city (>4 million population); * medium city (1–4 million); other: small city (0.25–1 million). The initial letters of the cities correspond to: London, Berlin, Rome, Bucharest, Stockholm, Barcelona, Nantes (all in Europe); Chicago, Philadelphia, San Diego, Ottawa, Edmonton, Portland (Oregon), Atlanta (all in North America); Mexico City, Santiago, Brasilia, Tegucigalpa, Iquitos (all in Latin America); Cairo, Nairobi, Bamako, East London (all in Africa); Beijing, Moscow, Seoul, Tehran, Sapporo, Ulaanbaatar, Kayseri, Cuttack, Kagoshima (all in Asia): Bangkok, Kuala Lumpur; Canberra (all in Australo-Indochina).

The second analysis, relating the number and length of greenspace wedges projecting into metropolitan areas to city size, provides quite different insights (Figure 11.4).

1. Metropolitan areas with long or very long greenspace wedges (Figure 11.4, top) probably have a high proportion of residents living near parkland, good access for residents to walk or bicycle for recreation outside the metro area, and good connectivity for outside species from rural and natural areas to populate city greenspaces.
2. The same benefits apply to many short greenspace wedges for medium-size cities, and many medium-length wedges for large cities (right).
3. Large cities with few short wedges (bottom left), in contrast, probably have little connectivity for recreational movement outward or species movement inward, and, unless separated parks in the metro area are adequate, many residents live far from greenspaces.
4. Most large cities have an intermediate number of short or medium-length wedges (center). In contrast, most small cities have few wedges which may vary from characteristically short to very long (left).

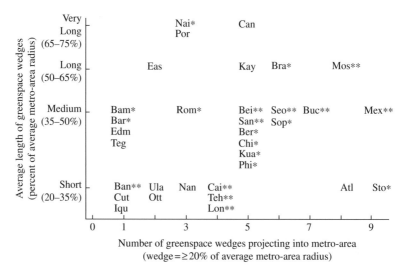

Fig. 11.4. Number and length of greenspace wedges for cities of different sizes. **Large city (>4 million population); *medium city (1–4 million); other, small city (0.25–1 million).

Numerous similar spatial analyses should provide a rich array of insights into how urban regions are structured and work, both ecologically and for associated human uses. Alternative urbanization models which simulate different expansion effects on the existing arrangement and flows could provide considerable insight for planning the future of an urban region.

In conclusion, the idea that urban planners or designers, architects, and landscape architects (with primary expertise on the city) could seriously collaborate with ecologists, conservationists, and natural resource planners (with primary expertise on land well beyond the city) has yet to gel. However, I think the place where it will soon happen is the urban region, including localities within it. Furthermore, the land mosaic perspective which has emerged from landscape ecology is the most likely and promising model available. Meshing nature and people on the land so they both thrive long-term is now within reach.

References

Acebillo, J. A. and Folch, R. (2000). *Atles Ambiental de l'Area de Barcelona: Balanc de recursos i problemes*. Barcelona: Ariel Ciencia y Barcelona Regional.

Bennett, A. F. (2001). *Linkages in the Landscape: The Role of Corridors and Connectivity in Wildlife Conservation*. Gland, Switzerland and Cambridge, UK: IUCN-The World Conservation Union.

Boada, M. and Capdevila, L. (2000). *Barcelona: Biodiversitat urbana*. Barcelona: Ajuntament de Barcelona.

Browder, J. O. and Godfrey, B. J. (1997). *Rainforest Cities: Urbanization, Development, and Globalization of the Brazilian Amazon*. New York: Columbia University Press.

Calthorpe, P. and Fulton, W. (2003). *The Regional City.* Washington DC: Island Press.

Cervero, R. (1998). *The Transit Metropolis: A Global Inquiry.* Washington DC: Island Press.

Ferguson, J., Connelly, M., Forman, R. *et al.* (1993). *Town of Concord 1992 Open Space Plan.* Concord, MA (USA): Concord Natural Resources Commission.

Forman, R. T. T. (1995). *Land Mosaics: The Ecology of Landscapes and Regions.* Cambridge and New York: Cambridge University Press.

Forman, R. T. T. (2002). Imaginar un mosaico terrestre en el que puedan desarrollarse tanto la naturaleza como las personas / Envisioning a land mosaic where both nature and people thrive. In *Jardines insurgentes: Arquitectura del paisaje en Europa 1996–2000.* Barcelona: Fundación Caja de Arquitectos, pp. 25–48.

Forman, R. T. T. (2004). *Mosaico territorial para la región metropolitana de Barcelona.* Barcelona: Editorial Gustavo Gili.

Forman, R. T. T. (2006). Good and bad places for roads: Effects of varying road and natural patterns on habitat loss, degradation, and fragmentation. In 2005 *Proceedings of the International Conference on Ecology and Transportation.* Raleigh: North Carolina, Center for Transportation and the Environment, North Carolina State University.

Forman, R. T. T., Sperling, D., Bissonette, J. A. *et al.* (2003). *Road Ecology: Science and Solutions.* Washington DC: Island Press.

Forman, R. T. T., Reeve, P., Beyer, H. *et al.* (2004). *Open Space and Recreation Plan 2004.* Concord, MA: Natural Resources Commission.

Greenberg, J. (2002). *A Natural History of the Chicago Region.* University of Chicago Press.

Groom, M. J., Meffe, G. K., Carroll, R. and Contributors. (2005). *Principles of Conservation Biology.* Sunderland, MA: Sinauer Associates.

Gross, R. A. (1976). *The Minutemen and Their World.* New York: Hill and Wang.

Gutzwiller, K. J. (ed.) (2002). *Applying Landscape Ecology in Biological Conservation.* New York: Springer.

Hobbs, R. J. (1995). Landscape ecology. *Encyclopedia of Environmental Biology,* **2**: 417–428.

Jongman, R. and Pungetti, G. (eds.) (2004). *Ecological Networks and Greenways: Concept, Design, Implementation.* Cambridge and New York: Cambridge University Press.

Kowarik, I. and Korner S. (eds.) (2005). *Wild Urban Woodlands: New Perspectives for Urban Forestry.* New York: Springer.

Maynard, W. B. (2004). *Walden Pond: A History.* Oxford and New York: Oxford University Press.

Opdam, P., Foppen, R. and Vos, C. C. (2002). Bridging the gap between ecology and spatial planning in landscape ecology. *Landscape Ecology,* **16**: 767–779.

Ozawa, C. P. (ed.) (2004). *The Portland Edge: Challenges and Successes in Growing Communities.* Washington DC: Island Press.

Patronat del Parc de Collserola (1990). *Parc de Collserola : Pla Especial d'Ordenació i de Protecció del Medi Natural, Realitzacions 1983–1989.* Barcelona: Area Metropolitana de Barcelona-Ajuntament de Barcelona.

Peterken, G. (2002). *Reversing the Habitat Fragmentation of British Woodlands.* Surrey, UK: WWF-Goldaming.

Pino, J., Roda, F., Ribas, J. and Pons, X. (2000). Landscape structure and bird species richness: Implications for conservation in rural areas between natural parks. *Landscape and Urban Planning,* **49**: 35–48.

Prat, N., Munné, A., Sola, C. *et al.* (2002). *La Qualitat Ecologica del Llobregat, El Besos, El Foix i la Tordera: Informe 2000.* Estudis de la Qualitat Ecologica dels rius no. 10. Barcelona: Diputació de Barcelona.

Roda, F., Retana, J., Gracia, C. A. and Bellot, J. (1999). *Ecology of Mediterranean Evergreen Oak Forests.* Berlin and New York: Springer-Verlag.

Rosell Pages, C. and Velasco Rivas, J. M. (1999). *Manual de prevenció i correcció dels impactes de les infrastructures viaries sobre la fauna.* No. 4. Barcelona: Departament de Medi Ambient, Generalitat de Catalunya.

Saunders, D. A. and Hobbs, R. J. (eds.) (1991). *Nature Conservation 2: The Role of Corridors.* Chipping Norton, Australia: Surrey Beatty.

Tess, G., Tess, B., Harms B., Smeets, P. and van der Valk, A. (eds.) (2004). *Planning Metropolitan Landscapes: Concepts, Demands, Approaches.* DELTA Series 4. Wageningen: IOS Press-Delft University Press.

12

The role of ecological modelling in ecosystem restoration

SVEN ERIK JØRGENSEN

12.1 Ecological modelling as a tool for ecological restoration

Ecological modelling is a powerful tool for planning and managing restoration projects. The idea behind the use of ecological management models is demonstrated in Figure 12.1. Urbanization and technological developments are having an increasing impact on the environment. Energy and pollutants are released into ecosystems, where they may cause more rapid growth of algae and/or bacteria, species damage, or alteration of the entire ecological structure of an ecosystem. An ecosystem is extremely complex, and so it is an overwhelming task to predict the environmental effects that an emission will have. It is here that the model comes into the picture. With sound ecological knowledge, it is possible to extract the features of the ecosystem that are involved in the pollution problem under consideration, and to construct the basis of an ecological model.

Figure 12.1 illustrates the idea behind the introduction of ecological modelling as a management tool. This introduction occurred around 1970. Today, environmental management and restoration of natural ecosystems and resources are more complex and have to apply environmental technologies, cleaner technologies and ecological engineering or use of ecotechnology (see also Jørgensen and Bendoricchio, 2001). Ecotechnology (Mitsch and Jørgensen, 1989; Mitsch and Jørgensen, 2003), which encompasses ecosystem restoration, design of constructed ecosystems (for instance wetlands), ecologically sound management of natural resources and ecosystems including the use of natural ecosystems for pollution abatement, is often applied to solve the problems of non-point or diffuse pollution, mainly originating from agriculture. The importance of non-point pollution was hardly acknowledged before about 1980. Furthermore, global environmental concerns play a more important role in environmental management today than twenty years ago. The abatement of the greenhouse effect and the depletion of the ozone layer are widely discussed and several international conferences have resulted in some governments taking the first

Ecological Restoration: A Global Challenge, ed. Francisco A. Comin. Published by Cambridge University Press.
© Cambridge University Press 2010.

246   *Sven Erik Jørgensen*

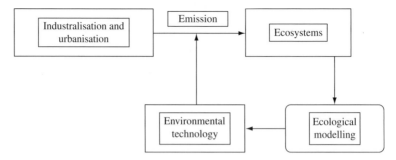

Fig. 12.1. Relationship between environmental science, ecology, ecological modelling and environment

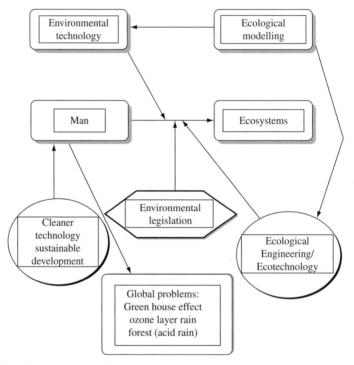

Fig. 12.2. The idea behind the use of environmental models in environmental management.

steps toward the use of international standards to solve these crucial problems. Figure 12.2 illustrates this more complex picture of environmental management today.

The application of models in ecological engineering is almost compulsory, because models are able to synthesize our knowledge about a problem and an ecosystem and to quantify the results of ecological management projects. They

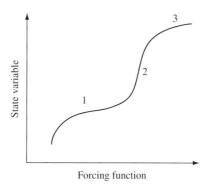

Fig. 12.3. The ecological buffer capacity is defined as the ratio change in state variables / change in forcing functions (impacts). The buffer capacity at point 1 and 3 is high but low at point 2. In ecological management it is important to assess when the buffer capacities are low and high, which is possible by the application of models.

can be applied to the design of constructed ecosystems or modifications of existing natural ecosystems. The results of different restoration projects can easily be compared by the use of ecological models, which of course facilitate the decision about which restoration method to select among a number of alternatives under given circumstances.

The reactions of the system might not necessarily be the sum of all the individual reactions (May, 1977); this implies that the properties of the ecosystem as a system cannot be revealed without the use of a model of the entire ecosystem. Moreover, the forcing functions or external variables describing the impact on the ecosystem and the state variables or internal variables describing the state of the ecosystems are rarely just linearly related. This implies that the buffer capacity or resistance capacity defined as changes in the impact state variable (Jørgensen, 2002) is not a constant; see Figure 12.3. A complex relationship between forcing functions and state variables can, however, be revealed by models, when they are based on a good knowledge of the problem and the ecosystem.

12.2 Models applied in ecosystem restoration

Structurally dynamic models (SDMs) that consider adaptation and shifts in species composition have as a goal, the identification of the properties that give the best survival. Eco-exergy, defined as the distance from thermodynamic equilibrium (Jørgensen and Padisak, 1996; Jørgensen and De Bernardi, 1998; Jørgensen *et al.*, 2000), has been applied several times successfully as a goal function. Eco-exergy can be found as:

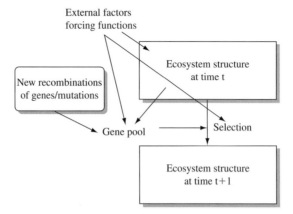

Fig. 12.4. Conceptualization of how external factors steadily change species composition. The possible shifts in species composition are determined by the gene pool, which is steadily changed due to mutations and new sexual re-combinations of genes. The development is, however, more complex. This is indicated by an arrows from "structure" to "external factors" and "selection" to account for the possibility that the species can modify their own environment (see below) and thereby their own selection pressure; 2) an arrow from "structure" to "gene pool" to account for the possibilities that the species can to a certain extent change their own gene pool.

$$\text{Eco} - \text{exergy} = \sum \beta_i \, c_i \ldots \text{detritus equivalent}$$

where β refers to a weighting factor that accounts for the complexity of the different organisms included in the model through the calculation of the information embodied, and c_i is the concentration of the ith component in the ecosystem.

The β value is for detritus that is used as a reference is 1.0. For the minimal cell β is about 4, for bacteria about 8, for crustaceans about 233, for fish about 499, for mammals about 2125 and for homo sapiens about 2170. A more complete list of β values can be found in Jørgensen *et al.* (2005). The experience of the application of structurally dynamic models is that the change in properties of the state variables are described by this approach according to the observations of the structural changes (see, for instance, Zhang *et al.*, 2003a; Jørgensen and Fath, 2004). The development of this type of model is very important for the application of models in ecological management, because the management strategy is to change the forcing functions, i.e., the conditions for the biological components in the ecosystem (Figure 12.4). It implies, of course, that the model developed under the previous conditions does not account for the changes in properties of the biological components due to the adaptation process or due to a shift in the species composition. Structurally dynamic models are therefore important when restoration methods are applied, because restoration implies that the conditions are changed and possibilities for adaptation to the new conditions and shifts in species composition may emerge.

12.3 Case studies

12.3.1 Pamolare II: a structurally dynamic model for the management of shallow lakes

This is a model prepared to help take decisions on alternative restoration methods for lakes subject to an accelerated process of eutrophication. So, major forcing functions are related to discharges of nutrients into the lake and to solar radiation as drivers of the trophic web relationships in the lake, and also to relationships between structural components of the lake. The model has been developed for the UNEP and can be downloaded from the Internet (www.unep.or.jp/ietc/pamolare/about_pamolare.asp).

The conceptual diagram of the model is shown in Figure 12.5. The model has twelve state variables: dissolved phosphorus (P), ammonium, nitrate, phytoplankton,

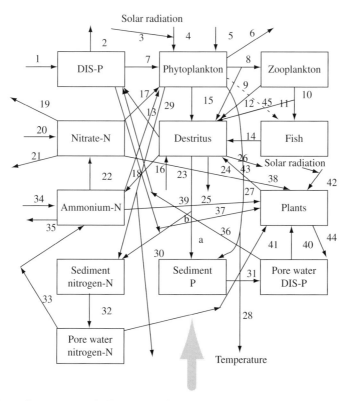

Fig. 12.5. The conceptual diagram of the shallow lake Pamolare model that has been applied to take decisions about the selection of lake restoration methods.

zooplankton, fish, submerged plants, nitrogen in sediment, phosphorus in sediment, nitrogen in pore water, phosphorus in pore water, detritus. The general structurally dynamic eutrophication model cannot be applied to shallow lakes when the competition between submerged plants and phytoplankton is an important process, or when submerged plants should be included as an important state variable. The processes and forcing functions included in the model are shown in Figure 12.5 and described in Table 12.1. The twelve state variables and differential equations for them, developed in accordance with the forty-five processes giving inputs and outputs to the state variables are shown in Table 12.2. These forty-five processes are described by forty-five different equations, which are presented in a help file in the model software.

The parameters of the model, which are listed in Table 12.3, can be established by four different methods:

1. By optimization of eco-exergy. The size of phytoplankton A and of zooplankton Z are changed by x percent after y days if the change yields more eco-exergy. This corresponds to survival of the fittest and this parameter estimation method is abbreviated EXOPT.
2. The parameters are so well known from the literature that they are not calibrated. These parameters are denoted LITT.
3. Parameters have only or mainly influence on one state variable, which can be followed when the parameters are changed. These parameters must however be recalibrated after an automatic calibration or after an eco-exergy optimization. These parameters are denoted ENKAL.
4. Four parameters have a major influence on several state variables at the same time. These parameters are denoted AUKAL. They can be found by an automatic calibration procedure included in the software.

Single calibrations can be carried out in accordance with the following guidelines:

1. When the automatic calibrations and eco-exergy optimization have been applied once, CC (the carrying capacity for zooplankton) can be calibrated to fit a good ratio between phytoplankton, A and zooplankton, Z. CC = 40 can be recommended as the initial value.
2. DNIC and NIC can be calibrated to give a good ratio between ammonium and nitrate after the automatic calibration and eco-exergy optimization have been applied once.
3. MIRSP and MIRSN can be calibrated to ensure the observed concentrations in the sediment when the automatic calibrations and eco-exergy have been applied once.

The use of the model for lake management requires that the calibrated and validated models are applied to set up prognoses, i.e., scenarios for different possible or expected changes of the forcing functions. By development of prognoses, the manual and automatically calibrated parameters obtained by the calibration are maintained while SDM size changes are presumed – this implies that SDM is activated. The number of days that the prognoses cover is selected in contrast to the calibration and validation, where 365 days = one year is always required. The results are presented in graphs and tables.

Table 12.1 *Description of the forcing functions and processes of the model Pamolare II.*

Forcing function (F) or process (P)	Description	Abbreviation
F	Discharge of inorganic phosphorus-P from various sources (wastewater, drainage water, precipitation)	TP, F
F	Outflow of dissolved phosphorus	UP
F	Solar radiation	RAD
PR	Growth of phytoplankton, determined by several factors included the forcing function and the temperature	
F	Discharge of phytoplankton to the lake from tributaries	
F	Outflow of phytoplankton	
PR	Uptake of phosphorus by phytoplankton	
PR	Grazing of zooplankton	GZ
PR	Faeces produced by grazing	FG
PR	Predation of zooplankton by fish	PF
PR	Faeces produced by predation	FP
PR	Zooplankton mortality	MZ
PR	Mineralization of detritus producing dissolved phosphorus	MIP
PR	Mortality of fish	MF
PR	Mortality of phytoplankton	MA
PR	Mineralization of detritus producing ammonium	MIA
PR	Uptake of nitrate-N by phytoplankton	ONI
PR	Uptake of ammonium-N by phytoplankton	ONA
F	Outflow of nitrate-N	UNI
F	Discharge of nitrate-N from various sources	TNI
PR	Denitrification, which implies that the nitrate is lost (dinitrogen is formed and transferred to the atmosphere)	DNI
PR	Nitrification	NIT
F	Discharge of detritus from various sources	TD
F	Outflow of detritus	UD
PR	Sedimentation of detritus 25a) covering the transfer of phosphorus, SIP, and 25b) covering the transfer of nitrogen	SIN
PR	Formation of carbon dioxide by mineralization. The carbon dioxide is transferred to the atmosphere	MIN
PR	Sedimentation of phytoplankton – transfer to exchangeable phosphorus in the sediment	SPP
PR	Settled phytoplankton – phosphorus that is non-exchangeable	NEP
PR	Sedimentation of phytoplankton – transfer to exchangeable nitrogen in the sediment	SPN
PR	Settled phytoplankton – nitrogen that is non-exchangeable	NEN
PR	Decomposition of phosphorus compounds in the sediment to dissolved phosphorus-P in the pore water	MISP
PR	Decomposition of nitrogen compounds to dissolved nitrogen (ammonium-N) in the pore water	MISN
PR	Ammonium-N transfer from pore water to lake water by diffusion	PON
F	Discharge of ammonium-N from various sources	TNA

Table 12.1 (*cont.*)

Forcing function (F) or process (P)	Description	Abbreviation
F	Outflow of ammonium-N	UNA
PR	Dissolved – P transfer from pore water to lake water	POP
PR	Uptake of phosphorus (P) from the water by submerged plants	PP
PR	Uptake of nitrate-N from the water by submerged plants	PNA
PR	Uptake of ammonium-N from the water by plants	UAM
PR	Uptake of phosphorus from the pore water by submerged plants	PNI
PR	Uptake of ammonium-N from the pore water by submerged plants	PNA
PR	Photosynthesis by submerged plants by uptake of carbon dioxide C from the water	FSP
PR	Mortality of submerged plants	MP
F	Removal of submerged plants from the lake by grazing (mainly birds)	FF
PR	Carps grazing on phytoplankton (if present)	FFP

Table 12.2 *State variables of the model Pamolare II indicating their symbols, units and differential equations.*

State variable	Symbol	Unit	Differential equation
Dissolved-P*	P	mg/l	$dP / dt = TP + MIP + POP/DYB - UP - OP - PP$
Phytoplankton	A	mg/l	$dA / dt = OP + VKA + ONI + ONA + TA - UA - GZ - MA - FG - SPP - NEP - SPN - NEN - FFP$
Zooplankton	Z	mg/l	$dZ / dt = GZ - PF - FP - MZ$
Nitrate-N	NO3	mg/l	$dNO3 /dt = TNI + NIT - ONI - PNI - UNI - DNI$
Detritus	D	mg/l	$dD / dt = MA + FG + MZ + FP + MF + MP + TD - MIA - MIP - UD\text{-} SIP - SIN \ MIN$
Fish	F	mg/l	$dF / dt = PF - MF + FFP$
Ammonium-N	NH4	mg/l	$dNH4 / dt = TNA + MIA + PON/DYB - UNA - ONA - PNA - NIT$
Sediment nitrogen-N	SNN	g/m^2	$dSNN/ dt = (SPN + SIN)*DYB - MISN$
Pore water nitrogen-N	PNN	g/m^2	$dPNN / dt = MISN - PON - PVN$
Sediment-P	SPP	g/m^2	$dSPP / dt = (SPP + SIP)*DYB - MISP$
Pore water-P	PPP	g/m^2	$dPPP / dt = MISP - POP - PVP$
Submerged plants/ macrophytes	M	g/m^2	$dM / dt = PVP + PVN + PNI*DYB + PNA*DYB + FSP +PP*DYB- MP - FF$

* P = phosphorus

Table 12.3 *The parameters of the model Pamolare II.*

Parameter	Abbreviation	Range	Unit	Method used
Max. growth rate A	GAMAX	0,5–5	1 / 24t	EXOPT
Optimum temp for A	opt	18–22	$^{\circ}$C	LIT: 20°C*
Water depth	DYB	?	m	LIT: 1.8m *
MM-constant	KP	0,01–0,2	g / m^3	EXOPT
MM-constant	KN	0,025–0,5	g / m^3	EXOPT
Ext coeff	EXT	0,05–0,5	m^3 / g m	AUKAL
Max. growth rate Z	GZMAX	0,1–1,0	1 / 24h	EXOPT
***MM-constant	KA	0,01–0,5	g / m^3	EXOPT
Carrying capacity	CC	5–100	g / m^3	ENKAL**
Temperature coeff.	TEK	1,02–1,10	–	LIT: 1,05*
Efficiency	0,60	0,5–0,7	–	LIT: 0,60*
Max growth rate F	GFMAX	0,01–0,12	1 / 24h	LIT: 0,05*
MM-constant	KZ	0,01–0,25	g / m^3	LIT: 0,05*
Mortality Z	ZMC	0,05–0,25	1 / 24h	EXOPT
Mineralization rate D	NDC	0,01–0,5	1 / 24h	AUKAL
Mortality F	MFC	0,005–0,05	1 / 24h	LIT: 0,02*
Mortality A	MAC	0,1–1,5	1 / 24h	EXOPT
Denitrification rate	DNIC	0–0,2	1 / 24h	ENKAL**
Nitrification rate	NIC	0,01–0,2	1 / 24h	ENKAL**
Sedimentation rate D	SDR	0,01–0,25	m / 24h	AUKAL
Exchangeable P in sed.	EXP	0,1'–'0,8	–	LIT: 17/29
Sedimentation rate A	SDRA	0,01–0,6	m / 24h	EXOPT
Exchangeable N in sed.	EXN	0,1–0,8	–	LIT: 0,15
Min.rate P-sediment	MIRSP	0,01–0,2	1 / 24h	ENKAL**
Min.rate N-sediment	MIRSN	0,02–0,5	1 / 24h	ENKAL**
Diffusion coeff.	DIFFC	0,01–0,25	1 / 24h	AUKAL
Max. growth rate M	GPMAX	0,02–0,5	1 / 24h	LIT: 0,2*
MM-constant M(P)	KPP	0,01–0,2	g / m^2	LIT: 0,05*
MM-constant M(N)	KPN	0,02–0,5	g / m^2	LIT: 0,35*
Max.growth rate M	GPOVMAX	0,01–0,5	1 / 24h	LIT: 0,155*
Mortality M	MM	0,005–0,1	1 / 24h	LIT: 0,041*

* Only valid for Glumsø. ** Covers single calibrations
*** MM denotes that a Michaelis-Menten equation is applied:, rate = rc. s / (s + k')

The model has been applied to set up prognoses for Lake Glumsø in Denmark, a shallow lake with a volume of about 0.5 million m^3. It has been possible by the use of the model to assess the effect on the water quality of constructing a surface wetland on the inflow side, after the waste water has undergone a mechanical-biological-chemical treatment and has been discharged down-stream. The model has also been applied on Lake Morgan, close to Ankara in Turkey, in order to assess the effect of increased submerged vegetation for the water quality. The increase of submerged vegetation was a result of reducing the water level to about 2 or 3 meters to give sufficient solar radiation. The application of a STELLA version of the model can be seen in Zhang *et al.* (2003*a*, 2003*b*). A two-layer version of the model has recently been applied to assess the effect of biomanipulation and aeration for Lake Fure, close to Copenhagen. In all these three cases, the prognoses have been partly validated and they gave a satisfactory result, in the order of 20–40 percent relative standard deviation for the most important state variables (phytoplankton, total P and total N). The model is also being applied to assess the effect of constructed wetland and biomanipulation for the Eastern Lake, close to Wuhan in China.

12.3.2 *Model of sub-surface wetland*

This model was developed for the purpose of designing a sub-surface wetland or modifiying an existing wetland to gain the capacity to treat waste water or drainage water, based on defined removal efficiencies in terms of BOD_5, nitrate, ammonium, organic nitrogen and phosphorus. To use this model, it is necessary to know: (1) the water flow; (2) the concentrations of the above-mentioned constituents in the water; and (3) the required removal efficiencies for these constituents (i.e., their concentrations in the treated water).

The sub-surface wetland being modelled consists of a constructed wetland area in which the "soil" is gravel, thereby allowing a good water flow through the wetland. The core design parameter is the volume of the wetland (denoted V). Wetland plants with an indicated density are planted in the gravel (see below under initial design parameters), and it is presumed the recommended plant density is followed.

The conceptual diagram of the model is presented in Figures 12.6 to 12.8. Figure 12.6 illustrates the organic matter sub-model, while Figure 12.7 illustrates the nitrogen sub-model, with three groups of nitrogen compounds (organic nitrogen, ammonium and nitrate), and Figure 12.8 illustrates the phosphorus sub-model.

The model state variables include BOD_5, organic nitrogen (ORN), ammonium (AMM), nitrate (NIT), and total phosphorus (TPO) in five successive boxes, making a total of twenty-five state variables (Table 12.4). As noted below, the sub-model

Table 12.4. *State variables of the model sub-surface wetland and their units.*

State variables	Unit
BOD5-A, BOD5-B, BOD5-C, BOD5-D, BOD5-E	mg O_2/L
NIT-A, NIT-B, NIT-C, NIT-D, NIT-E	mg N/l
AMM-A, AMM-B, AMM-C, AMM-D, AMM-E	mg N/L
TPO-A, TPO-B, TPO-C, TPO-D, TPO-E	mg P/L
ORN-A, ORN-B, ORN-C, ORN-D, ORN-E	mg N/L

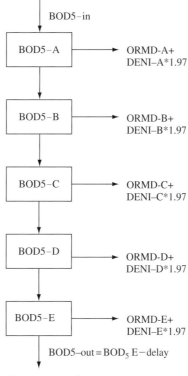

Fig. 12.6. The BOD$_5$ sub-model.

variables are expressed by three letters (e.g., NIT for nitrate), followed by IN, OUT or A, B, C, D or E, with the parameters using two letters.

The model has a number of forcing functions (Table 12.5), which the user must specify for a given model run.

A continuous transfer takes place from one state variable to another in the model simulations. This section identifies the processes that take place in the sub-surface wetland. It should be noted that the same processes take place in each box, although the concentrations are different in each box. Thus, the equations are repeated in the model program, with an indication of the concentrations in the five different model boxes – A, B, C, D and E. All the processes are expressed by four letters, followed by A, B, C, D or E, corresponding to the five boxes. The model expressions are the same for each box, although the applied concentrations of the modelled materials differ for each box. Exponent is expressed by the notation ($^$).

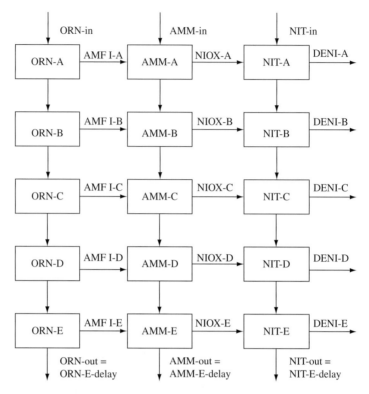

AMFI = oxidation of organic N to ammonium. NIOX = nitrification (ammonium –>
nitrate. DENI = denitrification (nitrate –> dinitrogen <)

Fig. 12.7. The nitrogen sub-model, illustrating the three nitrogen compounds
(organic–N, ammonium and nitrate).

A number of equations that represent processes taking place in the wetland are
repeated in each box, with an indication of the letters of the box (Table 12.6). See
also Figures 12.6, 12.7 and 12.8.

The model also calculates delay values (i.e., the concentrations of the five
constituents in the five boxes during one box-retention time (RTB) earlier). For
example,

$$\text{AMM-A-delay} = \text{AMM-A at time t- RTB, when t} > \text{RTB};$$

$$\text{if t} < \text{RTB, AMM-A-delay is 0.}$$

These equations are repeated for all five constituents in all five boxes, and the delay
concentrations are indicated with "-delay."

Furthermore, the simulated results are also used to determine the removal efficiencies, which are also shown on graphs. They are calculated as a function of time,
as follows:

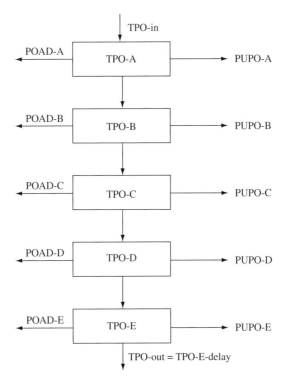

TPO = total phosphorus PUPO = plant uptake of phosphorus
POAD = adsorption of phosphorus to the gravel.

Fig. 12.8. The phosphorus sub-model.

- Efficiency of BOD_5 removal (percent) = 100*(BOD5-in – BOD5-out)/BOD5-in;
- Efficiency of nitrate removal (percent) = 100*(NIT-in – NIT-out)/NIT-in;
- Efficiency of ammonium removal (percent) = 100*(AMM-in – AMM-out)/AMM-in;
- Efficiency of organic-N removal (percent) = 100*(ORN-in – ORN-out)/ORN-in;
- Efficiency of nitrogen removal (percent) = 100* ((NIT-in + AMM-in + ORN-in) – (NIT-out + AMM-out + ORN-out))/(NIT-in + AMM-in + ORN-in); and
- Efficiency of phosphorus removal (percent) = 100*(TPO-in – TPO-out)/TPO-in.

The model parameters are indicated in Table 12.7.
The differential equations for the state variables of the model are as follows:

$$BOV*d\ BOD\text{-}5\text{-}A/dt = QIN*BOD\text{-}IN\text{-}\ QIN*(1\text{-}POM)*BOD\text{-}A\text{-}delay - BOV*ORMD\text{-}A\text{-}DENi\text{-}A*1.97$$

$$BOV*d\ BOD\text{-}5\text{-}B/dt = QIN*(1\text{-}POM)*BOD\text{-}A\text{-}delay - BOV*ORMD\text{-}B - QIN*BOD\text{-}B\text{-}delay - DENI\text{-}B*1.97$$

Table 12.5 *Forcing functions for the sub-surface wetland model.*

Forcing function	Abbreviation	Unit	Range	Default value
Volume of wetland		m3	10–10,000,000	
Flow of water	QIN	expressed as m^3/24 h	1'-'1,000,000	
Porosity	as fraction of POR	no unit	0–1	0.46
Input concentration of BOD5	BOD-IN	mg O_2/l	0–1000	
Input concentration of ammonium	AMM-IN	mg N/l	0–100	
Input concentration of nitrate	NIT-IN	mg N/l	0–100	
Input concentration of total phosphorus	TPO-IN	mg P/l	0–50	
Input concentration of organic nitrogen	ORN-IN	mg N/l	0–200	
Fraction of BOD5 as suspended matter	POM	no unit	0–1	
Fraction of organic-N matter as suspended matter	PON	no unit	0–1	
Fraction of phosphorus as suspended matter	POP	no unit	0–1	
Average oxygen concentration in Box A	AOX	mg/l	0–20	0.4 mg/l
Average oxygen concentration in Box B	BOX	mg/l	0–20	0.4 mg/l
Average oxygen concentration in Box C	COX	mg/l	0–20	0.4 mg/l
Average oxygen concentration in Box D	DOX	mg/l	0–20	0.4 mg/l
Average oxygen concentration in Box E	EOX	mg/l	0–20	0.4 mg/l
Retention time	RTT (= VOL*POR / Q; 24-hour)			
Retention time per box	RTB (= RTT / 5; 24-hour)			
Box volume	BOV (= VOL*POR / 5)			

Average temperature (TEMP; as function of time; daily average temperature is listed for the number of days to be simulated with the model). The length of model simulations must be indicated as number of days.

The final three forcing functions (RTT, RTB and BOV) are calculated

Table 12.6 *The equations representing the processes taking place in the sub-surface wetland model.*

Process	Equation
Ammonification	AMFI = ORN*AC* TA^ (TEMP-20)
Nitrification	NIOX = AMM*NC*INOX*TN ^ (TEMP-20)/ (AMM+MA)
Oxidation of BOD5	ORMD = BOD5*OC *INOO*TO ^ (TEMP-20)
Denitrification	DENI = NIT*DC*TD ^ (TEMP-20)/(NIT + MN)

INOX-A = AOX/(AOX + KO), and so on for boxes B, C, D and E, using the notations BOX, COX, DOX and EOX; however, KO is the same parameter for all five boxes

INOO-A = AOX/(AOX + OO), and so on for boxes B, C, D and E, using the notations BOX, COX, DOX and EOX; however, OO is the same parameter for all five boxes

Process	Equation
Plant uptake of ammonium	PUAM = AMM*PA
Plant uptake of nitrate	PUNI = NIT*PN
Plant uptake of phosphorus for box A	PUPO-A = TPO-A*PP*(1-POP)
Plant uptake of phosphorus for box B	PUPO-B = TPO-B*PP
Plant uptake of phosphorus for box C	PUPO-C = TPO-C*PP
Plant uptake of phosphorus for box D	PUPO-D = TPO-D*PP
Plant uptake of phosphorus for box E	PUPO-E = TPO-E*PP
Adsorption of phosphorus for box A	If POAD>0, POAD-A = TPO-A*(1-POP)* (POR) – AF*(1-POR); otherwise POAD = 0
Adsorption of phosphorus for boxes B, C, D and E	POAD = TPO*POR – AF*(1-POR)

Table 12.7 *The model parameters of the sub-surface wetland model*

Parameter	Range	Unit	Default value
AC	0.05–'0.8	1/24h	0.5
NC	0.1–'1.5	1/24h	0.8
OC	0.05–'0.8	1/24h	0.5
DC	0.25–5	1/24h	2.2
TA	1.02–'1.06	None	1.04
TN	1.02–'1.07	None	1.047
TO	1.02–'1.06	None	1.04
TD	1.05–'1.12	None	1.09
KO	0.1–2	mg/l	1.3
OO	0.1–2	mg/l	1.3
MA	0.05–2	mg/l	1
MN	0.01–1	mg/l	0.1
PA	0.00–1	1/24h	0.01
PN	0.00–1	1/24h	0.01
PP	0.00–1	1/24h	0.003
AF	0–100	None	1.0

$$BOV*d \; BOD\text{-}5\text{-}C/dt = QIN*BOD\text{-}B\text{-}delay - BOV*ORMD\text{-}C - QIN*BOD\text{-}C\text{-}delay - DENI\text{-}C*1.97$$

$$BOV*d \; BOD\text{-}5\text{-}D/dt = QIN*BOD\text{-}C\text{-}delay - BOV*ORMD\text{-}D - QIN*BOD\text{-}D\text{-}delay - DENI\text{-}D*1.97$$

$$BOV*d \; BOD\text{-}5\text{-}E/dt = QIN*BOD\text{-}D\text{-}delay - BOV*ORMD\text{-}E - QIN*BOD\text{-}E - delay\text{-} DENI\text{-}E*1.97$$

$QIN*BOD\text{-}E\text{-}delay$ indicates $BOD_5\text{-}OUT$, which appears on a graph together with measured values of the $BOD_5\text{-}OUT$, while $BOD_5\text{-}A$, B, C, D and E are shown in a table as functions of time.

$$BOV*dNIT\text{-}A/dt = QIN*NIT\text{-}IN - QIN*NIT\text{-}A\text{-}delay - BOV*DENI\text{-}A + BOV*NIOX\text{-}A - BOV*PUNI\text{-}A$$

$$BOV*dNIT\text{-}B/dt = QIN*NIT\text{-}A\text{-}delay - BOV*DENI\text{-}B + BOV*NIOX\text{-}B\text{-}BOV*PUNI\text{-}B - QIN*NIT\text{-}B\text{-}delay$$

$$BOV*dNIT\text{-}C/dt = QIN*NIT\text{-}B\text{-}delay - BOV*DENI\text{-}C + BOV*NIOX\text{-}C - BOV*PUNI\text{-}C - QIN*NIT\text{-}C\text{-}delay$$

$$BOV*dNIT\text{-}D/dt = QIN*NIT\text{-}C\text{-}delay - BOV*DENI\text{-}D + BOV*NIOX\text{-}D - BOV*PUNI\text{-}D - QIN*NIT\text{-}D\text{-}delay$$

$$BOV*dNIT\text{-}E/dt = QIN*NIT\text{-}D\text{-}delay - BOV*DENI\text{-}E + BOV*NIOX\text{-}E - BOV*PUNI\text{-}E - QIN*NIT\text{-}E\text{-}delay$$

$QIN*NIT\text{-}E\text{-}delay$ indicates NIT-OUT, which appears on a graph together with measured values of NIT-OUT, while NIT-A, NIT-B, NIT-C, NIT-D and NIT-E are all shown in a table as functions of time.

$$BOV*dAMM\text{-}A/dt = QIN*AMM\text{-}IN\text{-} QIN*AMM\text{-}A\text{-}delay - BOV*NIOX - A + BOV*AMFI\text{-}A\text{-} BOV*PUAM\text{-}A$$

$$BOV*dAMM\text{-}B/dt = QIN*AMM\text{-}A\text{-}delay + BOV*AMFI\text{-}B - BOV*NIOX\text{-}B - BOV*PUAM\text{-}B - QIN*AMM\text{-}B\text{-}delay$$

$$BOV*dAMM\text{-}C/dt = QIN*AMM\text{-}B\text{-}delay + BOV*AMFI\text{-}C - BOV*NIOX\text{-}C - BOV*PUAM\text{-}C - QIN*AMM\text{-}C\text{-}delay$$

$$BOV*dAMM\text{-}D/dt = QIN*AMM\text{-}C\text{-}delay + BOV*AMFI\text{-}D - BOV*NIOX\text{-}D - BOV*PUAM\text{-}D - QIN*AMM\text{-}D\text{-}delay$$

$$BOV*dAMM-E/dt = QIN*NIT-D-delay + BOV*AMFI-E -$$
$$BOV*NIOX-E - BOV*PUNI-E - QIN*AMM-E-delay$$

QIN*AMM-E-delay indicates AMM-OUT, which appears on a graph together with measured values of AMM-OUT, while AMM-A, AMM-B, AMM-C, AMM-D and AMM-E are all shown in a table as functions of time.

$$BOV*dORN-A/dt = QIN*ORN-IN - QIN*(1-PON)*ORN-A-delay -$$
$$BOV*AMFI-A$$

$$BOV*dORN-B/dt = QIN*ORN-A-delay - BOV*AMFI-B - QIN*ORN-$$
$$B-delay$$

$$BOV*dORN-C/dt = QIN*ORN-B-delay - BOV*AMFI-C - QIN*ORN-$$
$$C-delay$$

$$BOV*dORN-D/dt = QIN*ORN-C-delay - BOV*AMFI-D - QIN*ORN-$$
$$D-delay$$

$$BOV*dORN-E/dt = QIN*ORN-D-delay - BOV*AMFI-E - QIN*ORN-$$
$$E-delay$$

IN*ORN-E-delay indicates ORN-OUT, which appears on a graph together with measured values of ORN-OUT, while ORN-A, ORN-B, ORN-C, ORN-D and ORN-E are all shown in a table as functions of time

$$BOV*dTPO-A/dt = QIN-TPO-IN - QIN*(1-POP)*TPO-A-delay -$$
$$BOV*PUPO-A-BOV*POAD-A$$

$$BOV*dTPO-B/dt = QIN*(1-POP)*TPO-A-delay - BOV*PUPO-B-$$
$$BOV*POAD-B - QIN*TPO-B-delay$$

$$BOV*dTPO-C/dt = QIN*TPO-B-delay - BOV*PUPO-C-BOV*POAD-C -$$
$$QIN*TPO-C-delay$$

$$BOV*dTPO-D/dt = QIN*TPO-C-delay - BOV*PUPO-D-BOV*POAD-$$
$$D - QIN*TPO-D-delay$$

$$BOV*dTPO-E/dt = QIN*TPO-D-delay - BOV*PUPO-E-BOV*POAD-E -$$
$$QIN*TPO-E-delay$$

QIN*TPO-E-delay indicates TPO-OUT, which appears on a graph together with measured values of TPO-OUT, while TPO-A, TPO-B, TPO-C, TPO-D and TPO-E are all shown in a table as functions of time.

As mentioned above, the simulated values of BOD_5-out, nitrate-out (NIT-out), Ammonium-out (AMM-out), total phosphorus-out (TPO-out), and organic nitrogen-out (ORN-out) are shown in two ways: as tables and as graphs. If the measured values are available, they are shown on the same graphs, thus allowing a direct comparison.

The simulated results of the removal efficiencies are also shown on graphs, including:

- Efficiency of BOD_5-removal (percent);
- Efficiency of nitrate removal (percent);
- Efficiency of ammonium removal (percent);
- Efficiency of organic-N removal (percent);
- Efficiency of nitrogen removal (percent) and
- Efficiency of phosphorus removal (percent).

This model has been applied in the construction of nine sub-surface wetlands in Tanzania. UNEP has adopted the model to be applied for the construction of sub-surface wetlands in IRAQ for treatment of drainage water and waste water, with the objective to restore the wetlands between the Euphrates and the Tigris.

12.4 Conclusions

Models are powerful tools in ecological management and ecological engineering. Quantification, which is urgently needed in environmental management, is hardly possible without a model.

In ecological engineering models are widely used in two situations:

1. To answer the question: what is the effect of a proposed ecological engineering project, whether biomanipulation, construction of wetlands or aeration.
2. To design constructed wetlands or other constructed or modified natural ecosystems.

Two models that have been applied in ecological engineering both for ecosystem restoration and for the design of constructed ecosystems have been presented as examples and their practical application results have been mentioned briefly. From these applications, it is clear that models in ecological engineering are the powerful and useful tool that we expected from our general discussion on model applications in the introduction.

References

Jørgensen, S. E. (2002). *Integration of Ecosystem Theories: A Pattern*, 3rd ed. Dordrecht: Academic Publishers.
Jørgensen, S. E. and Padisak, J. (1996). Does the intermediate disturbance hypothesis comply with thermodynamics? *Hydrobiologia*, **323**: 9–21.

Jørgensen, S. E. and de Bernardi, R. (1998). The use of structural dynamic models to explain successes and failures of biomanipulation. *Hydrobiologia*, **359**: 1–12.

Jørgensen, S. E., Patten, B. C. and Straskraba, M. (2000). Ecosystems emerging: 4. growth. *Ecological Modelling*, **126**: 249–284.

Jørgensen, S. E. and Bendoricchio, G. (2001). *Fundamentals of Ecological Modelling*. Amsterdam: Elsevier.

Jørgensen, S. E. and Fath, B. (2004). Modelling the selective adaptation of Darwin's finches. *Ecological Modelling*, **176**: 409–418.

Jørgensen, S. E., Ladegaard, N., Debeljak, M. and Marques, J. C. (2005). Calculations of exergy for organisms. *Ecological Modelling*, **185**: 165–176.

May, R. M. (1977). *Stability and Complexity in Model Ecosystems*, 3rd ed. New York: Princeton University Press.

Mitsch, W. J. and S. E. Jørgensen. (1989). *Ecotechnology: An Introduction to Ecological Engineering*. New York: John Wiley and Sons.

Mitsch, W. J. and S. E. Jørgensen. (2003). *Ecological Engineering and Ecosystem Restoration*. New York: John Wiley and Sons.

Zhang, J., Jørgensen, S. E., Tan C. O. and Beklioglu, M. (2003a). A structurally dynamic modelling – Lake Mogan, Turkey as a case study. *Ecological Modelling*, **164**: 103–120.

Zhang, J., Jørgensen, S. E., Tan C. O. and Beklioglu, M. (2003b). Hysteresis in vegetation shift – Lake Mogan Prognoses. *Ecological Modelling*, **164**: 227–238.

13

Restoration as a bridge for cooperation and peace

AMOS BRANDEIS

13.1 Introduction

The most severe damage to our planet is usually caused in areas of conflicts, wars, poverty, growing population or increased development. The "real big issues" in these areas, in certain times, push the environmental problems aside. The urgent problems of conflicts regarding limited resources, the human desire to have more and more, or just hatred between people lead to unsustainable development and deterioration. The needs of today eclipse concern for the results tomorrow. These processes, together with others, including some on a global scale, such as climate change, cause environmental disasters today and store up even larger disasters for the future, which are created today and disregarded till they emerge, when it will be much more difficult to overcome them.

The restoration economy (Cunningham, 2002), is growing fast and more and more restoration projects are being initiated in order to deal with the problems that already exist. Also, more and more work is being done to avoid new problems. But most of the projects and actions are not being carried out in areas of conflicts, wars, poverty and growing population, because these are much more complicated and the chances of success are significantly lower. Most conflicts and wars are dealt with by the international community, either by diplomacy or by power. The war in Iraq with its historical rivers, and the former conflict between East and West Germany with the polluted Elbe River flowing between the countries, are just a few well-known examples. It seems that those involved in trying to resolve conflicts and serious problems seldom look into the common ground of the environment and the issue of restoration, as one of the means to reduce the conflicts and find a common language for collaboration. Just a few positive examples are known and recognized. One of them is Wangari Maathai, who received the 2004 Nobel Peace Prize for "her contribution to sustainable development, democracy, and peace." She was the founder of The Green Belt Movement, an organization that has mobilized thousands of Kenyans, mostly women, to plant more than 30 million trees across the country.

Ecological Restoration: A Global Challenge, ed. Francisco A. Comin. Published by Cambridge University Press.
© Cambridge University Press 2010.

Now a member of the Kenyan Parliament, she works to spread her message of peace through grass roots mobilization and by stressing that a healthy planet makes peace more possible (Maathai, 2003; Levitt, 2006).

This chapter will present and discuss two exceptional projects. The Alexander River Restoration Project (ARRP) became the bridge for cooperation and one of the features of the desired future peace between Israeli and Palestinian neighbors in an era and area where they fight each other. The Lake Bam Restoration Project (LBRP) demonstrates a collaboration among three continents, to restore a drying-out lake in Burkina Faso, one of the poorest countries in the world. Both projects are local projects, but global lessons might be learned from them and applied in other parts of the world.

13.2 The Alexander River Restoration Project

The Alexander River Restoration Project (ARRP), launched in 1995 and still underway, is a comprehensive and inter-disciplinary restoration project of a severely degraded cross-border river. Most of the actions taken so far are on the Israeli side. The cooperation between the Israelis and Palestinians regarding the elimination of pollution, building treatment plants, realization of the proposed "Peacepark" and the "Alexander River Basketball League" will be discussed in this chapter after the introduction to the river and the overall project on the Israeli side.

The Alexander River flows through agricultural land in the northern part of the Tel Aviv metropolitan area in Israel's central district. The river actually crosses borders, since it flows from the Palestinian city of Nablus into the Samarian Hills through the Palestinian city of Tul Karem up to the estuary on the Mediterranean Sea, situated north of the city of Netanya in Israel (Figure 13.1). Nature, scenery, several villages and extensive agricultural regions adorn its 32-km length. The only two cities in this catchment are Nablus and Tul Karem, both of them Palestinian cities. The river basin size is about 550 km^2. The quantity of fresh water flowing all year-round along about half its length is small. The water is less than 1.5 m deep, and in some sections no more than 20 cm. The width of the water surface is between 40 m and 2 m. In the winter floods, which occur a few times a year, the depth in certain sections can reach up to 8 m for a few hours. In the past, the river ecosystem was very rich, especially because of the relatively mild Mediterranean climate. The river is well known in Israel for the huge Nile soft-shelled turtles which live in the river. These turtles are up to 120 cm long and weigh about 50 kg. They used to live in all eastern Mediterranean rivers between Turkey and Egypt, but only in the Alexander River did one such community survive.

Over the last fifty years, the Alexander river basin and the river itself have been severely degraded (Plate 13.1). Fresh water was taken from the river and used for

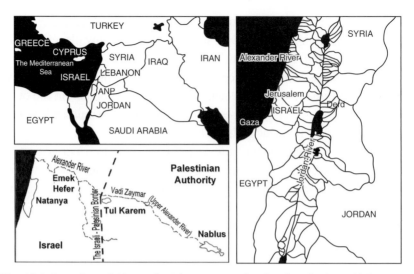

Fig. 13.1. Location of Alexander River, a cross-border river in the Middle East

domestic and agricultural use. Large quantities of raw sewage and effluent were dumped into the river. About a hundred sources of pollution have been identified. Some of the sources were in Israel, while others were upstream in the Palestinian territories, between Nablus and Tul Karem. The sewage from the Palestinian side continues to flow beneath the wall separating the Israelis and Palestinians, while the mosquitoes arising from the river fly over the wall. About half of the sewage penetrates into the ground water, and simply "disappears" in the mountain aquifer, which is the best and most important source of fresh water for the Israelis and Palestinians. The river has become a sewage channel. Most of the natural ecosystem disappeared and the water in the river became a black, stinky source of nuisance. The degraded river attracted mosquitoes and the river banks were sprayed to keep vegetation down. Flood damage was treated according to the most efficient and cheapest methods, disregarding all environmental aspects. A few meanders were shortened and concrete was used to build dams and to avoid erosion. Given its state, the local people, most of whom had learned to swim in the river in the recent past, turned their back on it. The river and its surroundings became the backyard of the area. Nobody wanted to get close to the waste along the sewage river. The river was unknown to the younger generation.

The Alexander River Restoration Administration (ARRA) was founded in 1995 with the aim of restoring the Alexander River and has been working tirelessly ever since. The administration consists of about twenty public and state entities at local, regional and national level. The leading entities of the administration are the Ministry for the Environment, The Jewish National Fund (JNF-KKL), the Emek Hefer Regional Council and the Sharon Regional Drainage Authority. The aim was to establish a voluntary, non-institutional and non-statutory administrative body. The method chosen to manage the degraded basin and the restoration project was to maintain a very simple, small and almost virtual organization, with no offices or properties, to serve as the coordinator between the existing entities. Cooperation between the administration and the Palestinian neighbors started in 1996, with the mediation and support of the German Government (BMZ Ministry working with the German Regional Development Bank, Kfw) since 1997.

The first phase of the Alexander River Restoration Project involved the preparation of a comprehensive and interdisciplinary master plan for the whole river basin (Brandeis, 2006). Three outline schemes for the various parts of the river were prepared in between 1997 and 2000. These plans were designed to establish the goals and principles of the master plan in statutory outline schemes, and to protect the open spaces along the river. Similar plans dealing differently with the land-use issue were prepared in Israel for the most polluted river in Israel – the Kishon River (Brandeis, 2001), and for the Yarkon River, which flows through the city of Tel Aviv (Rahamimoff, 1995).

Despite the ARRA's non-institutional character, it has run a large variety of projects since 1998, which are of a truly diverse nature and cover different sections of the river. The Demonstration Project is the *pièce de résistance* of the river restoration. The project involved the scenic, visual, ecological and drainage restoration of a section of the river of approximately 750 m. This part of the river was chosen because of its run-down condition prior to the project's development and its prime accessibility and exposure to the adjacent highway. The essence of the project was to develop an attractive river park intended to serve as an experimental project from the point of view of many aspects. The planning and implementation were based on international experience acquired from running projects of this type, such as the Skerne River Project in Denmark (Hansen, 1996) and the Cole in England (Vivash, 1999). The result attains a pre-agreed balance between drainage, ecological, scenic, tourist, maintenance, implementation and other considerations. A few more river parks have been developed over the last years, along different sections of the river and its main tributary. The nature of each development differs according to the setting and natural resources of each park (Plate 13.1).

The removal of pollutants is the main issue handled by the administration. Most sources have been eliminated since the administration was established. More projects have been carried out to protect the rare, huge Nile soft-shell turtles that live in the river and to rehabilitate the natural ecosystem (for example, the building of riffles and fish-ladders, based on experience from Denmark and Scotland). One component of the project was called "Saving Alex," referring to the turtles that are unique to this river. Basking areas and protected egg-laying areas were developed for the turtles.

Since the inception of the project, the principles of transparent and accountable planning have been among the top priorities. The importance of public participation as a major factor for success has been studied in many projects (Riley, 1998). The plans were designed in an intensive process of collaboration with the public. Children and students have taken part in various activities, such as planting trees along the river, documenting the restoration tasks, returning fish to the river, taking part in guided tours along the river, preparing assignments, etc. Each school in the area has a special curriculum for the students that covers the Alexander River Restoration process. Students have been given special lectures and tours, and they have prepared a wide variety of impressive papers, songs, etc. An annual small-scale river festival and river parade have been carried out in the last years. The river restoration has become a very important component of the vision of the region as a whole.

The Alexander River Restoration Project has won five awards, among them the 2003 International Thiess Riverprize, awarded annually in Brisbane, Australia. The main criterion for winning the award is the ability to demonstrate progress and achievements in affecting real river management outcomes (Nelson, 2004). The creation of the Lake Bam Twinning Restoration Project was a consequence of the Alexander River Restoration Project winning the award.

13.3 The Lake Bam Restoration Project

Lake Bam, a natural lake in Burkina Faso that is drying out, suffers from the impact of sedimentation, desertification, over-grazing, possible climate change, local human activity and increased use of its water by the fast-growing population.

Lake Bam is 115 km north of Ouagadougou, Burkina Faso's capital, in western Africa. The lake is part of the Nakanbe (Volta) river system, which flows through the Central Plateau of Burkina Faso. The plateau is about 300 m above sea level, and has a dry southern-Sahel climate. The mean annual rainfall is about 550 mm during the wet season, which lasts only three to four months. Only during this season does the river flow through the lake. The dry season is very hot (39 °C on average), and the annual evaporation is up to 1500 mm per dry season.

The length of the lake usually varies between 15 and 25 km, and the width is between 0.8 and 4 km. The lake is from 1.5 to 3.5 m deep during the wet season and only from 0 to 1 m deep during the dry season, and in extreme droughts it can dry out. This has happened four times in the last century. In recent years, large parts of the lake have dried out and divided the lake into a few separate lakes (Plate 13.2, Figure 13.2). The catchment area is about 2600 km^2. The volume of the water in the lake varies between 20 and 34 million m^3 during the wet season to between 0 and 9 million m^3 at the end of the dry season (Figure 13.3); between 15 and 20 million m^3 evaporate during the dry season while another 5 to 10 million m^3 are used for irrigation directly

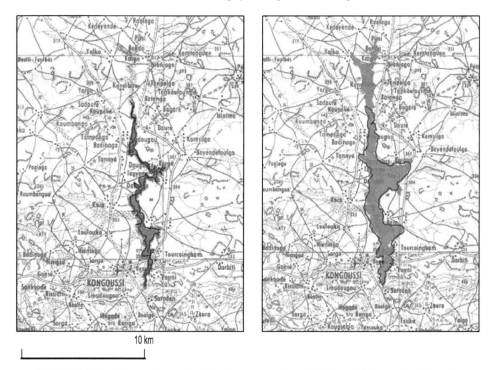

10 km

Fig.13.2. Lake Bam at the end of the dry season, June (left) and at the end of the wet season, September (right).

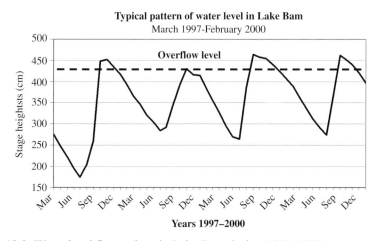

Typical pattern of water level in Lake Bam
March 1997-February 2000

Fig. 13.3. Water level fluctuations in Lake Bam during 1997–2000.

from the lake. Only less than 25 percent of the total volume remains in the lake at the end of the dry season.

The local population, about 60,000 people, live in mud houses in a town (Kongoussi) and in forty small villages around the lake (Plate 13.2). More than 45 percent live under the poverty line (US$ 0.5 a day). Almost all of the population subsist on sorghum millet and corn, which they grow in the wet season. Some families live on selling wood which they cut from wild trees and carry to the cities. About 560 families live on fishing in the lake. The largest source of income in the region is irrigated agriculture.

The lake is in a process of deterioration. For many hundreds of years the scattered population in the region lived in harmony with the shallow lake, in a sustainable environment. Natural sedimentation, high evaporation, and extreme droughts were natural factors that affected the lake. The constantly growing population (the population of the nearby town Kongoussi has grown from 1,100 residents to about 27,000 residents in the last 40 years), together with the global impacts of human activity, has caused over-exploitation of the natural resources (e.g., exploitation of water resources), the development of irrigated agriculture, sedimentation caused by agricultural activity around the lake and erosion of the fields, deforestation, over-grazing, an attempt to build a dirt dam through the lake, and more.

New sources of pollution, such as herbicides and pesticides, also threaten the lake. The problems of lack of organization and management and inadequate monitoring are also severe. Water is seen as a free public good and no water resource management exists at all. The results are that, on the one hand, the lake dries out more frequently in the dry season, often completely (a disaster for the population and the environment), and on the other hand, during the wet season the lake floods the nearby villages more frequently, as well as the agricultural fields. This causes damage, and restricts the farmers to one sowing cycle, which hinders attempts to reduce poverty.

The preparation of a feasibility study (Brandeis, 2006) regarding the restoration of Lake Bam, the Lake Bam Restoration Project (LBRP), started in the year 2006. The International River Foundation (IRF), based in Australia, has committed itself to fund this feasibility study, as a twinning project between the Alexander River Restoration Administration (winner of the 2003 International Riverprize) and the Ministry of Agriculture, Hydraulics and Fishery in Burkina Faso (Plate 13.2).

As part of the feasibility study, a few major conclusions were drawn from the analysis:

- Only 0.1 percent of the potential water (rainfall) is used by the population.
- The frequency of overflows during the wet season has increased in recent years.
- There is no correlation between rainfall and the overflows.
- There is a correlation between water use for agriculture and minimum water level (up to 30 percent of the water is used for irrigation during the dry season).
- Up to 60 percent of the water evaporates from the lake during the dry season.
- Large quantities of sediments are settled on the margins of the lake.
- The volume of the lake is shrinking, estimated at 3.3 million m^3 in forty years, equal to 9 percent of the total volume.
- There are frequent floods because of the sediments.

- The ecosystem (fauna and flora) has been severely damaged because of the deterioration of the lake.
- The water quality is deteriorating especially because of the herbicides and pesticides used in agriculture and because of livestock entering the lake.
- There is evidence of diseases in the population.

The multi-dimensional problems of the lake and their implications, need a comprehensive, multi-dimensional, integrated restoration strategy. The solutions that the feasibility study recommends are structured on three levels, which can and should support each other:

1. Physical and environmental works at catchment level, to reduce erosion and sedimentation.
2. A more efficient and sustainable use of natural resources, at catchment level.
3. Conditioning works in the lake and its surroundings, to secure water in the dry season and avoid floods in the wet season.

The projects and actions proposed by the feasibility study will hopefully be implemented in phases.

13.4 Lessons learned about how restoration projects become bridges for cooperation

The Alexander River Restoration Project (ARRP) and the Lake Bam Restoration Project (LBRP) are very different projects in very different situations. Thus the two projects provide case studies for cooperation on restoration projects from which restorers may learn lessons that they can use on other projects in different situations and locations. As restoration projects in general have many common features, some of the lessons from one project might be helpful elsewhere. The lessons must be learned carefully and adjusted to specific situations because, unfortunately, there are no prescriptions for planning and for cooperation between people. Here some results and lessons that can be learned from the work done on these two projects are presented and discussed.

13.4.1 *A disaster is usually the starting point for cooperation*

The first thing that brings people together to cooperate on an environmental issue is usually a shocking disaster. This is exactly what happened in the case of the Alexander River. For many years the pollution of the Upper Alexander River, called Vadi Zaymar, was almost ignored and bothered almost nobody. This area was ruled in the past by the Ottoman Empire, the British Mandate, Jordan, and Israel. With the establishment of the Palestinian Authority, this area became part of it. Sewage from the two Palestinian cities, the many Palestinian villages, refugee camps and industries penetrated directly into the ground water, or was dumped into the river channel. The sewage of the 250,000 people in this basin simply disappeared and it seems that no one really cared about it. No sewage treatment had been done at all throughout history. Until 1996 the Israeli regional drainage authority built a dirt dam on the border between the Palestinians and Israelis every year, and stopped the sewage that was running downstream (rather than penetrating into the ground water) from flowing into Israel. The seasonal "sewage lake" created there was used partly by the nearby farmers for irrigation. In the winter the floods would wash the sewage into the Mediterranean Sea, and it caused no actual damage to the Alexander River.

This so called "idyllic situation" changed at once in 1996. The peace process, which started in the early 1990s, was to be blamed for this sudden change. The beginning of this process raised high expectations for a better future in the Middle East. These expectations led to a building boom in the Palestinian cities. The new houses were connected to a central sewage system, which had no treatment facilities. The sewage was dumped in larger and larger quantities into the riverbed. In October 1996, the dirt dam on the border could not hold back the large quantities anymore; it collapsed and the black sewage flowed downstream. This was during the olive harvesting season, when more than twenty-five Palestinian olive mills dumped the concentrated waste of the olive mills into the riverbed. This is highly concentrated organic matter, with toxic ingredients (phenols and others). These events occurred at the end of the summer, when the quantity of water in the river naturally reaches it lowest point. The ecological disaster that followed can hardly be described. The whole river became totally black, with oil on its upper surface and white foam (stemming from detergents) and hundreds of thousands of dead fish floating on this liquid (Plate 13.1). The situation was shocking, especially since the Alexander River Restoration Project had started and there was already a master plan (Brandeis, 1996) defining the actions to be taken to restore the river.

The starting point of the Lake Bam restoration project was less shocking; it started as one of the twinning projects initiated by the International River Foundation (IRF). However, the almost dry lake at the end of the dry season, and the major floods at the end of the wet season were clear signs of an impending environmental and human crisis.

13.4.2 *A situation where the problem cannot be solved by one group alone leads either to conflict and war, or to collaboration*

It was clear from the outset that the major problem of the sewage contaminating the Alexander River could not be solved only by the Israeli side. Therefore, the immediate response of the Chairman of the Alexander River Restoration Administration (ARRA), Mr. Nachum Itzkovitz, who served also as the Mayor of the Emek Hefer Regional Council (1994–2006) was that cooperation with the Palestininian side was needed. War was out of the question. This was a time when everybody in the Middle East was hoping for peace and believing it to be a viable option. An immediate meeting was held in the Palestinian city of Tul Karem. After a few hours of talks, both sides signed a treaty, that covered the essential cooperation required to solve together the environmental and health problems caused by the sewage on both sides. This led to the preparation of a common survey and plan, prepared during one year of joint work by Palestinian and Israeli experts. The draft of this plan, together with a second treaty signed by both sides in 1998 in the Emek Hefer Municipality, was the basis for the German government to become the major mediator, coordinator and donor of this project (the German finance was only for the Palestinian side).

In Lake Bam in Burkina Faso, conflicts regarding the use of the precious water at the end of the dry season and protection against floods during the wet season actually began. There was no armed conflict, but as the the farmer cooperatives lost power, every farmer started to do whatever he thought would be best for him. This included the construction of terraces to protect their fields, or the opening of the outflow of the lake at the end of the wet season to lower the level of the water and reach specific fields. This action actually meant a loss of the precious water which was so badly needed at the end of the dry season. Therefore, it is quite clear that the establishment of a cooperative twinning project with the Alexander River Restoration Administration was a key element in the reduction of local conflicts and the resolution of problems through international cooperation.

13.4.3 Brave and powerful leaders are a pre-condition for success

The two restoration projects could never have happened if they had not been led by brave and powerful leaders. The leaders of the two projects have different characteristics and positions, but the leadership they have given has made a crucial contribution to the success of the projects.

The leader of the Alexander River Restoration Administration, the mayor of the Emek Hefer Regional Council, was a relatively young, ambitious, and well-educated leader. He knew how to bring all the stakeholders together to the table, reduce the conflicts between them and get decisions with wide consent. He invested a lot of time and effort on this project. Despite being the mayor for more than 30,000 inhabitants with many other issues requiring his attention, he was always there to solve problems and to initiate the next steps, and even to talk with the project manager (Amos Brandeis) usually more than once a day. His good relations with ministers, officials and high-ranked officers in the Israeli army were always useful for the restoration project.

On the Palestinian side, the first leader, the governor of Tul Karem and the second, the mayor of the city, are brave local leaders who share a vision. It was not always easy for them to cooperate with Israelis during times of fighting, curfews, suicide bombers, etc. in the Middle East in general and in this area specifically. Secret meetings held in neutral hospitals were a key for cooperation during times of heavy fighting. Thus the belief of local people in the project allowed this cooperation to continue even during the hardest times of the Intifada in the Middle East.

The Lake Bam project is led by very different leaders in a very different situation. There are actually three leaders, one for each participant. On the Australian side, Mr. Martin Albrecht, the chairman of the International River Foundation (IRF) is the man behind the idea of the twinning projects, and the one who succeeded in getting the budget for the feasibility study allocated, thanks to his passion for helping Burkina Faso and the environment. On the Burkina Faso side, the Senior Minister for Agriculture, Water Resources and Fish, is leading the process, making decisions, and trying to raise the money necessary for the implementation phase. On the Israeli side, Mr. Eitan Israeli, the Honorary Consul of Burkina Faso in Israel, is leading the cooperative effort. He is what is often called "Africa sick," (Golan and Ron, 2005) which means that he returns again and again to try to help the local people, in his case over a period of 45 years. He is the one who opens all the doors in Burkina Faso for the planning team. Without this extraordinary combination of leaders this project would have failed from the beginning, like many other projects do.

13.4.4 A clear vision is needed for cooperation

The leaders and professionals of both projects have clearly defined the vision of their project from the beginning. Such a clear vision is actually the guideline for the whole work throughout all the stages. In such complex situations it is easy to find oneself taking a detour which leads to a different direction. Therefore, a clear and simple vision is very important from the beginning.

In the case of the Alexander River the vision of the chairman of the Alexander River Restoration Administration was to restore the river and make it the artery of the life of the whole region, as it was when he learned to swim in the river as a child, about 45 years ago (Plate 13.1). The vision of the Palestinian leader was to improve quality of life and to reduce health risks for the population he represents. The common vision of both leaders was to demonstrate that Israelis and Palestinians can cooperate and live in peace as neighbors. Unfortunately this is so far one of the very few examples of such cooperation in the Middle East, and therefore the demonstration was badly needed.

In the case of Lake Bam, the vision was very different. The International River Foundation aims to promote global cooperation on river and water issues through twinning projects. The International and National Riverprize winners of each year are urged to find a twinning partner. The winner can choose his twin, and the IRF supports this cooperation, acting as a catalyst. So far, twinning projects are underway between Australia and Kenya, Canada and Argentina, the United States and Russia, France and Senegal, Australia and Thailand, and others. Each twinning project tackles a different situation with its own specific issues. The twinning projects are presented annually during the Riversymposium held in Brisbane each September.

The vision of the Senior Minister in Burkina Faso and the Alexander River team is to restore the dying lake in a sustainable way; it is expected that the project will be a case study for many other lakes and reservoirs in western Africa which suffer from similar problems.

13.4.5 A basic understanding of the other side is essential – listening and learning

Cooperation usually takes place between partners who are different. In the case of these two projects the differences are enormous. They are evident almost in any aspect one can think of. Israelis and Palestinians, have not only been in conflict for many decades, but also differ in culture, language, mentality, socioeconomic status, etc. For example, the Israeli GNP per capita is about ten times higher than the Palestinian one. Israel as a country of immigrants also represents differences in background and mentality (though smaller). The four people mainly involved in this project on the Israeli side have different origins: Israel, Iraq, Germany and Kurdistan. The Germans, as mediators, come also from a different background and situation.

The Palestinians speak Arabic among themselves, while the Israelis speak Hebrew and the Germans speak German. Although on every side there is somebody who speaks the languages of the others, all the meetings are held in a fourth language, English, which became the "neutral" language of the team.

On the Lake Bam project the differences are even bigger. The partners represent three different continents, Africa, Australia and Asia (a part that borders Europe and has strong American influences). While the GDP of Australia is 130 times bigger than that of Burkina Faso, the population is slightly less than twice. Life expectancy in Burkina Faso is about forty-seven years, while in Australia and in Israel is around eighty. The culture and mentality of these three countries could not be more different.

The team uses four languages during the planning process, and uses four translators. The African officials and team members speak French, the Israelis Hebrew. The material is written in French, Hebrew and English, and all that is in Hebrew is translated into English and French. At meetings with the local stakeholders by the lake, Amos Brandeis shows a presentation in French, speaks English and a translator explains in the local Mosse language.

There is no other way to overcome the differences between participants than to listen to each other very carefully, learn from each other, respect each other and understand each other. The spoken or written language should not be a barrier for having a real common language. The possession of a higher degree of education does not mean people are smarter or know more about the problems that must be solved. People anywhere have their knowledge and wisdom and their way of expressing it. The path for success relies on mutual understanding and cooperation as equal partners. Patronizing other people does not work.

13.4.6 *A definition of common interests is the basis for cooperation*

To understand each other and to learn from each other might be the first step for collaboration. But success depends mainly on the deep understanding and definition of the interests of each side and especially the mutual interests. In such complicated situations, the partners must have not only a vision regarding the restoration project, but also real interests which prove good reason to cooperate and which support the vision – interests the satisfaction of which will yield real achievements, satisfaction and pride.

Both partners in the ARRP deeply believe in improving the environment, solving the health problems and demonstrating that collaboration between Israelis and Palestinians on environmental issues is possible. But apart from this vision, both sides share also more concrete interests. Most of the sewage produced on the Palestinian side simply disappears by penetrating into the aquifer. This sewage is a "time bomb," which both sides know will explode soon. The best fresh water, an essential resource for both sides, is being polluted. Doing nothing to protect it means that the next generation will find it even harder to live in peace with a larger population and fewer resources. But even more urgent is the problem of diseases that stem from the open sewage flowing in the riverbed between houses on the Palestinian side. This sewage is the ideal ecosystem for mosquitoes to spread. Some of them carry the deadly West Nile Fever disease, which kills people on both sides every year. The construction of the security fence by the Israelis in 2004 in this area did not solve the problem. Unfortunately the mosquitoes can fly higher then the four-meter height of the fence, need no visa to cross the border and cannot been shot when they do so. Therefore the mutual interest in killing the mosquito larvae while still in the sewage led to the Israelis and Palestinians taking up the fight together against the common enemy – spraying them on both sides of the border, even during the hardest times of the conflict.

Economic interests play a role as well. The Palestinian economy is eager to encourage international investment which will bring money from outside and supply jobs to the poor population. The Palestinians, Israelis and Germans understood this desire and had the idea of holding international tenders for the rehabilitation of the Tul Karem sewage ponds. The companies who were successful used Palestinian contractors and the work boosted the local economy and supplied income to many Palestinian families.

The Israelis wanted to solve the environmental problem as soon as possible and to avoid security problems directly on the security fence. They found that collaboration between the Palestinians, Germans and Israelis enabled the work to ccontinue right next to the border, even during times when the Israeli army did not allow other Palestinians to enter this area.

The interests of the partners in Burkina Faso are totally different from those of the partners in the Middle East. The Australians want the work to be carried out scientifically and professionally so that they can secure the international funds for the implementation. They also want this restoration process to be a kind of template and case study for other twinning projects initiated by the IRF. The Senior Minister and his team want the work on site to start as soon as possible. The ARRA team tries to find the best way in between. It seems that all partners understand the interests of the others and have found a common language.

13.4.7 Mediators can help

In times of conflicts and wars, it is sometimes hard to talk with each other. In the case of the ARRP, the first contact was made by a mediator who arranged the first meeting between the Palestinians and Israelis in Tul Karem. The mediator was an Israeli Arab, who understands both sides well, and has rich experience as an ex-mayor of an Arab town in Israel. That first successful meeting and the general positive atmosphere of the time in the Middle East helped the partners to cooperate directly. Then when the situation deteriorated, the German government agency who came on board in 1998 was helped greatly by not only funding the project (on the Palestinian side), but by providing fair mediation. Both sides appreciate very much the sensitive work of the Germans in this respect, and their contribution to the direct dialog between the partners.

In the Lake Bam project no mediator was needed, but the Honorary Consul of Burkina Faso in Israel actually is a mediator in many situations. He knows all the partners very well and is the right person to help the partners to overcome differences and misunderstandings such as those that occurred during the planning process.

13.4.8 Direct talks are important to gain mutual trust

Mediation is important but nothing compares with direct talk and communication. After major differences were overcome, working together and respecting each other helped all partners to gain trust, to talk directly and openly with each other, and even to become real friends. Common meals and informal interaction led to a situation where everybody trusts each other. Mutual trust is a key element in working together. This process happened in both projects.

13.4.9 Cooperation must fit to culture and mentality

Trust is built not only by becoming friends and never letting one's partner down, but also by respecting the partner's culture and mentality. In both projects the partners are very different and come from very diverse cultures and backgrounds. Attempts to impose processes, ways of behavior or solutions which do not fit with the culture, values or mentality of the other side will probably fail.

13.4.10 Avoiding the "big issues" can help to solve the environmental problems

In most cases armed conflicts are not caused by environmental problems. This is certainly the case for the conflict between the Palestinians and the Israelis in the Middle East. Therefore, the method of "talking sewage only" was found very useful in this project. The potential for conflict in the meetings held between the Israelis and the Palestinians during the hardest time of the Intifada was enormous. Many "big issues," inherent differences and political issues could have caused a total failure of the talks. The environment is a much easier common ground for agreements and mutual understanding. But every decision has a direct impact on other political or "bigger" issues.

One of the most tense secret meetings held in a hospital during one of the worst times of the armed conflict began with a clear statement from both mayors: "We talk sewage, only sewage. Whoever will say anything related to political issues will have to leave the room immediately." This helped. Even the most sensitive issues discussed during this meeting were agreed as part of the rule "sewage only." One of the controversial questions was the question of who would treat the sewage, on which side of the border, and who would use the treated effluent. The Oslo agreements between the Palestinians and Israelis say that each side will be responsible for the treatment of its sewage and reuse. However, Oslo is far away from the sewage running in the riverbed in the Middle East, contaminating both the underground water and the river. While this principle is generally agreed by all, realistic short- or medium-term solutions must be implemented. Stages are required, and not everything can be done in the short term. The topography, land use, landownership and other engineering and economical aspects of the area round the Alexander River does not currently allow the full treatment of the Palestinian sewage on the Palestinian side. The option of doing nothing would maintain the catastrophic situation. Therefore, the partners decided in that meeting to avoid any discussion on the long-term policy, and the political issues surrounding it. The next stage, based on the engineering and environmental recommendations, was adopted and later implemented – the sewage ponds of the city of Tul Karem were rehabilitated (actually rebuilt), and connected to the treatment facilities on the Israeli side. This does not rule out any future solution in which the Palestinian sewage will be treated on the Palestinian side.

13.4.11 An integrated comprehensive approach is essential

To restore an ecosystem is not an easy task. There are many different facets which must be dealt with carefully by the partners. One of the most important, which must be put on the table right at the beginning, is the necessity for taking a comprehensive approach. All partners should understand that only an integrated comprehensive approach can really heal the environment, especially in areas of extreme conflicts, poverty or special problems. Easy answers that do not deal with all aspects of restoration may lead to failure. Partners who cooperate in such complex situations cannot take the risk of a failure. A failure means not only further deterioration of the environment, but also the end of hope for cooperation and peace.

The ARRP demonstrates a comprehensive approach to river restoration. It deals with many different aspects in an integrated manner. Issues of ecology, water quality, water quantity, landscape, land uses, drainage, soil preservation, economy, education, public participation and many others are addressed. Solving the sewage problem stemming from the Palestinian side is only one aspect of the overall project, though a very important one. But the project cannot focus only on the removal of the sewage. If park development, school programs and community events have to wait until the total removal of the sewage, nothing will actually happen. The different projects and actions, within the total framework of the integrated project, provide the synergy for the overall success of the restoration project. One can compare it to a Rubik's Cube (named after the man who invented this sophisticated toy, which was extremely popular in the 1980s). The "hands" that are needed to solve Rubik's cube are in this case the vision and the people.

The need for a comprehensive approach equally applies to the Lake Bam restoration project, which is in its initial stages. Solving only the immediate problem of the lake in engineering terms will not help to realize the overall vision of the restoration of a lake with a rich ecosystem, which is the source of life for the people,

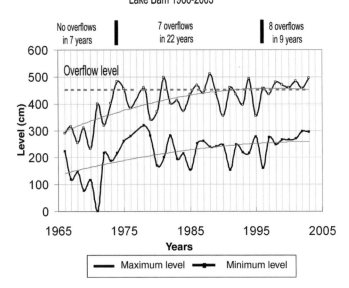

Fig. 13.4. Frequency of overflows of Lake Bam showing an increasing trend during recent years.

and supports the standard of living of the fast growing population (Plate 13.2). It is not enough just to solve the frequent flooding (Figure 13.4) and have more water in the lake during the dry season.

13.4.12 *A practical approach is useful to achieve real outcomes*

Environmental crises in an area of conflicts or severe poverty are not a theoretical issue, or just a subject for academic research. Usually, in these areas really urgent problems with dramatic impacts need urgent actions. It is also reasonable to predict that in these areas people are practical and want to cooperate, despite the situation, and this is because they understand that urgent actions are vital. This is at least the situation in the two projects which are examined here.

In both cases the local leaders wanted solutions on the ground. In the case of the Alexander River they wanted efficient treatment of the sewage and removal of the pollution from the riverbed. In the case of Lake Bam the immediate need was simply to avoid the drying-out of the deteriorating lake, and to reduce the risk of flooding. The aim was to secure water supply to the local population and to the ecosystem. In both cases the leaders wanted to see results on the ground as soon as possible. The results achieved so far in the ARRP act as an important catalyst to move on with the project.

13.4.12 Small successes are better than large failures

The question of scale and ambition is always a major one when dealing with restoration projects. On one hand, we would like to accomplish real success which will have a major impact. On the other hand, a large problem is much harder to solve. The case study of the ARRP is an excellent example for analysis with regard to this question.

Talks and cooperation between the Palestinians and their Israeli neighbors started in 1996 and the Germans came on board in 1998. Until the year 2000, all the talks concentrated on the overall, long-term solution to the sewage problem. The survey conducted by the Palestinian-Israeli team and more feasibility studies conducted by German experts were focused on the overall solution. Many reports were written and many meetings and workshops were held. Meanwhile, the pollution continued and the fish died.

In the year 2000 the ARRA decided to start the first stage of the project on the Israeli side of the border (financed by Israel). This was a few months before the second Intifada started, but it was not a result of the worsening situation in the region. It was due to the wish of the Mayor to move on and demonstrate real outcomes, which would facilitate the restoration of the river downstream and the development of parks on the Israeli side. This decision was followed by the construction of the Nablus Stream Emergency Project, also known as the Yad Hannah treatment plant. This treatment facility, which needs further improvement, partially purifies the sewage stemming from the Palestinian side of the border.

In 2003 a very important workshop took place in Jerusalem (Plate 13.1). In this workshop, for the first time, the partners agreed that the overall solutions should be put aside and that they should concentrate on the first stage of the implementation on the Palestinian side. An action plan was prepared and the timeframe was agreed. Within a short time a tender was published for the rehabilitation of the Tul Karem treatment ponds, and the project was carried out by Palestinian contractors. The funding was German, and a German engineer together with the Tul Karem munici-pality supervised the works. The Israeli partners helped by sorting out military and other organizational aspects.

The treatment ponds (Plate 13.1) were inaugurated in a ceremony held early in 2005 with about 75 participants from each side and Palestinian and Israeli journal-ists. The ceremony was organized by the project managers of both sides. The two mayors expressed their pride in this extraordinary achievement, which has helped further phases of the overall project to progress.

The same principle is being applied in the Lake Bam project. The question of implementing the overall project as a whole, implementation of phases, or starting with a pilot project, is under discussion.

13.4.13 Demonstration projects are an excellent starting point

After three years of planning, the decision was made in 1998 to start the implementation of the ARRP. The question of what to start with was raised, as this was actually the first river restoration project in Israel (together with the Yarqon River; see Rahamimoff, 1995). A demonstration project was chosen with the aim of demonstrating the potential of restoring the river. The most deteriorated section of the river, in the most accessible part (next to the main highway) was chosen. A 750-m long park was developed and called "the demonstration project." Every important guest visiting the region, including Ministers, officials from the Ministry of Finance, Parliament members, guests from other countries, etc. was brought to visit the demonstration project. The impact was very big; it helped to raise about fifteen times more money for the project than the demonstration project actually cost. This park is still the most attractive one along the river and more than 5,000 people come to visit it on nice weekends.

This concept of demonstration project as leverage for the overall project was used again when the Nablus Emergency Sewage Project was implemented and work to clean the riverbed on the Palestinian side began in 2006. It is planned that this section will become a "Peacepark", situated along the river on both sides of the border. To start with the Palestinian park and the Israeli park will be developed separately, but in the future, when the security fence is hopefully not needed anymore, the children will be able to play together on both sides of the border. The basic concept of the Peacepark was developed and implemented in South Africa. Hopefully the concept will, with some adjustments, be implemented on the Alexander River, and will act as a demonstration project for other cross-border rivers.

The planning team on the Lake Bam project have proposed the development of a demonstration pilot project near the city of Kongoussi in Burkina Faso. The idea is to start right away to build an internal reservoir, i.e., a water reservoir within the floodplain. A location near the city (where it is badly needed), and adjacent to the main transport junction (where it is most accessible and visible) was chosen. At this stage no decision has been taken regarding the implementation of the demonstration project.

13.4.14 Technical knowledge exchange is key for cooperation

Partners expect to learn from each other also as professionals. The idea that partners should exchange knowledge and help each other is very important and when they do this it benefits the project and the results.

A good example is the trash traps developed by the ARRA on the Israeli side. These traps, built across the riverbed, are aimed at stopping trash from flowing downstream during floods. The Israelis gave the concept, the plans and recommen-dations for improvements to their Palestinian colleagues, who implemented them on the Palestinian side as well.

The whole foundation of the twinning projects initiated by the IRF is the exchange of knowledge on an international level. This is certainly happening during the planning process of the Lake Bam restoration project.

13.4.15 Cooperation on a restoration project can be leverage for cooperation in other fields

The environment can be the starting point for cooperation in many other fields. Collaborating, gaining mutual trust and working together may lead to more colla-boration between the partners, maybe (but not necessarily) on environmental issues. The Alexander River project is again an interesting example in this respect. Thus even though this project is being implemented during a major conflict in the Middle East, new pathways for collaboration are being sought. One recent example is the establishment of the Alexander River Basketball League. The idea, supported by an American fund called "Peace Players," is to have a basketball league of children's teams from communities along the river. At this stage the plan is to have a Palestinian team from Tul Karem, an Israeli Arab team from a town in Israel and an Israeli Jewish team from a kibbutz. It is intended that the teams will practice then meet a few times a year for a tournament. The tournament will also be the opportunity for common social activities and for people to experience the ecology of the river, learn about it, and develop the shared responsibility for its health which all the people who live along it must share. This project was launched in January 2007. Friends of the Earth initiated a very succesful cooperation between children from both sides in a project called Good Water Neighbors.

13.4.16 Patience and belief are crucial

Projects in areas of conflicts, poverty and wars are not easy and need real detemina-tion and creativity. As the Mayor of Emek Hefer always used to say: "Where there is a door, open it. Where is no door look for a window, and if there is no window look for one till you find it." Belief and patience are major keys for success in these complex situations.

13.5 Conclusions

The environment can be good common ground for cooperation anywhere, but especially in areas of conflicts, wars, poverty and growing populations. The impact

Table 13.1. *Comparison of major characteristics and conclusions from the Alexander River Restoration Project and Lake Bam Restoration Project case studies.*

Feature	Alexander River Restoration Project	Lake Bam Restoration Project	Common facts or features
Type of project	River restoration	Lake restoration (a river pool).	Part of a river system. Affected by the whole basin.
Main issue	Pollution	Lack of water and floods.	
Major figures	550 km² catchment. 32 km long. 2–40 m wide. 0.2–1.5 m deep.	2,600 km² catchment. 15–25 km long. 0–4,000 m wide. 0–3.5 m deep.	Quite similar dimensions (except from one being a flowing river and the other a river pool).
Location	Asia – Middle East	Africa – north west.	
Year launched	1995	2004	
Starting point	Ecological disaster – heavy pollution (sewage).	Initiative to start a twinning project with the Alexander River Restoration Project, after it won the 2003 International Riverprize.	
Initiator	Local Regional Council, with the help of the Ministry for the Environment, an environmental NGO (JNF), and others.	International Riverfoundation (IRF) urged the Alexander River Restoration Project to seek a twinning project in a developing country.	
Participating entities	On the Israeli side, Alexander River Restoration Administration. On the Palestinian side, the town of Tul Karem.	Africa – The Ministry for Agriculture, Water Resources and Fish in Burkina Faso. Asia – The Alexander River Restoration Administration. Australia – IRF.	Very different entities cooperate.
Source of funding	On the Israeli side, Government, NGOs, local authorities and private donors. On the Palestinian side, the German government (through Kfw).	International River Foundation, Australia.	International funds

Planning team	Interdisciplinary planning team headed by architect Amos Brandeis.	Interdisciplinary planning team headed by architect Amos Brandeis.	Interdisciplinary planning team with professionals from all partners.
Main problems	The Middle East conflict. A cross-border river. Population growth. Heavy pollution from many sources.	Deterioration of catchment. Desertification and possible climate change. Population growth. Extreme poverty and no resources.	"Big issues" such as wars, desertification, poverty, etc. are challenging factors that must be dealt with on the local level.
Vision	Clean river. Healthy and sustainable environment. Better quality of life for all people in the region. Demonstration.	Reasonable stable water levels in the lake (no drying out and no major floods). Healthy and sustainable environment. Food supply to the local people. Demonstration.	Improving quality of life of all people in the region. Sustainable environment. Better resource management. Demonstration for other projects.
Methodology	Master plan, outline scheme, implementation.	Feasibility study then implementation in stages.	Integrated comprehensive approach. Practice-oriented approach.
Leadership	Local mayors.	Minister in Burkina Faso and his general manager, chairman and manager of the ARRA, chairman of the IRF.	Extremely enthusiastic, devoted, brave and target-oriented leaders.
Cooperation	Strong initial efforts to agree and cooperate. "Talking business."	Internationally stimulated cooperation, focused on real outcomes.	Avoid "big issues". Focus on practical environmental objective with on-the-ground results.
Understanding of all parties	Much effort of both sides to understand each other and work towards meeting the interests of each side.	Every participant tried to understand the others and overcome the gaps between mentalities.	Mutual understanding. Much effort to meet the interests of all partners. Recognizing the differences between mentalities.
Common interests of partners	Improving the environment. Demonstrating that cooperation in the Middle East is possible. Improving the environment.	Solving health problems.	Helping the local people. Improving the environment. Using the environmental issues as a bridge for cooperation.

Table 13.1. (*cont.*)

Feature	Alexander River Restoration Project	Lake Bam Restoration Project	Common facts or features
Mediators	German Government (BMZ Ministry through Kfw).	Honorary consul of Burkina Faso in Israel.	Very helpful international mediators, crucial for success.
Public involvement	On the Israeli side only.	Local stakeholders around the lake.	Very successful and fruitful where it was done.
Outcomes	On the ground results. A huge improvement in the environmental situation.	At this stage a feasibility study as a basis for raising money for the implementation stage.	Excellent cooperation between partners led to outcomes, according to stage of each project.
Demonstration project	Park on Israeli side. Treatment ponds and river cleanup on Palestinian side.	First stage near the town is planned as a demonstration project.	An excellent tool for further support.
Success	Huge success. Five awards so far including the 2003 International Riverprize.	Too early to assess.	
Cooperation in other fields	Basketball tournament –the Alexander River League.	Not yet.	
Prospects for the future	Continuous cooperation to solve the remaining problems, and to demonstrate that Israelis and Palestinians can live together in peace on one small piece of land.	Continuous cooperation to implement the plans and restore the lake.	Environmental projects are long term projects. Much still has to be done. Patience and belief are crucial

of an environmental crisis can usually be well understood by the people who live in these areas and by their leaders. In most cases, conflicts are not caused by the environmental problems, but are a result of them, or of neglecting them because of the "big issues." Therefore, if people can find a common language and start to cooperate on environmental issues which are not directly part of the conflict, they might find a resolution of the overall conflict. Even if this does not happen, at least the environment and the local people will benefit.

The Alexander River Restoration Project, with its thirteen-year history of cooperation between Israelis and Palestinians at a local level, certainly raises hopes for a better future in the Middle East. Even if this does not come in the near future, at least the environment has been improved and some friendships have been created. In the case of the Lake Bam Restoration Project no results can yet be analyzed. However, expectations are high that the approaching environmental crisis can be solved. Cooperation between the two "twins" from Burkina Faso and Israel, their "mother" (IRF from Australia), and hopefully a "father" (an international aid agency which will help with the funding of the implementation) seems to be the only practical route to a more sustainable future for the lake and its surroundings.

At first sight one could think that these two projects do not have much in common, beside the fact that they were planned and managed by the same person. But as can be learned from the summary in Table 13.1, the projects have many similar features that demonstrate how restoration projects in very different locations and circumstances can be a bridge for cooperation and peace.

The two projects analyzed here are exceptional examples of cooperation on restoration projects in different parts of the world. Unfortunately, not many projects of this kind have been initiated or actually carried out. Many lessons can be drawn from them, but they should be used very carefully in other places, because the circumstances are always different. Particularly in areas of conflicts a very sensitive and site-specific approach must be used. It is always much easier to understand the turtles and the silent fish in the river, than the whole complexity of the human system in the area that needs restoration.

Wars on water have been held throughout history. It might be the right time to have peace on water, instead of another "Water War", a "Water Peace."

References

Brandeis, A. (1996). *Alexander River Restoration Master Plan. Alexander River Restoration 1996 Report*. Hod-Hasharon, Israel.(www.restorationplanning.com).

Brandeis, A. (2001). *Kishon River Master Plan*. Kishon, Israel: Kishon River Authority. (www.kishon.org.il).

Brandeis, A. (2006). *Alexander River Outline Scheme. Alexander River Restoration 2000 Report*. Hod-Hasharon, Israel. (www.restorationplanning.com).

Cuningham, S. (2002). *The Restoration Economy, The Greatest New Growth Frontier*. San Francisco: Berrett-Koehler Publishers.

Golan, T. and Ron, T. (2005). *Gorillas and Diplomacy*. Tel Aviv: Am Oved Press.

Hansen, O. H. (ed.) (1996). *River Restoration–Danish Experience and Examples*. Copenhagen: National Environment Research Institute, Denmark.

Levitt, J. L. (2006). Illegal peace? An inquiry into the legality of power-sharing with warlords and rebels in Africa. *Michigan Journal of International Law*, **27**: 495–568. (www.drjeremylevitt.com/images/uploads).

Maathai, W. (2003). *The Greenbelt Movement: Sharing the Approach and the Experience*. New York: Lantern Books.

Nelson, S. (2004). Science and celebrations. The riverfestival experience "downunder." *Proceedings of the 4th Canadian River Heritage Conference*. Guelph, Ontario: International Riverfoundation. (www.grandriver.ca/RiverConferenceProceedings).

Rahamimoff, A. (1995). *Yarqon River Master Plan. Yarqon River Authority*. (www.yarqon.org.il).

Riley, A. L. (1998). *Restoring Streams in Cities*. Washington DC: Island Press.

Vivash, R., Murphy, D., Janes, M., Holmes, N. and Haycock, N. (eds.) (1999). *Manual of River Restoration Technologies*. Silsoe Campus: The River Restoration Center (RRC).

Index

acid test, 16
acidification, 200
afforestation, 21, 38, 39, 49, 57, 67, 68
agricultural land, 49, 57, 63, 83, 115, 126, 128,
   130, 265
agriculture, i, xvi, 9, 23, 36, 57, 59, 63, 118, 119, 122,
   142, 145, 150, 153, 154, 157, 158, 160, 178, 209,
   210, 225, 233, 234, 245, 270
agroforestry, 118, 122, 131, 132
aquaculture, 195, 205, 208, 210, 213
aridification, 145
autotrophic respiration, 24, 25, 28

beach, 13, 192, 193, 194, 200, 202, 203, 204,
   211, 234
biocarbon fund, 51
biodiversity
   benefit, 64, 65, 71, 120, 122, 134, 234
   conservation, xvi, 39, 68, 100, 103, 117, 118,
     119, 121, 123, 124, 131, 161, 204, 214, 228,
     231, 234
   loss, 5, 8, 9, 13, 53, 58, 115, 151, 164, 190, 198, 200,
     205, 230, 240
   service, 55, 61, 133, 144, 152
   value, 4, 55, 61, 67, 143, 185, 207
bio-energy, 55
biofuels, 55
biogeochemical
   cycles, 4, 7, 8, 32, 190, 202, 205
   models, 26, 31
   processes, xix, 10, 205, 214, 216
   transformations, 205
biological
   diversity, xvi, 21, 95, 96, 116, 161, 197
   invasions, 95
   populations, 190
   productivity, 153
   structure, xix, 190, 215
biomass, 6, 21, 23, 24, 25, 28, 31, 49, 53, 58, 59, 60, 69,
   159, 195
bioremediation, 69
biosphere, 4, 10, 11, 83

carbon
   balance, 25, 27, 30, 31, 32, 33, 34, 36, 62, 69
   bottom-up model, 32
   budget, 21, 24, 28, 30
   credits, 38, 47, 48, 49, 50, 56, 57, 66, 71, 73
   cycle, xx, 21, 22, 23, 24, 26, 27, 30, 31, 32, 34, 39
   dioxide, 22, 31, 36, 54, 196
   finance, 46, 48, 55, 56, 57, 64, 69, 70, 71, 73
   flow, 24
   forestry, 46, 50, 52, 53, 55, 56, 57, 58, 60, 61, 63, 64,
     65, 66, 67, 69, 71, 72, 73
   market, 38, 45, 46, 48, 53, 60, 66, 68, 69, 70, 71,
     72, 73
   neutral, 48
   offsets, 48, 51
   pool, 25, 28
   sequestration, 21, 36, 39, 55, 57, 58, 60, 62, 66, 67,
     72, 73
   sink, 23, 25, 60, 195, 196, 202
certified emission reductions, 47
challenge, xx, 3, 4, 5, 6, 11, 13, 15, 17, 46, 64, 66, 70,
   73, 93, 94, 95, 104, 141, 161, 162, 165, 187, 189,
   190, 191, 200, 206, 212, 213, 216
Clean Development Mechanism, 39, 46, 47, 48, 49, 55,
   67, 71, 72
climate
   change, i, xvi, 8, 10, 13, 16, 23, 37, 38, 47, 48, 53, 54,
     56, 65, 68, 70, 71, 72, 73, 88, 95, 96, 98, 107, 141,
     150, 163, 182, 189, 194, 196, 198, 199, 201, 204,
     206, 264
   warming, 145, 146, 150
Climate, Community and Biodiversity standard, 64
coastal
   erosion, 196, 199, 204, 206
   lagoons, 190, 197, 198
   management, 193, 206, 207
   protection, 195, 200, 202
   resilience, 206, 212
   squeeze, 205
connectivity, 16, 124, 125, 132, 144, 186, 194, 213,
   215, 231, 238, 241
Convention on Biological Diversity, 71

cooperation, xix, xx, 16, 161, 162, 165, 206, 264, 265, 271, 273, 274, 275, 276, 278, 279, 281, 282, 283
coral reefs, 193, 194, 199, 200, 202, 207
corporate social responsibility, 48
corridors, 10, 104, 124, 125, 214, 226, 230, 238, 239
cultivated land, 25, 26, 150
cultural diversity, 12, 95, 96
cyclone, 199

dam, 164, 175, 179, 185, 204, 211, 270, 272
deforestation, 8, 9, 21, 23, 26, 38, 39, 45, 46, 49, 52, 53, 57, 61, 65, 73, 129, 134, 197, 270
degradation, 46, 53, 73
delta, 175, 176, 182, 183, 184, 191, 196, 205, 206, 207, 211, 233
denitrification, 185
desertification, 21, 39, 53, 65, 69, 70, 71, 140, 141, 144, 147, 151, 152, 153, 160, 161, 164, 165, 268
diversity, 91, 92, 94, 95, 98, 116, 131, 135, 193, 197, 212, 231

Earth, xv, xvi, xvii, xviii, xx, 3, 4, 5, 21, 23, 78, 83, 84, 88, 89, 92, 106, 107, 151, 177, 189, 193, 201
ecological threshold area, 135
economic value, xvi, 89, 140, 160, 177, 198
ecosystem services, 10, 12, 55, 61, 71, 79, 94, 95, 152, 153, 198
ecotechnology, 245
ecotone, 165, 189
eddy covariance, 29, 31, 32, 36, 37
emissions trading, 48, 51
energy, 6, 11, 23, 24, 32, 47, 49, 55, 66, 72, 82, 98, 106, 141, 176, 179, 189, 193, 197, 201, 202, 203, 204, 205, 209, 214
erosion, 25, 45, 65, 67, 115, 125, 128, 131, 132, 142, 147, 149, 151, 153, 158, 160, 194, 197, 199, 200, 202, 211, 212, 214, 233, 266, 270, 271
ethics, 13, 16, 100, 101, 102, 105, 108
eutrophication, 153, 194, 195, 198, 207, 208, 209, 210, 249, 250
exergy, 247, 250

farmland, 124, 128, 141, 144, 149, 154, 164, 165, 184, 225, 226, 228, 231, 232, 233, 239
fast-growing species, 62
fish harvest, 202
fisheries, xvi, 5, 180, 195, 196, 200, 201, 202, 205, 207, 209, 213, 215, 240
focal restoration, 91, 97, 98
food, 4, 5, 8, 11, 22, 64, 78, 79, 96, 130, 141, 144, 153, 176, 190, 196, 197, 198, 226, 230, 232, 233, 239
    web, 144, 153, 196
forest
    conservation, 49, 73
    cover, 50, 53, 73, 121, 127, 129, 130, 134
    ecosystems, 28, 29, 37, 116, 117, 141, 142, 143, 155
    management, 25, 38, 39, 49, 52, 56, 68, 142, 152
    productivity, 35
    restoration, 45, 53, 60, 61, 63, 67, 71, 115, 117
forestry offsets, 48
fragmentation, 11, 130, 143, 164, 181, 202, 208, 210, 230

glacial state, 22
global
    climate, i
    ecological restoration, xviii, 15, 61, 73
    scale, i, v, xviii, xix, xx, 3, 6, 7, 10, 13, 31, 45, 54, 71, 73, 79, 83, 107, 113, 186, 189, 202, 210, 215
grassland, 6, 25, 26, 29, 59, 68, 125, 141, 142, 143, 145, 147, 149, 152, 153, 154, 156, 159, 160, 164
grazing, 38, 59, 63, 64, 149, 153, 156, 159, 176, 268, 270
green edge strip, 239
greenhouse, 160, 233
    emissions, 23
    fertilizers, 234
    gases, 8, 26, 38, 46, 49, 52, 64, 73, 202
greenspace strips, 237
gross primary production, 24, 25, 31, 36, 37

habitat loss, 210
heterotrophic respiration, 25, 28
hurricane, 183, 191, 193, 199, 201, 214

indigenous people, xvi, 107, 108
interglacial state, 22
Intergovernmental Panel on Climate Change, 23
inverse modelling, 33

joint implementation, 48

Kyoto Protocol, xvi, 38, 39, 46, 47, 48, 50, 61, 72

lagoons, 197, 217
lake, 148, 153, 155, 164, 175, 181, 182, 240, 249, 265, 268, 271, 277, 279
land
    degradation, 39, 45, 61, 62
    mosaic, 225, 226, 242
land–atmosphere carbon flux, 26
landscape
    ecology, 125, 225, 226, 231, 242
    restoration, 124, 125, 127, 132, 133, 134, 135
land-use, 45, 49, 53, 56, 57, 63, 70, 71, 72, 73, 142, 184, 226, 228, 230
land-use, land-use change, and forestry, 49
Latin America, 51
leaf area index, 30
leakage, 49, 56, 57, 59, 60, 63, 70
Loess Plateau, 147, 152, 158, 160, 164

mangrove, 6, 83, 94, 142, 144, 156, 176, 186, 195, 196, 200, 201, 207, 210, 211, 212
market, 46, 47, 48
Marsh Arabs, 175, 185, 186
marshes, 176, 177, 180, 185, 190, 196, 201, 203, 205, 210, 211, 215, 234
    marshland, 186
meadows, 95, 195, 207
Mediterranean, 35, 211, 215, 234, 265, 272
metropolitan, 225, 226, 232, 236, 238, 239, 265
mine, 157
mire, 177

model
   bottom-up, 31, 32, 33, 34
   top-down, 34
monoculture plantations, 121
mosquitoes, 178, 266, 276
mud flats, 196, 197
multi-issue solutions, 234
multi-purpose species, 122
multi-species plantations, 117, 123

native species, 61, 62, 69, 72, 95, 101, 121, 122, 123, 160
net biome productivity, 25
net ecosystem production, 25
net primary production, 24, 25, 31, 37, 80
normalized difference vegetation index, 30
nurse tree, 62, 68
nursery, 195, 212

ocean–atmosphere carbon flux, 8, 26, 32
Official Development Assistance, 53
oil
   removal, 208
   spills, 197, 208, 209, 216
overgrazing, 65, 159
overpasses, 231, 237, 238

participation, xv, xvii, 9, 68, 91, 97, 107, 161, 208, 216,
   268, 279
pastures, 63, 104, 156
peace, 264, 265, 272, 273, 274, 276, 279, 287
peatlands, 176, 177, 178, 186
photosynthetically absorbed active radiation, 30
pollination, 80
pollutants, 149, 208, 212, 234, 236, 245, 266, 267
post-Kyoto regime, 50, 53, 68, 73
poverty, xvi, 16, 53, 122, 152, 161, 264, 269, 270, 279,
   280, 283
prairie, 104, 177

radar, 31
radiation use efficiency, 31, 33
recreation, 79, 200, 216, 226, 227, 228, 240
reduced emissions, 46, 52, 73
reference
   area, 127
   site, 16, 104
   system, 15, 16
reforestation, 38
rehabilitation, xvi, 65, 117, 123, 125, 131, 141, 156,
   161, 163, 164, 166, 198, 213, 277, 281
remote sensing, 30, 31, 32, 33, 36
resilience, 63, 88, 135, 190, 211, 212
rice, 125, 175, 177, 211
riparian
   buffers, 184, 185
   strips, 125
river, 26, 142, 144, 146, 148, 154, 163, 175, 179, 187,
   189, 194, 201, 204, 205, 211, 226, 264, 265, 278,
   282, 283, 284

rocky shores, 197
roots, 24, 25, 28, 59, 60, 265
root shoot ratio, 60

salinization, 142, 195
savannas, 25, 68
sea-level rise, 189, 190, 196, 198, 204, 206,
   215, 218
seagrass, 194, 195
secondary vegetation, 62
sediment balance, 204, 206
seedbanks, 61
sequestration rate, 59, 67
sewage, 208, 234, 266, 271, 272, 276
shrimp farming, 210
shrubland, 233
soil respiration, 25, 27
spectral reflectance, 30
stakeholders, 16, 49, 64, 72, 125, 126, 129, 134, 161,
   165, 166, 273, 276
stepping stones, 88, 125
stream, 15, 62, 104, 125, 175, 178, 184, 226, 230, 231,
   234, 238, 239, 253
succession, xix, 16, 62, 68, 69, 115, 158
swamp, 96, 144, 177, 182, 186

technology, 11, 52, 84, 95, 162, 225
   technological restoration, 91, 95, 96, 97
tide gauge, 198
topsoil, 68
total ecosystem respiration, 25, 37
town planning, 226
traditional ecological knowledge, 108
transplantation, 194, 195
trash trap, 282
tree
   farming, 122
   planting, 64, 122
tundra, 186
turtle, 265, 267, 287

underpass, 231, 238
United Nations Convention to Combat
   Desertification and land
   Degradation, 71
United Nations Framework Convention on Climate
   Change, 23, 38, 47
urban planning, 206, 214
urbanization, 8, 144, 149, 190, 194, 197, 200, 208, 210,
   215, 231, 239, 240

voluntary carbon standard, 57, 58
voluntary market, 51, 61, 70
Vostok ice core, 22

war, xvii, xix, 140, 264, 272
wetland ecosystems, 144, 154
wetlanders, 175
World Bank, 51, 61